Drug Delivery Across Physiological Barriers

Drug Delivery Across Physiological Barriers

Editor

Jakub Sarto

scitus
academics

Drug Delivery Across Physiological Barriers

Edited by **Jakub Sarto**

ISBN: 978-1-68117-694-9
Library of Congress Control Number: 2016958856

SCITUS Academics LLC,
616, Corporate Way, Suite 2, 4766,
Valley Cottage, NY 10989

www.scitusacademics.com

Notice

Printed in United States of America on Acid Free Paper

Preface

Drug delivery is the method or process of administering a pharmaceutical compound to achieve a therapeutic effect in humans or animals. For the treatment of human diseases, nasal and pulmonary routes of drug delivery are gaining increasing importance. These routes provide promising alternatives to parenteral drug delivery particularly for peptide and protein therapeutics. Drugs have long been used to improve health and extend lives. The practice of drug delivery has changed dramatically in the last few decades and even greater changes are anticipated in the near future. Biomedical engineers have not only contributed substantially to our understanding of the physiological barriers to efficient drug delivery—such as transport in the circulatory system and drug movement through cells and tissues—they have contributed to the development of a number of new modes of drug delivery that have entered clinical practice.

Drug Delivery Across Physiological Barriers covers the physiological barriers controlling penetration and transport of substances in the body, strategies for drug penetration across tissue compartments, and drug transport into cells and their compartments. It provides an overview of the current advances in this field, including considerations on the biological regulation and natural mechanisms overcoming these barriers, as well as drug delivery strategies assisting the transport of drugs and their carriers at the tissue, cell, and subcellular levels. The recent progress of nano drug delivery systems in breaking the physiological barriers for more effective in vivo anti-tumor effect, in order to shed a new perspective on the development of tumor-targeted nano drug delivery systems are reviewed.

For any therapeutic agent to be effective, it must accumulate in target cells in optimal concentrations for a required duration of time.

Unfortunately, physicochemical and physiological barriers can lead to heterogeneous accumulation of various therapeutic molecules, particles, and cells in solid tumors. In a disease such as cancer, failure to treat a small fraction of cells can result in tumor regrowth. Thus, it is crucial to know which cells have been successfully treated and which have not. To determine this, high-resolution imaging of drug distribution and its physiological determinants is essential. Only with this knowledge can we improve drug delivery to all regions of a tumor..

Table of Contents

Chapter 1

Strategies for Enhanced Drug Delivery to the Central Nervous System

V. S. N. M. Dwibhashyam[1] and A. N. Nagappa[*]

[1]TherDose Pharma (P) Ltd., Plot No: 30-32, Survey No. 400, Prasanth Nagar, IE, Hyderabad-500 072, India

ABSTRACT

Treating central nervous system diseases is very challenging because of the presence of a variety of formidable obstacles that impede drug delivery. Physiological barriers like the blood-brain barrier and blood-cerebrospinal fluid barrier as well as various efflux transporter proteins make the entry of drugs into the central nervous system very difficult. The present review provides a brief account of the blood brain barrier, the P-glycoprotein efflux and various strategies for enhancing drug delivery to the central nervous system.

The brain is a delicate organ. The brain-microvascular endothelial cells of the blood-brain barrier protect this organ from exogenous substances. Certain efflux transporters such as P- glycoprotein also perform the same function. However, these protective barriers restrict the entry to the brain from the periphery of compounds that might be of therapeutic value in the treatment of fatal central nervous system (CNS) diseases, such as brain tumors, HIV encephalopathy, epilepsy, cerebrovascular disease and neurodegenerative disorders, and of other pathologies. Delivery of drugs to the brain is a challenging task because of the presence of efficient protective mechanisms. The existing conventional drug delivery systems have been inefficient in selectively targeting drugs to the CNS. In response to this, aggressive research efforts have focused on the development of new strategies to more effectively deliver drug molecules to the brain [1].

THE BLOOD BRAIN BARRIER (BBB) AND P-GLYCOPROTEIN (P-GP)

The BBB is one of the most challenging barriers in the body. It is created by the way the blood vessels in the brain are organized. Brain capillaries are different from capillaries of other parts of the body in that normal brain endothelia have fewer pinocytic vesicles, more mitochondria, no fenestrations and adjacent cells are maintained in close apposition by tight junctions [2–4]. The shielding effect of the BBB is further strengthened by the presence of certain efflux transporters such as P-gp in the luminal membrane of the cerebral capillary endothelium. (Schematic depiction of comparison between brain capillaries and capillaries in general is given in fig. 1). P-gp is a brain microvascular endothelial cell protein, which possesses several essential pharmacological functions of drug portage and expulsion [5]. It is present in high concentration on the apical surface of these endothelial cells. P-gp is an active drug efflux transporter protein. While P-gp is involved in protecting the brain exposure to a variety of pharmacologically active hydrophobic agents, it is an impediment to the treatment of various CNS diseases such as primary brain tumors and cerebral human immuno deficiency virus (HIV) infection [5]. P-gp has affinity for a broad range of structurally unrelated large hydrophobic compounds including Vinca alkaloids, epipodophyllotoxins, anthracyclines, cyclosporine A, digoxin, and various HIV protease inhibitors [6–9]. Various approaches are being tried to bypass P-gp efflux. Reversal agents such as R-verapamil, PSC 833 (cyclosporine analog), and biricodar, inhibit P-gp mediated drug transport and increase the influx of therapeutic agents they are co-administered with [10]. Certain non-ionic surfactants like Tween-80 and Cremophor EL have also been found to have the reversal activity [11,12]. However, most of these agents have been found to be pharmacologically active and elicit significant toxicity at doses required for P-gp inhibition.

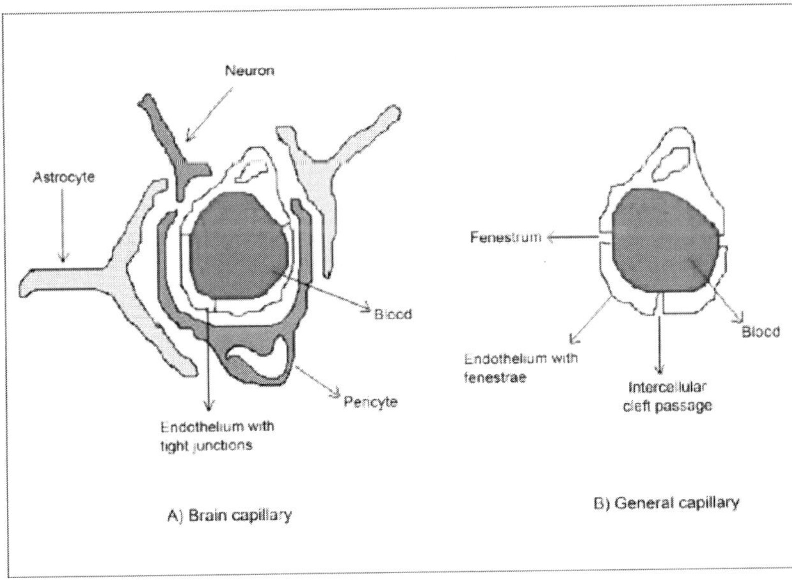

Figure 1. Schematic comparison between brain and general capillaries.

STRATEGIES FOR ENHANCED CNS DRUG DELIVERY

Various strategies have been studied to circumvent the multitude of barriers inhibiting brain penetration by therapeutic agents. These strategies generally fall into one or more of the following three broad categories viz., manipulating drugs, disrupting the BBB, and exploiting alternative routes for drug delivery

ENHANCING BBB PERMEABILITY BY DRUG MANIPULATION

Lipophilic analogs:

Drug penetration through the BBB is favored by lipophilicity. Because a drug's lipophilicity correlates so strongly with cerebrovascular permeability, hydrophobic analogs of small hydrophilic drugs ought to more readily penetrate the BBB. However, the drug molecule should have an optimum octanol-water partition coefficient with Log P value of approximately 1.5 to 2.5 to be efficacious when delivered via the circulatory system [13].

Prodrugs:

A prodrug consists of a drug covalently attached to an unrelated chemical moiety that improves the drug's pharmacokinetic properties. The prodrug itself is inactive but becomes active when the attached moiety is cleaved *in vivo* by enzymatic or hydrolytic processes. To enhance a drug's penetration through the BBB, prodrugs are often designed by attaching chemical moieties that increase the drug's lipophilicity. The best example of this approach is the series of analogs of morphine. Morphine does not readily cross the BBB. Latentiation via acetylation of both hydroxyl groups yields the hallucinogenic heroin, which readily traverses the BBB, and subsequent hydrolytic cleavage of the acetyl groups yields high concentrations of morphine trapped in the brain due to its hydrophilicity [2].

Chemical drug delivery systems (CDS):

The chemical delivery systems cross the BBB by smuggling compounds across as their lipophilic precursors. The CDS approach is a part of retro metabolic drug design approach. These are inactive chemical derivatives of a drug obtained by one or more chemical modifications that provide site-specific or site enhanced delivery through multi-step enzymatic and/or chemical transformations. They include two types of bio-removable moieties: a targeter (T) responsible for targeting, site-specificity, and lock-in; and modifier functions (F_1-F_n) that serve as lipophilisers, protect certain functions, or fine tune the necessary molecular properties to prevent premature, unwanted, metabolic conversions. Targeting is achieved by design: CDSs will undergo sequential metabolic conversions, disengaging the modifier functions and finally the targeter, after they fulfill their site- or organ-targeting role.

There are a wide variety of CDSs possible both theoretically and in practice. For convenience, the major CDSs can be divided into three classes. They are a) enzymatic physico-chemical CDSs that exploit site-specific traffic properties by sequential metabolic conversions resulting in altered transport properties. b) Site-specific enzyme activated CDSs that exploit specific enzymes found primarily, exclusively, or at higher activity at the site of action and c) Receptor-based CDSs that enhance selectivity and activity through transient, reversible binding to target receptors.

Brain targeting CDSs based on altered transport properties are the most studied ones, that are based on the idea that if a lipophilic compound that enters the brain is converted there into a lipid insoluble molecule, it will become 'locked in'. Bodor *et al,* developed a creative approach to exploit the permeability characteristics of the BBB to achieve site-specific and/or sustained release of drugs to the brain [14–16]. Targeting is assisted because the same conversion-taking place in the rest of the body accelerates elimination.

In principle, many targeter moieties are possible, but the one based on the 1,4-dihydrotrigonelline-Trigonelline system, where the lipophilic 1,4-dihydro form (T) is converted *in-vivo* to the hydrophilic quaternary form ($T^{'}$), proved the most useful. This conversion takes place easily everywhere in the body because it is closely related to the ubiquitous NADH-NAD^{+} coenzyme system associated with cell respiration. Oxidation takes place with direct hydride transfer, without generating highly active or reactive radical intermediates, providing a non-toxic targeter system. The resulting T^{+} is locked in the brain (fig. 2), but is easily eliminated from the body due to acquired positive charge, which enhances water solubility. After a relatively short time, the delivered drug D (as the inactive, locked-in T^{+}--D) will be present essentially only in the brain, providing sustained, brain-specific release of the active drug. These systems have been tried for a wide range of drug classes such as steroid hormones, antiinfective agents, anticancer agents, and anti retroviral agents among these, oestradiol-CDS is the most advanced and is currently undergoing Phase 1 and 2 clinical trials.

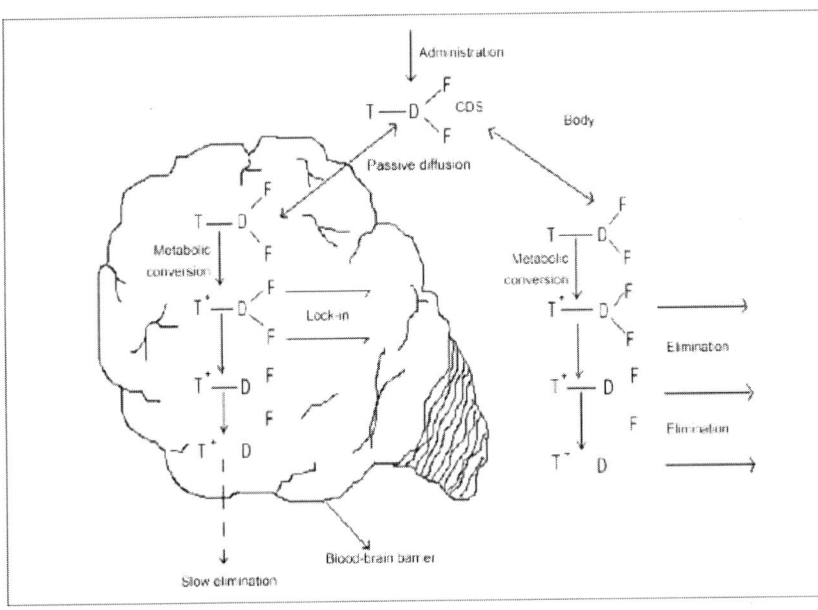

Figure 2. The lock-in mechanism for a brain chemical drug delivery system (CDDS).

Molecular packaging:

Delivering the peptides like enkephalin, TRH (thyrotropin-releasing hormone), and Kyotorphin analogs through the BBB is an even more complex problem because they can be rapidly inactivated by ubiquitous peptidases [17–20]. Three issues are to be solved simultaneously to enhance penetration through BBB. They are, to enhance passive transport by increasing the lipophilicity, assure enzymatic stability to prevent premature degradation, and exploit the lock-in mechanism to provide targeting. This complex approach is known as molecular packaging strategy, where the peptide unit is part of a bulky molecule, dominated by groups that direct BBB penetration and prevent recognition by peptidases. In general, a brain targeter packaged peptide delivery system contains a red-ox targeter (T), a spacer function (S), consisting of strategically used amino acids to ensure timely removal of the charged targeter from the peptide, the peptide itself (P) and a bulky lipophilic moiety (L) attached through an ester bond or sometimes through a C-terminal adjuster (A) at the carboxy terminal to enhance lipid solubility and to disguise the peptide nature of the molecule.

The first successful delivery with a package was for Tyr-D-Ala-Gly-Phe-D-Leu (DADLE), an analogue of leucine enkephalin, a naturally occurring linear pentapeptide (Tyr-Gly-Gly-Phe-Leu) that binds to opioid receptors. A similar strategy was used to deliver a thyrotropin-releasing hormone (TRH) analogue to the CNS [17–18]. These analogues are potential agents for treating neurodegenerative disorders such as Alzheimer's disease.

Carrier-mediated drug delivery:

This strategy takes advantage of the facilitative transport systems present in the brain endothelium. Active transport systems have been found for various substrates such as monosaccharides, monocarboxylic acids, amines, vitamins, hormones, purines, acidic and basic amino acids.

The cerebro vascular membranes are rich in facilitative carrier systems for both glucose and large neutral amino acids (LNAAs). The carrier system for glucose is highly specific for monosaccharides, rendering it useless for drug delivery. However, the LNAA carrier system is capable of transporting numerous endogenous as well as exogenous LNAAs with great structural variety [2,21]. This characteristic has made exploitation of the LNAA carrier system an attractive strategy for CNS drug delivery.

The LNAA carrier system has been exploited to deliver L-3,4-dihydroxyphenalanine (levodopa), an endogenous precursor to dopamine, to the brain. Unlike dopamine, levodopa has a high affinity for the LNAA carrier system. After traversing the anti-luminal membrane of the cerebral endothelium, levodopa is decarboxylated to yield dopamine. The LNAA carrier system has also been exploited to deliver the antineoplastic agent melphalan into the brain.

Receptor/vector-mediated drug delivery:

This strategy exploits several specific transcytosis systems that are actually meant for the extravasation of important nutrients and signaling molecules that cannot diffuse through the cerebromicrovasculature. These include systems for the transport of insulin, transferrin, and insulin like growth factor. Chimeric peptide delivery is one of such strategies, the principle of which lies in the coupling of a non-transportable peptide pharmaceutical to a transportable peptide or protein, which undergoes receptor mediated or absorptive mediated transcytosis through the BBB [22–23]. For example, the non-transportable protein β-endorphin was linked to the transportable protein-cationized albumin via a disulfide linkage. This chimeric peptide was successfully transcytosed into the brain and enzymatically cleaved in the parenchyma by thiol reductase.

Binding of the vector to its receptor on the luminal surface of brain capillary endothelial cells initiates endocytosis. Following exocytosis at the abluminal plasma membrane and release into brain interstitial space, the pharmacologically active moiety of the chimeric peptide may be released by enzymatic cleavage if a cleavable linkage between the vector and the drug is employed. The free peptide drug would then be able to interact with its specific target receptor on brain cells. A covalent conjugate of cationised albumin and the opioid peptide D-Ala-β-endorphin (DABE) was the first example of a chimeric opioid peptide to be investigated *in vitro*[24] and *in vivo*[25] with regard to its transport at the BBB. This chimeric peptide was linked by the disulfide based cross-linking agent, N-succinimidyl-3-(2-pyridyl-dithio) propionate (SPDP). The same technology has been applied to the vasoactive intestinal peptide analog (VIPa) (nontransportable pharmaceutical)[26–28]. This chimeric peptide consists of VIPa and a covalent conjugate of an antitransferrin receptor monoclonal antibody (mouse Mab OX26) and avidin vector (fig. 3).

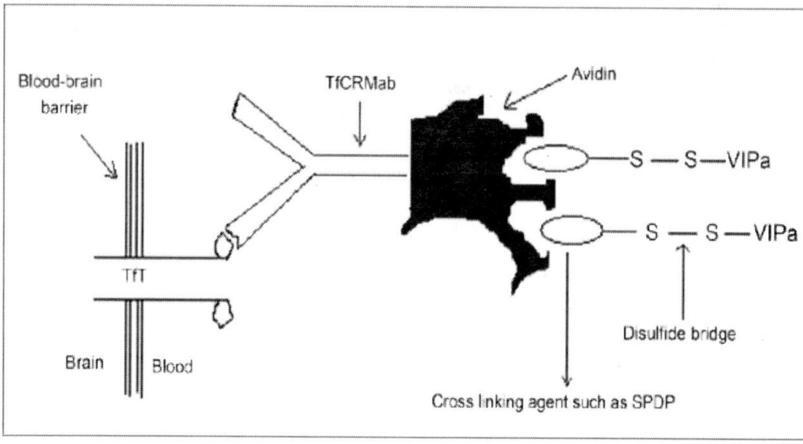

Figure 3. Transferrin transporter-based chemical peptide delivery.

The chimeric peptide technology has been applied to PNAs to achieve intracranial delivery of Polyamide (peptide) nucleic acids (PNAs) [29]. This PNA can successfully be delivered to the murine brain parenchyma when covalently coupled to the OX-26 monoclonal antibody. This technique has been expanded to successfully deliver drug-loaded immunoliposomes to the brain [30]. Immunoliposomes consist of liposomes coated with the inert polymer PEG conjugated to the OX-26 antitransferrin-receptor monoclonal antibody. Liposomes and proteins can be used as a cargo of membrane-permeable proteins for the improvement of their intracellular permeability[31].

Nanoparticles have been used as cargo for peptides. PLGA nanoparticles are the most studied ones in this regard. These nanoparticles consist of a colloidal polymer particle of poly-butyl cyanoacrylate (PBCA) with the desired peptide adsorbed onto the surface and then coated with Polysorbate 80. Nanoparticles have been used as a vector for delivery of hexapeptide dalargin (an enkephalin analog)[32–33]. Drugs that have successfully been transported across the BBB with the nanoparticles include loperamide, tubocurarine and doxorubicin [34,35]. The mechanism of nanoparticle transport has not yet been fully elucidated. The most probable transport pathway seems to be endocytosis by the blood capillary endothelial cells following adsorption of blood plasma components, most likely apolipoprotein E (apoE), after intravenous injection. Alternatively, transport may occur by transcytosis of the nanoparticles with drug across the endothelial cells. Percoating of nanoparticles with polysorbate led to adsorption of apoE and possibly other plasma components, which seem to be able to interact with the LDL receptors on the brain endothelial cells that could lead to their endocytosis. In addition to these processes, polysorbates seem to be able to inhibit efflux pump. This inhibition could contribute to the brain transport properties of the nanoparticles [36]. Luca *et al*, found an increased permeability of PLGA nanoparticles when conjugated with five short peptides [37]. Permeability of zidovudine and lamivudine across the BBB has been found to increase by 8-20 and 10-18 folds, respectively when administered as PBCA (polybutylcyanoacrylate) nanoparticles [38]

DISRUPTING THE BBB

This invasive technique for enhanced CNS drug delivery involves the systemic administration of drugs in conjunction with transient BBB Disruption (BBBD). A variety of techniques that transiently disrupt the BBB have been investigated. However, many of these have been proved to be toxic and are not clinically useful. These include the infusion of solvents such as dimethyl sulfoxide or ethanol and metals such as aluminum, X-irradiation, and the induction of pathological conditions including hypertension, hypercapnia, hypoxia, or ischemia [3,39–41].

Osmotic BBBD:

The most frequently applied clinical technique for achieving BBBD is the intracranial infusion of a hyperosmolar solution of mannitol [42,43]. By infusing the solution directly into an artery that feeds the target area of the brain; it is somewhat possible to achieve localized BBBD and/or drug delivery [44]. As the hyperosmolar solution flows through the cerebral capillaries, acute dehydration of endothelial cells results in cell shrinkage, which in turn widens the tight junctions connecting adjacent membranes. Subsequent rehydration in the presence of normal plasma leads to complete restoration of the BBB about 4 h. following treatment. An unfavourable toxic/therapeutic ratio often is observed with hyper osmotic BBBD. One reason for this is that this methodology results in a 25% increase in the permeability of the tumor microvasculature, in contrast to a 10-fold increase in the permeability of normal brain endothelium [45]. Therefore the normal brain is left vulnerable to potentially neurotoxic chemotherapeutic agents.

Biochemical BBBD:

These safer biochemical techniques selectively disrupt the intratumoral BBB while minimally altering the normal BBB [45]. Selective opening of brain tumor capillaries, by the intracarotid infusion of leukotriene C4 was achieved without concomitant alteration of the adjacent BBB [46]. It has been demonstrated in experimental animals that bradykinin, histamine and the synthetic bradykinin analog RMP-7 (receptor-mediated permeabilizer) infusion also selectively open the blood-tumor barrier. The biochemical mechanism has yet to be elucidated, but it has been established that the effect of the RMP-7 is mediated specifically through bradykinin B2 receptors.

Ultrasound-induced disruption:

The possibility of enhanced drug delivery to the CNS by inducing hyperthermia has been investigated recently [47]. Ultrasound induced mild hyperthermia, which can be controlled and localized to a small volume within the tissue, may offer promise [48]. This technology is still in its infancy.

ALTERNATIVE ROUTES FOR CNS DRUG DELIVERY

Changes at molecular levels and the mode of drug delivery may not always help in an increased penetration of drugs into the brain parenchyma. Certain methodologies have been proposed to enhance drug penetration into the brain, which are based on drug administration by alternative routes bypassing the cardiovascular system.

Olfactory and Trigeminal pathways to the CNS:

The neural pathway between the nasal mucosa and the brain provide a unique pathway for noninvasive delivery of therapeutic agents to the CNS [49–51] (fig. 4). The olfactory neural pathway provides both intraneuronal and extra neuronal pathways in to the brain. The intraneuronal pathway involves axonal transport and requires hours to days for drugs to reach different brain regions [52–56]. The extra neuronal pathway probably relies on bulk transport through perineural channels, which deliver drug directly to the brain parenchymal tissue, to the cerebro spinal fluid (CSF), or to both. This extra neuronal pathway allows therapeutic agents to reach the CNS with in minutes. Intranasal delivery of agents to the CSF is not surprising as CSF normally drains along the olfactory axon bundles as they traverse the cribriform plate of the skull and approach the olfactory submucosa in the roof of the nasal cavity where the CSF is then diverted in to the nasal lymphatics [57,58]

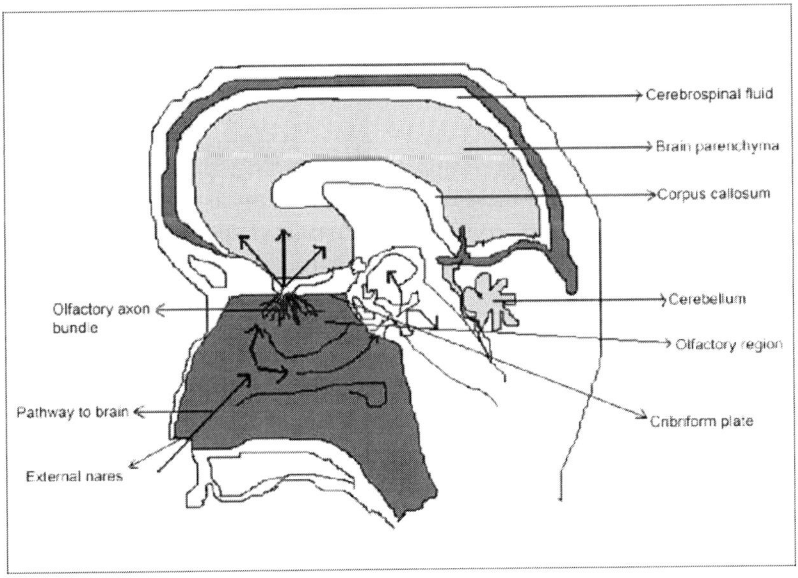

Figure 4. The Olfactory pathway between nose and central nervous system.

The Trigeminal neural pathway also may be involved in rapidly delivering protein therapeutic agents, such as insulin like growth factor-1 to the brain following intranasal administration. The trigeminal nerves innervating areas of the nasal cavity are responsible for most chemo-perception other than olfaction and sense diverse stimuli, including hot spices, formaldehyde, and other chemicals [59]. The properties of small molecules, including size and lipophilicity, have been reported to affect delivery to the CNS following intranasal administration [60,61]. The lower molecular weight and higher lipophilicity are favourable characteristics for rapid intranasal uptake of small molecules into the CNS.

There are limitations on the use of intranasal delivery as a means to bypass the BBB, including limitation on the concentrations achievable in different regions of the brain and spinal cord, which will vary with each agent. Delivery is expected to decrease with increasing molecular weight of the drug. Additionally, some therapeutic agents may be susceptible to partial degradation in the nasal mucosa or may cause irritation to the mucosa. Finally, nasal congestion from colds or allergies may interfere with this method of delivery.

Intraventricular/intrathecal delivery:

Drugs can be infused intraventricularly using an Ommaya reservoir, a plastic reservoir implanted subcutaneously in the scalp and connected to the ventricles within the brain via an outlet catheter. Drug solutions can be subcutaneously injected into the implanted reservoir and delivered to the ventricles by manual compression of the reservoir through the scalp. This strategy has several advantages compared to vascular drug delivery. This route bypasses the BBB and results in high CSF drug concentrations. A smaller dose is enough and systemic toxicity can be greatly reduced [21]. Drugs in the CSF face minimized protein binding and decreased enzymatic activity relative to drugs in plasma, leading to longer drug half-life in the CSF [62]. However, this strategy has certain disadvantages, including a slow rate of drug distribution within the CSF and increase in intracranial pressure associated with fluid injection or infusion into small ventricular volumes, resulting in to high clinical incidence of haemorrhage, CSF leaks, neurotoxicity and CNS infections[22]. CSF-brain barrier also limits the success of this approach.

Interstitial delivery:

This route of administration bypasses BBB. High CNS drug concentrations can be obtained with minimal systemic exposure and toxicity. Intracranial drug concentrations can be sustained, which is crucial in the treatment of many neurodegenerative disorders and for the antitumor efficacy of many chemotherapeutic agents. Ommaya reservoir, infusaid pump, MiniMed PIMS system and Medtronic SynchroMed system are some of the systems, which have been developed for delivering drugs directly to the brain interstitium. Until recently the most widely used method has been the interstitial injection or infusion of drugs using an ommaya reservoir or implantable pump. The adaptation of the ommaya reservoir to achieve interstitial drug delivery simply involves placing the outlet catheter directly in the intracranial target area. This technique has often been applied to neurooncological patients in whom the outlet catheter is placed in the resection cavity following surgical de-bulking of a brain tumor. Chemotherapeutic agents can be periodically injected into the subcutaneous reservoir and then delivered directly to the tumor bed. This technique, however, does not achieve truly continuous drug delivery.

The ommaya reservoir or infusion pumps have thus far been used in various clinical trials with brain tumor patients to interstitially deliver the chemotherapeutic agents BCNU or its analogs, methotrexate, adriamycin, bleomycin, fluodeoxyuridine, cisplatin, and interleukin 2(IL-2). In most of these studies the intratumoral drug concentrations were often high, and the side effects of the therapy were mild. The success of these techniques is limited by catheter clogging or blocking by tissue debris, inadequate distribution throughout the tumor, and a high degree of burden to the patient.

Biotechnological approaches:

Achieving interstitial drug delivery using biological tissues is another promising technique. It involves implanting into the brain, a tissue that naturally secretes a desired therapeutic agent. Transplantation of embryonic dopamine-releasing neurons is an attractive therapeutic strategy because these cells demonstrate good post-transplantation survival and growth characteristics in animal models [63]. However, lack of neovascular innervation limits this strategy. With out rapid neovascularization, implanted solid grafts undergo irreparable ischemic injury, leading to cell death.

Gene therapy has also been attempted to deliver drugs to the CNS. Prior to implantation, cells will be genetically modified to synthesize and release specific therapeutic agents. The therapeutic potential of this technique in the treatment of brain tumor was demonstrated. The utility of non-neuronal cells for therapeutic protein delivery to the CNS has been reviewed recently [64]. The survival of foreign tissue grafts may be improved by advancements in techniques for culturing distinct cell types. Co-grafted cells engineered to release neurotropic factors with cells engineered to release therapeutic proteins may enhance the survival and development of foreign tissue [65]. Direct application of protein-based therapeutics to the brain could soon include variations of diphtheria toxin to combat refractory gliobastomas and engineered anti-apoptotic factor (FNK) with powerful cytoprotective activity, to protect against ischemia [66,67]. As for neurodegeneration, one seemingly attractive new therapy has been the use of growth factors, such as glial-derived neurotrophic factor (GDNF) as a potential means of reducing the depletion of certain key population of cells lost in Alzheimer's or Parkinson's diseases [68,69].

Antisense drug delivery is another recent technology in CNS drug delivery. Peptide nucleic acids (PNAs) (fig. 5) are antisense oligonucleotides containing a polypeptide backbone. Receptor mediated transcytosis has been exploited to promote PNA delivery to the CNS. For example, the attachment of PNAs to the anti-transferrin (OX26) receptor antibodies has been shown to increase the brain uptake of the PNAs, with out loss of the ability of the PNAs to hybridize to target mRN [70].

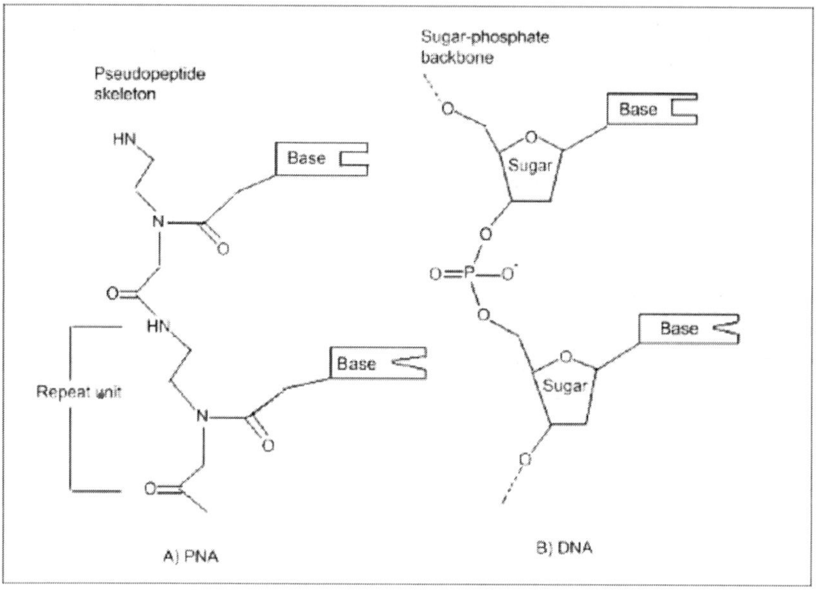

Figure 5. Structural Similarity between PNA and DNA oligomers.

REFERENCES

1. Martin B, Robert L, Henry B, Mathiowitz E. Encyclopedia of controlled drug delivery. Vol. 1. New York: John Wiley and Sons Inc; 1999. pp. 184–210.

2. Pardridge WM. Recent advances in blood-brain barrier transport. Annu Rev Pharmacol Toxicol.1988;28:25–39.

3. Tamargo RJ, Brem H. Transporting therapeutics across the blood-brain barrier. Neurosurg. 1992;2:259–79.

4. Abbott NJ, Romero IA. Transporting therapeutics across the blood-brain barrier. Mol Med Today.1996;3:106–13.

5. Gelperina SE, Khalansky AS, Skidan IN, Smirnova ZS, Bobruskin AI, Severin SE, et al. Toxicological studies of doxorubicin bound to polysorbate 80-coated poly (butyl cyanoacrylate) nanoparticles in healthy rats and rats with intracranial glioblastoma. Toxicol Lett. 2002;126:131–41.

6. Pankajavalli R. The Role of P-glycoprotein in the Blood-Brain Barrier. Einstein Quart J Biol Med.2003;19:160–5.

7. Dwibhashyam VS, Vijayaratna J. The Permeability Glycoprotein (P-gp): A Threat to Effective Drug Therapy? Indian Drugs. 2006;43:609–18.

8. Tsuji A, Tamai I. Blood-brain barrier function of P-glycoprotein. Adv Drug Deliv Rev. 1997;25:287–98.

9. Masereeuw R, Jaehde UM, Langemeijer W, DeBoer AG, Breimer DD. In vitro and in vivo transport of zidovudine (AZT) across the blood brain barrier and the effect of transport inhibitors. Pharm Res.1994;11:324–30.

10. Varma , MV , Yasvanth A, Chinmoy SD, Ramesh P. P-glycoprotein inhibitors and their screening: A perspective from bioavailability enhancement. Pharmacol Res. 2003;48:347–59.

11. Bhagvant D, Joseph PYK, James PE. Effects of nonionic surfactants on membrane transporters in Caco-2 cell monolayers. Eur J Pharm Sci. 2002;16:237–8.

12. Yu-li Lo. Relationships between the hydrophilic-lipophilic balance values of pharmaceutical excipients and their multidrug resistance modulating effect in Caco-2 cells and rat intestines. J Control Release.2003;90:37–48.

13. Madrid Y, Langer LF, Brem H, Langer R. New directions in the delivery of drugs and other substances to the central nervous system. Adv Pharmacol. 1991;22:299–324.

14. Bodor N, Buchwald P. Drug targeting via retrometabolic approaches. Pharmacol Therp. 1997;76:1–27.

15. Bodor N, Farag HH, Brewster ME. Site-specific, sustained release of drugs to the brain. Science.1981;214:1370–2.

16. Omar FA, Farag HH, Bodor N. Synthesis and evaluation of a redox chemical delivery system for brain-enhanced dopamine-containing an activated carbamate-type ester. J Drug Target. 1994;2:309–16.

17. Bodor N. A strategy for delivering peptides into the central nervous system by sequential metabolism.Science. 1992;257:1698–700.

18. Prokai L. Chemical delivery system to transport a pyroglutamyl peptide amide to the central nervous system. J Am Chem Soc. 1994;116:2643–44.

19. Prokai K, Tatrai PL, Bodor N. Brain-targeted delivery of a leucine-enkephalin analogue by retrometabolic design. J Med chem. 1996;39:4775.

20. Brownlees J, Williams CH. Peptidases peptides and the mammalian blood-brain barrier. J Neurochem.1993;60:793.

21. Greig NH. Optimizing drug delivery to brain tumors. Cancer Treat Rev. 1987;14:1–28.

22. Scheld WM. Drug delivery to the central nervous system: General principles and relevance to therapy for infections of the central nervous system. Rev Infect Dis. 1989;2:1669–90.

23. Bickel U, Yoshikawa T, Pardridge WM. Pharmacokinetics and saturable blood-brain barrier transport of biotin bound to a conjugate of avidin and a monoclonal antibody to the transferrin receptor. Adv Drug Del Rev. 1993;10:205–45.

24. Kumagai AK, Eisenberg J, Pardridge WM. Absorptive-mediated endocytosis of cationized albumin and a beta-endorphin-cationized albumin chimeric peptide by isolated brain capillaries: Model system of blood-brain barrier transport. J Biol chem. 1987;262:15214–9.

25. Pardridge WM, Triguero D, Buciak JL. Beta-endorphin chimeric peptides: Transport through the blood-brain barrier in vivo and cleavage of disulfide linkage by brain. J Endocrinol. 1990;126:977–84.

26. Jefferies WA, Brandon MR, Hunt SV, Williams AF, Gatter KC, Mason DY. Transferrin receptor on endothelium of brain capillaries. Nature. 1984;3:162–3.

27. Yoshikawa T, Pardridge WM. Biotin delivery to brain with a covalent conjugate of avidin and a monoclonal antibody to the transferrin receptor. J Pharmacol Exp Ther. 1992;263:897–903.

28. Friden PM, Walus LR, Watson P, Doctrow SR, Kozarich JW, Backman C, et al. Blood-brain barrier penetration and in vivo activity of an NGF conjugate. Science. 1993;259:373–7.

29. Pardridge WM, Boado RJ, Kang YS. Vector-mediated delivery of a polyamide ("peptide") nucleic acid analogue through the blood-brain barrier in vivo. Proc Natl Acad Sci USA. 1995;92:5592–6.[PMC free article]

30. Huwyler J, Wu D, Pardridge WM. Brain drug delivery of small molecules using immunoliposomes. Proc Natl Acad Sci USA. 1996;93:14164–9. [PMC free article]

31. Lindgren M, Hallbrink M, Prochiantz A, Langel U. Cell penetrating peptides. Trends Pharmacol Sci.2000;21:99–103.

32. Misra A, Jain NK. Progress in controlled and novel drug delivery systems. 1st ed. New Delhi: CBS Publishers and Distributors; 2004. pp. 23–47.

33. Kreuter J, Alyautdin RN, Kharkevich DA, Ivanor AA. Passage of peptides through the blood- brain barrier with colloidal polymer particles nanoparticles. Brain Res. 1995;674:171–4.

34. Kreuter J. Nanoparticlate systems for brain delivery of drugs. Adv Drug Deliv Rev. 2001;47:65–81.

35. Kreuter J. Transport of drugs across the blood-brain barrier by nanoparticles. Curr Med Chem.2002;2:241–9.

36. Zordan-Nudo T, Ling , Liv Z, Georges E. Effect of nonionic detergents on P-glycoprotein drug binding and reversal of multidrug resistance. Cancer Res. 1993;53:5994–6000.

37. Luca C, Francesca G, Giovanni T, Francesco R, Vandelli MA, Flavio F. Peptide-derivatized biodegradable nanoparticles able to cross the blood-brain barrier. J Control Release. 2005;108:84–96.

38. Yung-Chih K, Hung-Hao C. Effect of nanoparticulate polybutylcyanoacrylate and methylmethacrylate-sulfopropylmethacrylate on the permeability of zidovudine and lamivudine across the in vitro blood-brain barrier. Pharm Nanotechnol. 2006;327:160–9.

39. Gulati A, Nath C, Shanker K, Srimal , RC , Dhawan KN, Bhargava KP. Effect of alcohols on the permeability of blood-brain barrier. Pharmacol Res Commun. 1985;1:85–93.

40. Borisenko SA. Effects of drugs on blood-brain barrier permeability in rats chronically intoxicated by ethanol. Ann 1st Super Sanita. 1990;26:39–42.

41. Mayordomo F, Renau-Piqueras J, Megias L, Guerri C, Iborra FJ, Azorin I, et al. Cytochemical and stereological analysis of rat cortical astrocytes during development in primary culture: Effect of prenatal exposure to ethanol. Int J Dev Biol. 1992;36:311–21.

42. Kroll RA, Neuwelt EA. Outwitting the blood-brain barrier for therapeutic purposes: Osmotic opening and other means. Neurosurgery. 1998;42:1083–99.

43. Kroll RA, Pagel MA, Muldoon LL, Roman-Goldstein S, Fiamengo SA, Neuwelt EA. Improving drug delivery to intra cerebral tumor and surrounding brain in a rodent model: A comparison of osmotic versus Bradykinin modification of the blood-brain and/or blood-tumor barriers. Neurosurgery. 1998;43:879–86.

44. Bullard DE, Bigner SH, Bigner DD. Comparison of intravenous versus intracarotid therapy with 1,3-bis(2-chloroethyl)-1-nitrosourea in a rat brain tumor model. Cancer Res. 1985;45:5240–5.

45. Inamura T, Nomura T, Bartus RT, Black KL. Intracarotid infusion of RMP-7, a bradykinin analog: A method for drug delivery to brain tumors. J Neurosurg. 1994;81:752–8.

46. Chio CC, Baba T, Black KL. Selective blood-tumor barrier disruption by leukotrienes. J Neurosurg.1992;77:407–10.

47. Shivers RR, Wijsman JA. Blood-brain barrier permeability during hyperthermia. Prog Brain Res.1998;115:413–24.

48. Cho C W, Lin Y, Cobb WN. Ultrasound-induced mild hyperthermia as a novel approach to increase drug uptake in brain micro vessel endothelial cells. Pharm Res. 2002;115:413–24.

49. Thorne RG, Frey WH. Delivery of neurotrophic factors into the central nervous system. Clin Pharmacokinet. 2001;40:907–46.

50. Illum L. Transport of drugs from the nasal cavity to the central nervous system. Eur J Pharm Sci.2000;11:1–18.

51. Mathison S, Nagilla R, Bhaskar KU. Nasal route for direct delivery of solutes to the central nervous system: Fact or fiction? J Drug Target. 1998;5:415–41.

52. Thorne RG, Emory CR, Ala TA, Frey WH. Brain, Quantitative analysis of the olfactory pathway for drug delivery to the brain. Brain Res. 1995;692:278–82.

53. Balin BJ, Broadwell RD, Salcman M, El-Kalliny M. Avenues for entry of peripherally administered protein to the central nervous system in mouse, rat, and squirrel monkey. Comp Neurol. 1986;251:260–80.

54. Baker H, Spencer RF. Transneuronal transport of peroxidase-conjugated wheat germ agglutinin (WGA-HRP) from the olfactory epithelium to the brain of the adult rat. Exp Brain Res. 1986;63:461–73.

55. Broadwell RD, Balin BJ. Endocytic and exocytic pathways of the neuronal secretory process and trans-synaptic transfer of wheat germ agglutinin-horseradish peroxidase in vivo. J Comp Neurol. 1985;242:632–50.

56. Shipley MT. Transport of molecules from nose to brain. Brain Res Bull. 1985;15:129–42.

57. Kida S, Pantazis A, Weller RO. CSF drains directly from the sub arachnoid space into nasal lymphatics in the rat -Anatomy, histology and immunological significance. Neuropathol Appl Neurobiol. 1993;19:480–8.

58. Foldi M. The brain and the lymphatic system. Lymphol. 1996;29:1–9.

59. Feron VJ, Arts JH, Kuper CF, Slootweg PJ, Woutersen RA. Health risks associated with inhaled nasal toxicants. Crit Rev Toxicol. 2001;31:313–47.

60. Sakane T, Akizuki M, Yamashita S, Sezaki H, Nadai T. Direct drug transport from the rat nasal cavity to the cerebrospinal fluid: The relation to the dissociation of the drug. J Pharm Pharmacol. 1994;46:378–9.

61. Sakane T, Akizuki M, Taki Y, Yamashita S, Sezaki H, Nadai T. Direct drug transport from the rat nasal cavity to the cerebrospinal fluid: The relation to the molecular weight of drugs. J Pharm Pharmacol.1995;47:379–81.

62. Harbaugh RE, Saunders RL, Reedert RF. Use of implantable pumps for central nervous system drug infusions to treat neurological disease. Neurosurgery. 1988;23:693–8.

63. Sladek J, Gash D. Nerve-cell grafting in Parkinson's disease. J Neurosurg. 1988;68:337–51.

64. Snyder EY, Senut MC. The use of non neuronal cells for gene delivery. Nerobiol Dis. 1997;4:69–102.

65. Yurek DM, Sladek JR. Dopamine cell replacement: Parkinson's disease. Annu Rev Neurosci.1990;13:415–40.

66. Cohen KA, Liu T, Bissonette R, Puri RK, Frankel AE. DAB389EGF fusion protein therapy of refractory Glioblastoma Multiforme. Curr Pharm Biotechnol. 2003;4:39–49.

67. Asoh S, Ohsawa I, Mori T, Katsura K, Hiraide T, Katayama Y, et al. Protection against ischemic brain injury by protein therapeutics. Proc Natl Acad Sci USA. 2002;99:17107–12. [PMC free article]

68. Alexi T, Venero JL, Hefti F. Protective effects of neurotrophin-4/5 and transforming growth factor-alpha on striatal neuronal phenotypic degeneration after excitotoxic lesioning with quinolinic acid. Neuroscience.1997;78:73–86.

69. Susan A. Biotechnology, brain and future. Trends Biotechnol. 2005;23:34–41.

70. Banks WA. Oligonucleotides targeting prion diseases. J Pharmacol Exp Ther. 2001;3:1113–21.

Chapter 2

Ocular Drug Delivery Systems: An Overview

Ashaben Patel, Kishore Cholkar, Vibhuti Agrahari, and Ashim K Mitra

[1] *Division of Pharmaceutical Sciences, School of Pharmacy, University of Missouri-Kansas City, Kansas City, Missouri-64108, United States*

ABSTRACT

The major challenge faced by today's pharmacologist and formulation scientist is ocular drug delivery. Topical eye drop is the most convenient and patient compliant route of drug administration, especially for the treatment of anterior segment diseases. Delivery of drugs to the targeted ocular tissues is restricted by various precorneal, dynamic and static ocular barriers. Also, therapeutic drug levels are not maintained for longer duration in target tissues. In the past two decades, ocular drug delivery research acceleratedly advanced towards developing a novel, safe and patient compliant formulation and drug delivery devices/techniques, which may surpass these barriers and maintain drug levels in tissues. Anterior segment drug delivery advances are witnessed by modulation of conventional topical solutions with permeation and viscosity enhancers. Also, it includes development of conventional topical formulations such as suspensions, emulsions and ointments. Various nanoformulations have also been introduced for anterior segment ocular drug delivery. On the other hand, for posterior ocular delivery, research has been immensely focused towards development of drug releasing devices and nanoformulations for treating chronic vitreoretinal diseases. These novel devices and/or formulations may help to surpass ocular barriers and associated side effects with conventional topical drops. Also, these novel devices and/or formulations are easy to formulate, no/negligibly irritating, possess high precorneal residence time, sustain the drug release, and enhance ocular bioavailability of therapeutics. An update of current research advancement in ocular drug delivery necessitates and helps drug delivery scientists to modulate their think process and develop novel and safe drug delivery strategies. Current review intends to summarize the existing conventional formulations for ocular delivery and their advancements followed by current nanotechnology based formulation developments. Also, recent developments with other ocular drug delivery strategies employinggels, implants, contact lens and microneedles have been discussed.

Keywords: Anatomy and physiology; Cornea; Contact lens; Drug delivery; Eye; Emulsions; Formulations; Implants; Liposomes; Nanomicelles; Ointments; Retina; Suspensions

INTRODUCTION

The eye is a complex organ with an unique anatomy and physiology. The structure of eye can be divided into two main parts: anterior segment and posterior segment (Figure 1). Anterior segment of the eye occupies approximately one-third while the remaining portion is occupied by the posterior segment. Tissues such as cornea, conjunctiva, aqueous humor, iris, ciliary body and lens make up the anterior portion. Back of the eye or posterior segment of the eye include sclera, choroid, retinal pigment epithelium, neural retina, optic nerve and vitreous humor. The anterior and posterior segment of eye is affected by various vision threatening diseases. Diseases affecting anterior segment include, but not limited to glaucoma, allergic conjunctivitis, anterior uveitis and cataract. While, age-related macular degeneration (AMD) and diabetic retinopathy are the most prevalent diseases affecting posterior segment of the eye.

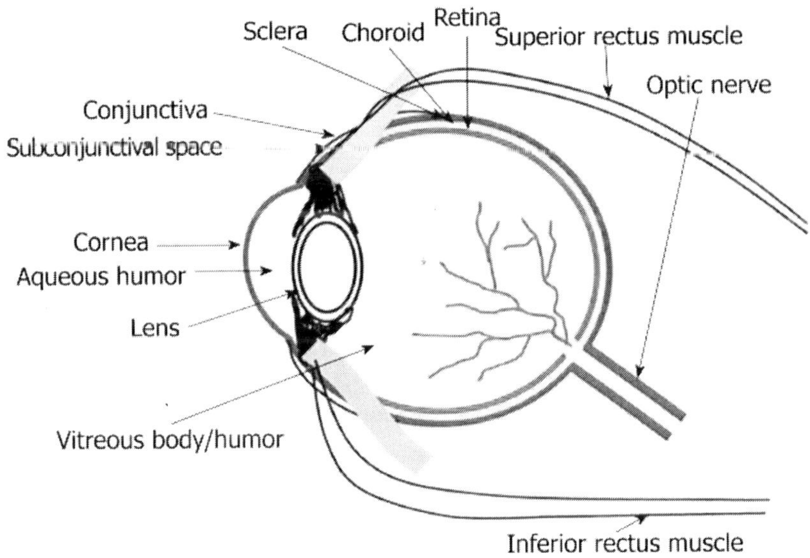

Figure 1. Structure of the eye.

Topical instillation is the most widely preferred non-invasive route of drug administration to treat diseases affecting the anterior segment. Conventional dosage forms such as eye drops account for 90% of the marketed ophthalmic formulations. The reason may be attributed to ease of administration and patient compliance[1,2]. Nonetheless, the ocular bioavailability is very low with topical drop administration. Numerous anatomical and physiological constraints such as tear turnover, nasolachrymal drainage, reflex blinking, and ocular static and dynamic barriers pose a challenge and impede deeper ocular drug permeation[3]. Hence, less than 5% of topically applied dose reaches to deeper ocular tissues[4]. Also, it is difficult to achieve therapeutic drug concentration into posterior segment ocular tissues following topical eye drops instillation because of the above mentioned barriers. The drug can be delivered to the posterior segment ocular tissues by different mode of administrations such as intravitreal injections, periocular injections, and systemic administration. However, small volume of eye compared to whole body and presence of blood retinal barriers; makes systemic administration an impractical approach. Intravitreal injection is the most common and widely recommended route of drug administration to treat posterior ocular diseases. Though, the need of repeated eye puncture with intravitreal injections causes several side effects such as endophthalmitis, hemorrhage, retinal detachment and poor patient tolerance[5]. The transscleral drug delivery with periocular administration route is evolved as an alternative mode of drug delivery to the posterior ocular tissues. Although transscleral delivery is comparatively easy, less invasive and patient compliant, drug permeation is compromised by ocular static and dynamic barriers. Ocular barriers to transscleral drug delivery include: static barriers *i.e.,* sclera, choroid and retinal pigment epithelium (RPE), and dynamic barriers, *i.e.,*lymphatic flow in the conjunctiva and episclera, and the blood flow in conjunctiva and choroid[6,7].

To overcome the ocular drug delivery barriers and improve ocular bioavailability, various conventional and novel drug delivery systems have been developed such as emulsion, ointments, suspensions, aqueous gels, nanomicelles, nanoparticles, liposomes, dendrimers, implants, contact lenses, nanosuspensions, microneedles, and *in situ* thermosensitive gels for the earlier mention ocular diseases. This review will provide an overview on various conventional and novel ophthalmic drug delivery systems developed to deliver drug to diseased ocular tissues for the treatment of ocular diseases.

CONVENTIONAL OCULAR DRUG DELIVERY SYSTEMS

Topical drop instillation into the lower precorneal pocket is a patient compliant and widely recommended route of drug administration. However, most of the topically administered dose is lost due to reflux blinking and only 20% (~7 μL) of instilled dose is retained in the precorneal pocket[8]. Concentration of drug available in the precorneal area acts as a driving force for its passive diffusion across cornea. However, for efficient ocular drug delivery with eye drops, high corneal permeation with longer drug cornea contact time is required. Several efforts have been made toward improving precorneal residence time and corneal penetration. To improve corneal permeation iontophoresis, prodrugs, ion-pair forming agents and cyclodextrins are employed[9–13]. There is a wide range of ophthalmic products available in the market out of which around 70% of prescriptions include conventional eye drops. The reasons may be due to ease of bulk scale manufacturing, high patient acceptability, drug product efficacy, stability and cost effectiveness.

Topical liquid/solution eye drops

Topical drops are the most convenient, safe, immediately active, patient compliant and non-invasive mode of ocular drug administration. An eye drop solution provides a pulse drug permeation post topical drop instillation, after which its concentration rapidly declines. The kinetics of drug concentration decline may follow an approximate first order. Therefore, to improve drug contact time, permeation and ocular bioavailability; various additives may be added to topical eye drops such as viscosity enhancers, permeation enhancers and cyclodextrins. Viscosity enhancers improve precorneal residence time and bioavailability upon topical drop administration by enhancing formulation viscosity. Examples of viscosity enhancers include hydroxy methyl cellulose, hydroxy ethyl cellulose, sodium carboxy methyl cellulose, hydroxypropyl methyl cellulose and polyalcohol[14–16].

Permeation enhancers improve corneal uptake by modifying the corneal integrity. Other additives such as chelating agents, preservatives, surface active agents and bile salts were studied as possible permeation enhancers. Benzalkonium chloride, polyoxyethylene glycol ethers (lauryl, stearyl and oleyl), ethylenediaminetetra acetic acid sodium salt, sodium taurocholate, saponins and cremophor EL are the examples of permeation enhancers investigated for improving ocular delivery[17–19]. Addition of permeation enhancers to ocular solutions improves ocular drug bioavailability but few studies revealed a local toxicity with permeation enhancers[20]. Hence, research is still being conducted to modify the effect of permeation enhancers and evaluate their safety on corneal tissues. Hornof *et al*[21] evidenced that polycarbophil-cysteine as an excipient did not damage the corneal tissue integrity and suggested that it could be safe for ocular formulations. Cyclodextrins act as carriers for hydrophobic drug molecules in aqueous solution. This helps to deliver drugs to the surface of biological membrane. Highly lipophilic biological membrane has much lower affinity towards hydrophilic cyclodextrins. Therefore, cyclodextrins remain in aqueous solution and the hydrophobic drug is absorbed by the biological membrane. Optimal bioavailability was achieved for eye drops with cyclodextrins concentration of < 15%[22]. Other applications of cyclodextrins in eye drop formulation were recently reviewed and described in detail elsewhere by Cholkar *et al*[23].

Among these approaches, viscosity enhancers and cyclodextrins suffer from the disadvantage of precorneal loss. In the case of penetration enhancers, care should be taken in the selection due to high sensitivity of ocular tissues. Hence, it leads to development of other conventional formulations approaches with inert carrier systems for ocular delivery of therapeutics. Conventional ocular formulations such as emulsions, suspensions, and ointments are developed to improve solubility, precorneal residence time and ocular bioavailability of drugs. In the current era of nanotechnology, these conventional formulations still retain their place, importance and capture the market at large. However, these formulations are associated with various side effects such as ocular irritation, redness, inflammation, vision interference and stability issues[24]. Currently, research is being conducted to improve *in vivo* performance of these carrier systems and to minimize their side effects[25]. Several attempts are also being made to deliver drugs to posterior ocular tissues with conventional formulations. In the following sections, attempts have been made to describe the recent efforts made to improve *in vivo* performance of conventional ocular formulation and reduce their side effects.

Emulsions

An emulsion based formulation approach offers an advantage to improve both solubility and bioavailability of drugs. There are two types of emulsions which are commercially exploited as vehicles for active pharmaceuticals: oil in water (o/w) and water in oil (w/o) emulsion systems[26]. For ophthalmic drug delivery, o/w emulsion is common and widely preferred over w/o system. The reasons include less irritation and better ocular tolerance of o/w emulsion. Restasise™, Refresh Endura® (a non-medicated emulsion for eye lubrication) and AzaSite® are the examples of currently marketed ocular emulsions in the United States. Several studies have demonstrated applicability of emulsions in improving precorneal residence time, drug corneal permeation, providing sustain drug release and thereby enhancing ocular bioavailability[27].

In a recent study, Tajika *et al*[28] demonstrated improved anti-inflammatory activity of prednisolone derivative, 0.05% [^{3}H] difluprednate, with emulsion as vehicle. Results confirmed that in the rabbit eye, emulsion could deliver drug to the anterior ocular tissues with small amount of drug reaching posterior tissues following single and multiple topical drop instillation. Single and multiple topical drop instillation studies revealed highest radioactivity in cornea followed by iris-ciliary body > retina-choroid > conjunctiva > sclera > aqueous humor > lens > and vitreous humor. Post single drop administration, T_{max} for cornea, conjunctiva, lens, iris-ciliary body, aqueous and vitreous humor was 0.5 h while for retina-choroid was 1 h. Negligible amount of drug was quantified in systemic circulation. With repeated dose instillation, T_{max} for lens and retina-choroid was 8 and 0.5 h, respectively. After 168 h, a total dose of approximately 99.5% of radioactivity was excreted in urine and feces. This study suggests difluprednate emulsion as a potential candidate for treating anterior ocular inflammations.

Emulsions with lipid additives such as soyabean lecithin, stearylamine were evaluated as carrier systems for azithromycin to demonstrate better ocular performance and bioavailability[29]. A comparative study for azithromycin solution *vs* emulsion at different doses (3, 5 and 10 mg/mL azithromycin) was studied for tear elimination characteristics. *In vivo* studies were conducted in rabbits with topical drop administration. Emulsion, not only observed to behave as a vehicle for azithromycin but also slowed drug release, improved its chemical stability and precorneal residence time. Additionally, emulsion formulation improved the chemical stability ($t_{1/2}$) of azithromycin at pH 5.0 and 7.0 relative to aqueous solutions. Altogether, results suggest that lipid emulsion could be a promising vehicle for ocular drug delivery.

Similarly, another novel approach is to derivatize active pharmaceutical ingredients (API), and improve its ocular bioavailability with an emulsion as carrier system. This strategy may help to reduce ocular irritancy and improve the effect of API. To test this hypothesis, Shen *et al*[25] made attempts to improve emulsion biocompatibility for the flurbiprofen. In this study, a derivative of flurbiprofen, flurbiprofen axetil, with castor oil and tween-80 was used to prepare emulsion[30]. Four different emulsions with varying ratios of castor oil (0.1 wt%–2.5 wt%) and tween 80 (0.08 wt%–4 wt%) were prepared and labeled as F1, F2, F3 and F4 respectively. *In vivo* studies were conducted in male New Zealand albino rabbits with a topical drop instillation. Aqueous humor pharmacokinetic studies showed F2 emulsion (castor oil to tween 80 wt% ratio of 0.5:0.4) to be better relative to other emulsion formulations and solution. The F2 emulsion translocated high drug concentrations into aqueous humor, post topical drop administration, relative to 0.03% flurbiprofen sodium eye drops (Figure 2). Similarly, ocular irritation studies with F2 emulsion demonstrated better biocompatibility relative to other emulsions (F1, F3 and F4).

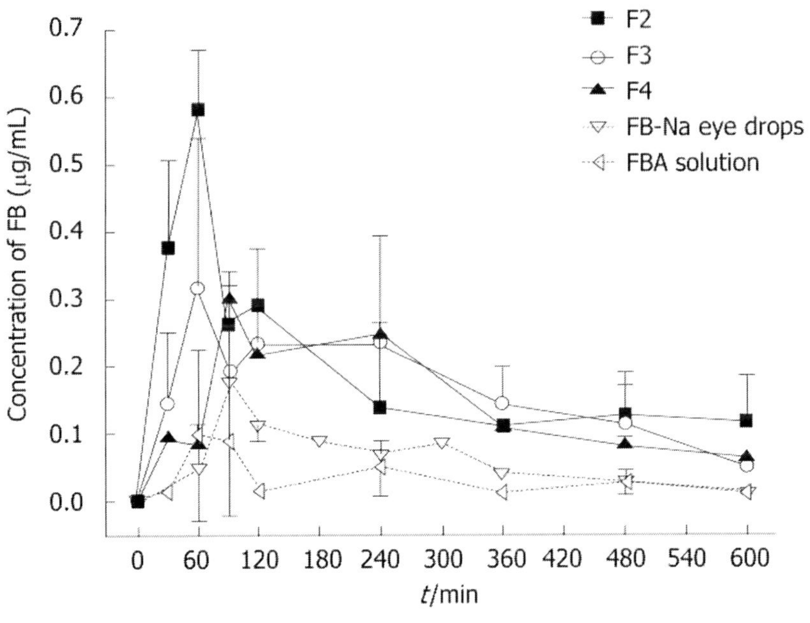

Figure 2. Concentration-time profiles of flurbiprofen (in the aqueous humor after instillation of flurbiprofen axetil emulsion F2-F4, FB-Na eye drops and flurbiprofen axetil-oil solution in rabbits

Several researchers have introduced mucoadhesive polymers such as chitosan and hydroxypropyl methyl cellulose ether for emulsion coating. Studies concluded that chitosan surface coating improves precorneal residence time of

API and thereby ocular bioavailability. Indomethacin loaded o/w emulsion was prepared employing castor oil and polysorbate-80 and the resultant emulsion was surface coated by chitosan[31]. A comparative *in vivo* study for chitosan coated *vs* non-coated indomethacin emulsions were conducted in male albino rabbits with topical drop instillation. Tear fluid pharmacokinetic study showed that emulsion surface coating with chitosan improves emulsion mean residence time and also half-life by 1.5 and 1.8 times, respectively relative to non-coated emulsion. Indomethacin concentrations were quantified in cornea, conjunctiva and aqueous humor, post 1 h of emulsion instillation. Indomethacin concentrations with emulsion system were found to be about 5.3 and 8.2 times higher in cornea relative to conjunctiva and aqueous humor.

Suspensions

Suspensions are another class of non-invasive ocular topical drop drug carrier systems. Suspension may be defined as dispersion of finely divided insoluble API in an aqueous solvent consisting of a suitable suspending and dispersing agent. In other words, the carrier solvent system is a saturated solution of API. Suspension particles retain in precorneal pocket and thereby improve drug contact time and duration of action relative to drug solution. Duration of drug action for suspension is particle size dependent. Smaller size particle replenishes the drug absorbed into ocular tissues from precorneal pocket. While on the other hand, larger particle size helps retain particles for longer time and slow drug dissolution[32]. Thus, an optimal particle size is expected to result in optimum drug activity. Several suspension formulations are marketed worldwide to treat ocular bacterial infections. TobraDex® suspension is one of the widely recommended commercial products for subjects responding to steroid therapy. TobraDex® is a combination product of antibiotic, tobramycin (0.3%), and steroid, dexamethasone (0.1%). The major drawback of this commercial product is high viscosity. Recently, Scoper *et al*[33] made attempts to reduce the viscosity of TobraDex® and to improve its *in vivo* pharmacokinetics along with bactericidal activity. The rationale behind developing this formulation was to improve the suspension formulation characteristics such as quality, tear film kinetics and tissue permeation. The new suspension (TobraDex ST®) consists of tobramycin (0.3%), and steroid, dexamethasone (0.05%). Suspension settling studies showed that new formulation had very low settling over 24 h (3%) relative to marketed Tobra-Dex® (66%). Ocular distribution studies showed higher tissues concentrations of dexamethasone and tobramycin in rabbits treated with TobraDex ST® relative to Tobra-Dex®. New suspension formulation was found to be more effective than TobraDex® against Staphylococcus aureus and Pseudomonas aeruginosa. Clinical studies in human subjects showed high dexamethasone concentrations in aqueous humor than TobraDex®. These results suggest that new suspension formulation to be an alternative to marketed suspension. This is because the new suspension possesses better formulation characteristics, pharmacokinetics, bactericidal characteristic and patient compliance than marketed TobraDex® suspension.

In another study, to treat dry eye, 4 wk, randomized, double masked, multicenter phase II clinical trials were conducted with rebamipide (OPC-12759) suspension[34]. Suspension formulation at two different doses, *i.e.,*1% and 2% rebamipide were employed for this study, where placebo served as control. The efficacy and safety of suspension formulation were determined in human subjects following topical instillation. A dose dependent response was observed for placebo, 1% and 2% rebamipide suspension for both fluorescein corneal staining and Lissamine green conjunctival staining studies at 2 and 4 wk. Tear production showed no significant difference from baseline from day 1 to week 4. But, the tear film break up time showed significant change in 1% and 2% rebamipide relative to placebo. All the subjects receiving treatment with suspension rebamipide formulation reported improvement of 64.1% and 54.9% respectively than subjects receiving placebo. Dysgeusia, ocular irritation and nasopharyngitis adverse events were frequently observed in 27.2%, 29.1% and 30.4% patients receiving placebo, 1% and 2% suspension, respectively. Drug induced adverse effects such as eye irritation was observed in 3.9%, 2.9% and 2.0% subjects receiving placebo, 1% rebamipide and 2% rebamipide respectively. All these adverse effects were found to recover without any additional treatment. This 4 wk studies revealed that suspension formulations were well tolerated and both formulations were effective in treating dry eye. In some measures, of the two formulations, 2% rebamipide suspension was found to be more effective relative to 1% suspension.

Ointments

Ophthalmic ointments are another class of carrier systems developed for topical application. Ocular ointment comprises of mixture of semisolid and a solid hydrocarbon (paraffin) that has a melting point at physiological ocular temperature (34 °C). The choice of hydrocarbon is dependent on biocompatibility. Ointments help to improve ocular bioavailability and sustain the drug release[35].

Vancomycin HCl (VCM) is a glycopeptides antibiotic with an excellent activity against aerobic and anaerobic gram positive bacteria and methicillin and cephem resistant *Staphylococcus aureus* (MRSA). Inspite of better activity of VCM, no appropriate topical formulation was available in the market. Better ocular tissue permeability of VCM was not expected in a normal eye but few clinical effects of VCM solution were reported in ocular disease treatment. The reason for the observed effects was hypothesized due to broken ocular barrier system, which might have improved drug permeation. Fukuda *et al*[36] studied the intraocular dynamics of vancomycin hydrochloride ophthalmic ointments in rabbits. Thus, authors made attempts to demonstrate ocular dynamics of VCM ophthalmic ointment (TN-011) with indications limited to extraocular MRSA infections. The minimum growth inhibitory concentration to treat MRSA bacterial infections was found to be 1.56 µg/g. *In vivo* studies were conducted in rabbits [normal *vs Bacillus subtilis*(BS) group]. The BS group was developed in cornea by injecting BS solution into the central portion of parenchyma. Treatment was by topical ocular ointment (1% VCM) administration to normal and BS group rabbit eye. In normal group, after 15 min, VCM concentration in cornea of 12.04 ± 4.73 µg/g was attained at 30 min which was decreased to 0.49 ± 0.97 µg/g at 120 min. On the other hand, VCM concentrations in BS group cornea was 25.60 ± 11.01 µg/g after 15 min and 3.68 ± 1.38 µg/g after 240 min of administration. The concentrations of VCM were maintained above MIC levels, in MRSA infection induced BS group, a considerable benefit to the patients from TN-011 is expected.

In another study by Eguchi *et al*[37], four different ointment formulation of vancomycin with varying concentrations (0.03%, 0.10%, 0.30% and 1.00%) were prepared in 1:4 mixtures of liquid paraffin and vaseline. The efficacy of formulations was evaluated in rabbit model of MRSA keratitis infection after topical application. It was observed that at low drug concentrations, *i.e.,* 0.03% and 0.10%, numerous infiltrates were found in corneas with abscesses. On the other hand, animals treated with 0.3% formulation showed no recurrence of keratitis in any eye over 14 d study period. Therefore, 0.3% vancomycin ointment was suggested to be adequate and effective to resolve corneal MRSA keratitis.

Though considerable effort is being put into research to improve efficacy, still there is a need to overcome certain drawbacks associated with conventional formulations. The above mentioned formulations: emulsion, suspension, and ointment are known to cause ocular adverse effects such as irritation, redness of eye and interference with vision. Also, chronic administration may increase systemic API availability which may lead to severe systemic complications[38–40]. Formulations with preservatives also induce adverse reactions upon systemic absorption[41,42]. Therefore, to overcome formulation based adverse effects and to deliver therapeutic amounts of drug in ocular tissues, research is now being focused on exploring and developing other novel strategies of ocular drug delivery. In the following sections, we have discussed about the recent developments made in nanotechnology and controlled release devices in past decade to improve ocular drug delivery.

NOVEL OCULAR DRUG DELIVERY SYSTEMS

Nanotechnology based ocular drug delivery

In a last few decades, many approaches have been utilized for the treatment of ocular diseases. Nanotechnology based ophthalmic formulations are one of the approaches which is currently being pursued for both anterior, as well as posterior segment drug delivery. Nanotechnology based systems with an appropriate particle size can be designed to ensure low irritation, adequate bioavailability, and ocular tissue compatibility. Several nanocarriers, such as nanoparticles, nanosuspensions, liposomes, nanomicelles and dendrimers have been developed for ocular drug delivery (Figure 3). Some of them have shown promising results for improving ocular bioavailability.

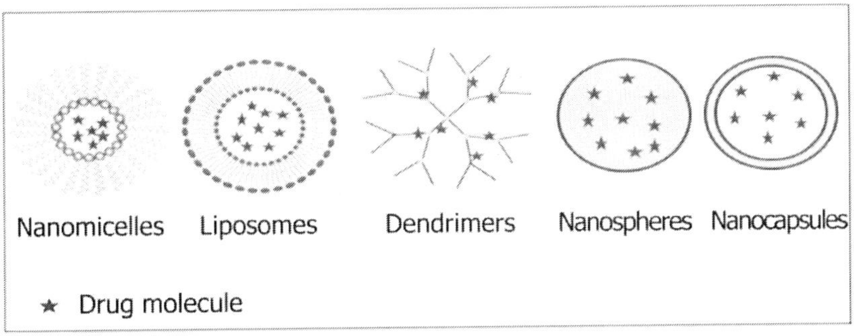

Figure 3. Nanocarriers for ocular drug delivery.

Nanomicelles

Nanomicelles are the most commonly used carrier systems to formulate therapeutic agents in to clear aqueous solutions. In general, these nanomicelles are made with amphiphilic molecules. These molecules may be surfactant or polymeric in nature. Recently, Cholkar *et al*[43] have reviewed in detail about ocular barriers and application of nanomicelles based technology in ocular drug delivery.

Currently, tremendous interest is being shown towards development of nanomicellar formulation based technology for ocular drug delivery. The reasons may be attributed due to their high drug encapsulation capability, ease of preparation, small size, and hydrophilic nanomicellar corona generating aqueous solution. In addition, micellar formulation can enhance the bioavailability of the therapeutic drugs in ocular tissues, suggesting better therapeutic outcomes. So far, several proofs of concept studies have been conducted to investigate the applicability of nanomicelles in ocular drug delivery. For instance, Civiale *et al*[44] developed dexamethasone loaded nanomicelles by employing copolymers of polyhydroxyethylaspartamide [PHEAC(16)] and pegylated PHEAC(16) for anterior segment delivery. *In vivo* dexamethasone concentration time profiles were studied and determined in rabbits with aqueous humor sampling. Results showed that dexamethasone loaded PHEA micelles have higher ocular bioavailability relative to dexamethasone suspension. The area under the curve for dexamethasone micellar formulation was 40% higher than that of control suspension. Results suggest that nanomicellar formulations are a viable option for topical ocular delivery of small molecules. Researchers have also utilized nanomicelles for ocular gene delivery. In a study, Liaw *et al*[45] made attempts to deliver genes by topical drop administration to cornea. Copolymer, poly (ethylene oxide)-poly (propylene oxide)-poly (ethylene oxide) (PEO-PPO-PEO) was used to develop micelles as a vehicle for gene delivery. This polymeric system efficiently transferred plasmid DNA with *LacZ* gene in rabbit and mice ocular tissues. Results were promising and indicated the potential application of copolymers in DNA transfer. Further studies were conducted with the copolymer to deliver two cornea specific promoters, *i.e.,* keratin 12 (K12) and keratocan. Transgene expression was quantified with β-Gal activity. Significant elevated levels were quantified following six doses of eye drop of pK12-Lac Z-PM three times a day in both mouse and rabbit corneas. The probable mechanism of transfection was endocytosis and particle size dependent paracellular transport of polymeric micelles[46].

Several attempts are also being made to utilize nanomicelles for the posterior ocular drug delivery. Recently, the authors have made a significant stride to deliver therapeutic drugs to the posterior ocular tissues with the aid of topical drops of mixed nanomicellar formulations. To bolster the hypothesis that the nanomicelles can deliver the drug to the posterior ocular tissues, *in vivo* studies were carried out in rabbits using voclosporin loaded nanomicelles[43]. Interestingly, the nanomicelle formulations were able to efficiently traverse ocular tissues and deliver drug to back of the eye tissues. Ocular tolerability of nanomicelles was evaluated against Restasis® as control in New Zealand White (NZW) rabbits. A detailed 72 h study with Hackett-McDonald scoring with microscopic ocular examination was included for two voclosporin (0.02% and 0.2%) micellar and Restasis® formulations. Post 1 h-topical drop administration of Restasis® highest ocular irritation was observed relative to two micellar voclosporin formulations. It was demonstrated that the novel mixed nanomicellar formulations were well tolerated and induced markedly low irritation than Restasis®. Further, authors also prepared dexamethasone and rapamycin mixed nanomicellar formulations at a concentration of 0.1 and 0.2 wt%, respectively. Ocular tissue distribution studies with single drop instillation showed that nanomicellar formulation encapsulating voclosporin, dexamethasone and rapamycin was able to deliver therapeutic concentrations of drug to back of the eye tissues post topical drop instillation. These studies suggest that small size, hydrophilic nanomicellar corona help to evade ocular barriers and deliver drug cargo to posterior ocular tissues. A non-corneal pathway of drug delivery has been hypothesized for back of the eye drug delivery. Ideta *et al*[47] made attempts to deliver fluorescein isothiocyanate-labeled poly-L-lysine [FITC-P(Lys)] to back of the eye tissues *via* intravenous drug administration to treat back of the eye tissue neovascularization. *In vivo* studies with unformulated FITC-P(Lys) resulted in death of animals post 1 h of administration. On the contrary encapsulating the FITC-P(Lys) in polyehthylene glycol-block-poly-α,β-aspartic acid micelles resulted in no death. This indicates no free drug was available in nanomicellar formulation. Micellar formulation showed a C_{max} at 4 h in retina-choroid and drug was detected up to 7 d following single intravenous administration. Prolonged micellar circulation was achieved by controlling polymer to drug charge ratios. Authors speculated that longer systemic micellar circulation may aid in enhanced permeation and retention (EPR) effect at neovascularization site. Micellar constructs were observed to selectively accumulate at the pathologic neovascular site to a greater extent than in normal tissues.

In another study, Ideta *et al*[48] made attempts to encapsulate dendritic photosensitizer (DP) in PEG-b-P(Lys) micellar construct for the treatment of exudative AMD with photodynamic therapy. *In vitro* cytotoxicity studies were performed under dark and light irradiation for DP alone and DP loaded polyionic complex (PIC) micelles to be more cytotoxic in light irradiated conditions. This higher cytotoxic effect of polymeric ion complex micelles under light irradiation was utilized for the treatment of exudative AMD. Photocoagulation was induced in rat eye. DP loaded PIC micelles were administered by intravenous injection and DP accumulation in choroidal neovascular site was observed. Application of mild laser light treatment destroyed/choked the abnormal vasculature. This new technology prevents further drug leakage. Histological studies revealed accumulation of PIC micelles at ocular lesion site. Reason may be attributed due to EPR effect. Administered free DP was eliminated within 24 h. On the other hand, PIC micelles encapsulated DP were detected after 24 h indicating micellar construct accumulation at lesion site with slow cell uptake. A reduction in fluorescence was observed post 25 min intravenous administration of DP loaded PIC micelles, due to chocking of abnormal vasculature. Hypofluorescence of DP micelles was increasing with time indicating increased vascular chocking. Normal endothelial cell destruction was not observed, possibly due to lower DP accumulation. Results suggest that small size and hydrophilic negatively charged micellar corona resulted in considerable EPR effect. This resulted in selective drug accumulation in the choroidal neovascular tissues with minimal/no drug induced adverse effects on normal cells.

Ocular research is currently focused to non-invasively deliver therapeutic levels of drugs to both anterior and posterior ocular segments. Advent of nanomicellar technology to delivery drugs in a non-invasive route, topical drop, is gaining interest. Due to their extremely small size and hydrophilic corona, nanomicelles may be retained in systemic circulation for longer time and accumulate at the diseased tissue *via* EPR effect. Thereby, non-specific drug accumulation in to normal tissues may be minimized. Proper selection of surfactant/polymer and engineering technique may aid in delivery of drugs to both anterior and posterior eye segments.

Nanoparticles

Nanoparticles are colloidal carriers with a size range of 10 to 1000 nm. For ophthalmic delivery, nanoparticles are generally composed of lipids, proteins, natural or synthetic polymers such as albumin, sodium alginate, chitosan, poly (lactide-co-glycolide) (PLGA), polylactic acid (PLA) and polycaprolactone. Drug loaded nanoparticles can be nanocapsules or nanospheres (Figure 3). In nanocapsules, drug is enclosed inside the polymeric shell while in nanospheres; drug is uniformly distributed throughout polymeric matrix. From past few decades, nanoparticles have gained attention for ocular drug delivery and several researchers have made attempts to develop drug loaded nanoparticles for delivery to both anterior and posterior ocular tissues (Table 1)[49–58].

Nanoparticles represents a promising candidate for ocular drug delivery because of small size leading to low irritation and sustained release property avoiding frequent administration. However, like aqueous solutions, nanoparticles may be eliminated rapidly from precorneal pocket. Hence, for topical administration nanoparticles with mucoadhesive properties have been developed to improve precorneal residence time[59]. Polyethylene glycol (PEG), chitosan and hyaluronic acid are commonly employed to improve precorneal residence time of nanoparticles.

Chitosan coating is most widely explored for improving precorneal residence of nanoparticles. The chitosan is positively charged and hence it binds to negatively charged corneal surface and thereby improves precorneal residence and decreases clearance. For instance, natamycin loaded chitosan/lecithin nanoparticles exhibited high ocular bioavailability at reduced dose and dosing frequency in rabbit eye compared to marketed suspension. Following topical administration, the concentration-time curve (AUC) (0–8) was increased up to

Table 1. Summary of recent developments with nanoparticles as ocular drug delivery vehicles

Drug	Polymer	Features
Carboplatin	CH, SA	Carboplatin loaded NPs demonstrated elevated and sustained anti-proliferative activity in a retinoblastoma cell line (Y-79), with IC_{50} of 0.56 and 0.001 μg/mL for free carboplatin and carboplatin loaded NPs, respectively[49]
5-FU	CH, SA	CH coated SA-CH nanoparticles (CH-SA-CH NPs) loaded with 5-FU showed significantly higher concentration of 5-FU in aqueous humor as compared to SA-CH 5-FU loaded NPs and 5-FU solution. The higher C_{max} was achieved in case of CH-SA-CH NPs (24.67 μg/mL) compared to 5-FU solution (6.14 μg/mL)[50]
Sparfloxacin	PLGA	After topical application, sparfloxacin-loaded nanoparticles were retained for a longer duration on the corneal surface as compared to an aqueous solution, which was drained rapidly from the corneal surface. Also, *in vitro* release studies revealed an extended release of sparfloxacin[51]
BT	Sodium alginate	BT-loaded nanoparticles provided prolong drug release over a period of 8 h after topical instillation to albino rabbits[52]
Levofloxacin	PLGA	The nanosuspensions was retained for the longer time on rabbit eye surface and drained out slowly compared to marketed formulation. Results of *ex-vivo* transcorneal permeation study across excised goat cornea revealed that levofloxacin from the marketed formulation was permeated 36.9% in 4 h whereas levofloxacin from PLGA nanoparticles was permeated 47.43% in 4 h across cornea[53]
DS	PLGA	An extended DS release was observed from the nanoparticles under *in vitro* conditions. The developed polymer nanoparticles formulation was non-irritant to cornea, iris, and conjunctiva for as long as 24 h after application[54]
Pilocarpine	PLGA	The *in vivo* miosis studies showed that the duration of miotic response increased by 40% for the nanoparticles compared to the eye drops[55]
Gatifloxacin/Prednisolone	Eudragit RS 100 and RL 100, coating with hyaluronic acid	*In vitro* release studies revealed prolonged drug release compared to the free drugs with no burst effect Nanoparticles formulation showed better bioavailability of gatifloxacin in rabbit eye with 1.76 fold increase in C_{max} of gatifloxacin in the aqueous humor in comparison to the eye drops[56]
Cloricromene (AD6)	Eudragit	Nanosuspension enhanced stability of the ester drug for several months as compared to an AD6 aqueous solution[57]
Brimonidine Tartrate	Eudragit RS 100 Eudragit RL 100	The AUC (ΔIOP *vs* time) for the selected nanoparticles formulations were about seven times higher than that of eye drop formulations in rabbit eye[58]

CH: Chitosan; SA: Sodium alginate; 5-FU: 5-Fluorouracil; PLGA: Poly (lactide-co-glycolide); IOP: Intraocular pressure; AUC: Area under the curve; BT: Brimonidine tartrate; DS: Diclofenac sodium.

1.47 fold and clearance was decreased up to 7.40 fold in case of chitosan/lecithin nanoparticles compared to marketed suspension[60]. In another study, Musumeci *et al*[61] reported that melatonin loaded PLGA-PEG nanoparticles were most effective and demonstrated significant intraocular pressure (IOP) lowering effect compared with melatonin loaded PLGA nanoparticles and aqueous solution of equivalent concentration in the rabbit eye (Figure 4). It was speculated that the reduced zeta potential of nanoparticles fabricated from PLGA-PEG than the PLGA allowed better and longer interaction between the nanoparticles and eye surface leading to higher hypotensive effect for prolonged period.

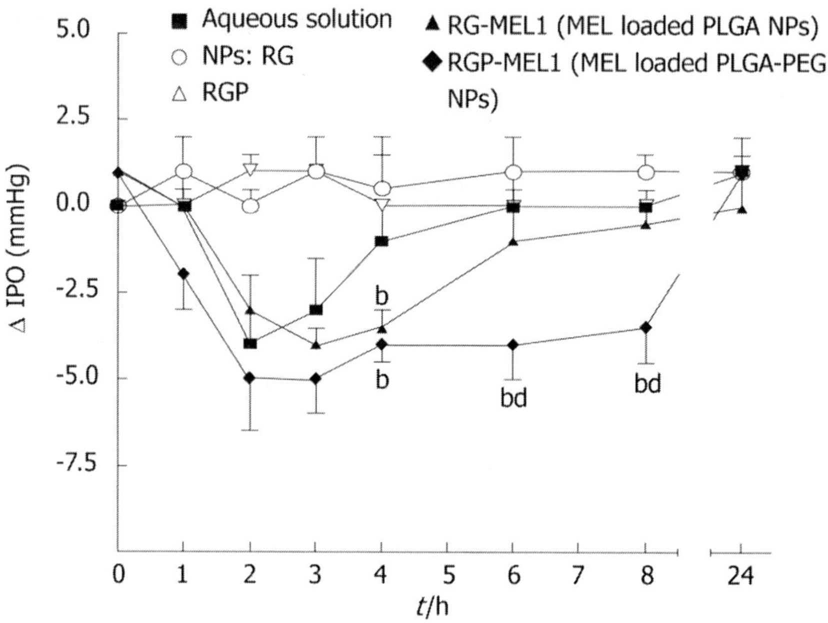

Figure 4. Intraocular pressure in normotensive rabbit eyes after topical instillation of melatonin (MEL)

Nanoparticles have also been successfully employed as an alternative strategy for long term drug delivery to the posterior segment ocular tissues. For posterior segment delivery, disposition of nanoparticles depends on the size and surface property. Following, periocular administration in to Sprague-Dawley rats, 20 nm particles were cleared rapidly from periocular tissues. The rapid clearance can be due to removal by conjunctival, episcleral or other periocular circulatory systems. On the other hand, particles in the range of 200–2000 nm were retained at the site of administration for at least two months. Moreover, due to the rapid clearance and fast drug release, small size nanoparticles could not sustain retinal drug level. Therefore, it can be concluded that for prolonged transscleral drug delivery to the back of the eye, nanoparticles with slow drug release and low clearance by blood and lymphatic circulations are suitable drug delivery candidates[62,63].

Following intravitreal injection, nanoparticles migrate through the retinal layers and tend to accumulate in the RPE cells. The PLA nanoparticles were present in rat RPE tissues up to 4 mo following single intravitreal injection which suggest that nanoparticles have great potential for achieving steady and continuous delivery to the back of the eye. Zhang *et al*[64] investigated the pharmacokinetics and tolerance of dexamethasone (DEX) loaded PLGA nanoparticles in rabbits following intravitreal injection. Authors concluded that DEX when encapsulated in nanoparticles exhibited sustained release for 50 d. The constant DEX levels were maintained in vitreous over 30 d with a mean concentration of 3.85 mg/L. Contrary, only trace amounts of DEX being detected on the 7th day after injection of DEX solution. These results imply that intravitreal injection of dexamethasone nanoparticles may be employed for sustained delivery of drugs for the treatment of posterior segment eye diseases.

The surface property of nanoparticles is a key factor affecting their distribution from vitreous humor to retinal layers[65]. Koo *et al*[66] studied correlation between surface properties of the nanoparticles and their distribution in the vitreous and retina after intravitreal injection. Heterogeneous polyethyleneimine/glycol chitosan (PEI/GC), human serum albumin (HSA)/GC, and HSA/hyaluronic acid (HA) nanoparticles were prepared by blending two polymers. The value of zeta potential of these nanoparticles were 20.7 ± 3.2, −1.9 ± 4.1 and −23.3 ± 4.4 for PEI/GC, HSA/GC, and HSA/HA nanoparticles, respectively. The nanoparticles were injected into vitreous cavity of Long Evans rats and vitreous/retinal distribution was evaluated by confocal microscopy. Figure 5 shows vitreal and retinal distribution of intravitreally administered heterogeneous nanoparticles. It can be depicted from the Figure 5 that PEI/GC nanoparticles easily penetrated the vitreal barrier and reached at the inner limiting membrane. However, PEI/GC nanoparticles did not penetrate through the physical pores of inner limiting membrane into the deeper retinal layers and also some aggregates were observed in vitreous. Similar to PEI/GC nanoparticles, HSA/GC nanoparticles reached to inner limiting membrane but could not penetrate to the deeper retinal layers which might be due to inhibition of the interaction between HSA and the Müller cells in retina by GC. On the other hand, negatively charged HSA/HA nanoparticles, could penetrate the whole retina structures and reach the outer retinal layers such as the photoreceptor layer and RPE which was attributed to interaction between anionic surface and Müller cells. In another study, HSA-NPs penetrated the whole retina and localized inside the RPE of the normal retina after intravitreal injection in rat eyes. Furthermore, in the laser photocoagulated retina, HSA-NPs were observed to reach the choroid through the disruption site of the RPE and Bruch's membrane. Therefore, the anionic HSA-NP could be promising drug delivery carrier for the treatment of AMD which required drug distribution to the choroid region in order to inhibit choroidal neovascularization[67].

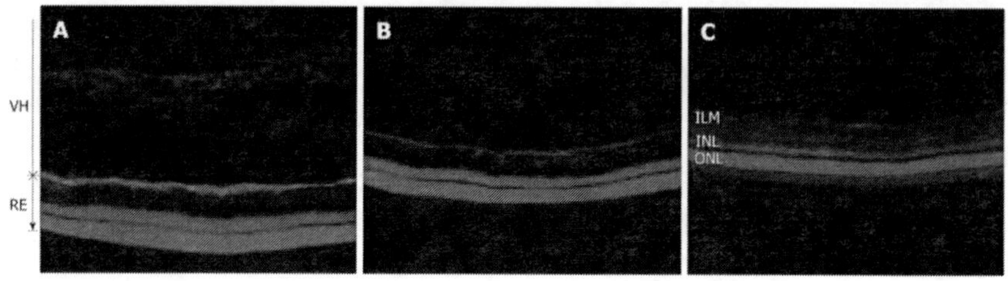

Figure 5. Vitreal and retinal distribution of intravitreally administered

Nanosuspensions

Nanosuspensions are colloidal dispersion of submicron drug particles stabilized by polymer(s) or surfactant(s). It is emerged as promising strategy for delivery of hydrophobic drugs. For ocular delivery, it provides several advantages such as sterilization, ease of eye drop formulation, less irritation, increase precorneal residence time and enhancement in ocular bioavailability of drugs which are insoluble in tear fluid[68]. The efficacy of nanosuspensions in improving ocular bioavailability of glucocorticoids has been demonstrated in several research studies.

Glucocorticoids such as prednisolone, dexamethasone and hydrocortisone are widely recommended for the treatment of inflammatory conditions affecting anterior segment ocular tissues. The current therapy with these drugs requires frequent administration at higher doses which induce cataract formation, glaucoma, and damage optic nerve. Efforts have been made toward improving ocular bioavailability of glucocorticoids by formulating as nanosuspensions. For instance, Kassem *et al*[69] compared ocular bioavailability of various glucocorticoids (prednisolone, dexamethasone and hydrocortisone) from nanosuspensions, solutions and microcrystalline suspensions. The formulations were instilled into the lower cul-de-sac of the rabbit eye and intraocular pressure (IOP) was measured at frequent time intervals up to 12 h. The area under percentage increase in IOP *vs* time curve (AUC) values for all the suspensions were higher than that for the respective drug solutions. In addition, higher extent of drug absorption and more intense drug effects were observed for all steroids form nanosuspensions compared with solutions. In another study, Ali *et al*[70] compared ocular bioavailability of hydrocortisone (Hc) nanosuspensions prepared by precipitation and milling method with HC solution in rabbits post topical instillation. Nanosuspensions prepared by both the precipitation and milling method achieved significantly higher AUC (0–9 h) values of 28.06 ± 4.08 and 30.95 ± 2.2 µg/mL than that of HC solution (15.86 ± 2.7 µg/mL). A sustained drug action which was represented in terms of changes in intraocular pressure was maintained up to 9 h for the nanosuspensions compared to 5 h for the drug solution (Figure 6).

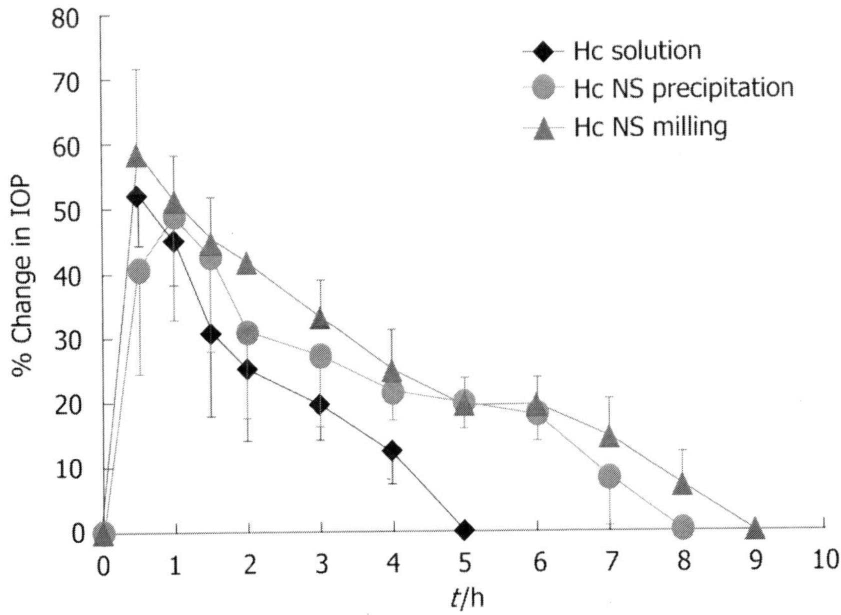

Figure 6. Changes in intraocular pressure of rabbits eyes following administration of hydrocortisone solution and nanosuspensions produced by milling and precipitation

From the results of above research studies, it can be concluded that nanosuspensions could be an efficient ophthalmic drug delivery system for delivery of poorly soluble drugs. In addition, nanosuspension can also be incorporated into hydrogels or ocular inserts for achieving sustained drug release for stipulated time period.

Liposomes

Liposomes are lipid vesicles with one or more phospholipid bilayers enclosing an aqueous core (Figure 3). The size of liposomes usually range from 0.08 to 10.00 µm and based on the size and phospholipid bilayers, liposomes can be classified as small unilamellar vesicles (10–100 nm), large unilamellar vesicles (100–300 nm) and multilamellar vesicles (contains more than one bilayer)[71].

For ophthalmic applications, liposomes represent ideal delivery systems due to excellent biocompatibility, cell membrane like structure and ability to encapsulate both hydrophilic and hydrophobic drugs. Liposomes have demonstrated good effectiveness for both anterior and posterior segment ocular delivery in several research studies. Recent advancements in liposomal ocular drug delivery are summarized in Table 2[72–81]. In a recent study, for delivery of latanoprost to anterior segment ocular tissues, liposomal formulation was developed by Natarajan *et al*[82]. The single subconjunctival injection of latanoprost/liposomal formulation in rabbit eye produced sustained IOP lowering effect over a period of 50 d with IOP reduction comparable to daily eye drop administration. For drug delivery to anterior segment of the eye, efforts are mainly put toward improving precorneal residence time by incorporating positively charged lipids or mucoadhesive polymer in liposomes. The positively charged liposomes *i.e.,* cationic liposomes have exhibited better efficacy in ocular delivery than negatively charged and neutral liposomes due to binding with negatively charges of corneal surface. Didodecyldimethylammonium bromide, stearylamine, and *N*-[1-(2,3-dioleoyloxy)propyl]-*N,N,N*-trimethylammonium chloride are commonly employed for fabricating cationic liposomes.

Table 2. Recent advancements in liposomal ocular drug delivery

Drug	Type of Liposomes	Result
Acetazolamide	Multilamellar, unilamellar	Multilamellar liposomes produced a more significant lowering in IOP in comparison with REVs liposomes[72]
Ciprofloxacin	Multilamellar	The mean residence time of ciprofloxacin was three fold higher for the CS-coated liposomes (3.85 h) compared to commercially available eye drops Ciprocin® (1.39 h)[73]
Cytochrome C		The cytochrome C loaded freeze-dried liposomes exhibited significant efficacy in retarding the onset and progression of cataract formation in rat eye[74]
VIP	Pegylated liposomes	After intravitreal injection, VIP concentration in ocular fluids was 15 times higher for liposomal formulation (155 ± 65 ng/mL) than the solution (10 ± 1 ng/mL), at 24 h[75]
Coumarin-6	Multilamellar	After topical administration in mice, the intensity of coumarin-6 in the retina was much higher with PLL modified liposomes[76]
Bevacizumab (Avastin)		Vitreous concentration of bevacizumab after 42 d of administration was 16 and 3.3 µg/mL in the eyes for liposomal and non-liposomal bevacizumab, respectively. The AUC (conc *vs* time) for liposomal bevacizumab was 1.5 fold higher compared with non-liposomal bevacizumab[77]
Fluorescence probe (coumarin-6)	Submicron-sized liposomes (ssLips) and multilamellar	After topical instillation of submicron-sized liposomes (ssLips), drug was delivered to the posterior segment ocular tissues including retina[78]
Fluconazole		Antifungal activity of fluconazole in liposomal formulation was better than that of fluconazole solution[79]
Edaravone	Submicron-sized liposomes	Topical administration of edaravone-loaded ssLips protected retina against light-induced dysfunction in mice eye while there was no marked protection found in the group treated with free edaravone[80]
Diclofenac	Multilamellar	Topical administration of diclofenac loaded PVA-R modified liposomes lead to improved retinal delivery in rabbit eye. Concentration of diclofenac in the retina–choroid was enhanced by 1.8 fold in case of drug loaded PVA-R modified liposome compared to that of the diclofenac solution[81]

REVs: Reverse phase evaporation; PLL: Poly-L-lysine; VIP: Vasoactive intestinal peptide, PVA: Polyvinyl alcohol; IOP: Intraocular pressure; AUC: Area under the curve.

Acyclovir loaded cationic and anionic liposomes were prepared by incorporating stearylamine and dicetylphosphate (DP), as cationic and anionic charge-inducing agents, respectively. In rabbit eyes, the acyclovir concentration in the cornea at 2.5 h after topical administration of positively charged liposomes was greater than those of negatively charged liposomes and free acyclovir. ACV concentrations in cornea were 253.3 ± 72.0, 1093.3 ± 279.7 and 571.7 ± 105.3 ng/g for ACV solution, ACV loaded positively and negatively charged liposomes, respectively. Also, the extent of ACV absorption through cornea was higher from positively charged liposomes which can be observed from ACV concentrations in aqueous humor at 2.5 h after instillation (Figure 7). The suggested reason was the higher binding of positively charged liposomes with negatively charged corneal surface *via* electrostatic interaction which ultimately lead to an increase of residence time and increase in acyclovir absorption[83]. In another study, when Coenzyme Q_{10}(CoQ_{10}) loaded liposomes was coated with mucoadhesive trimethyl chitosan, there was a 4.8 fold increase in the precorneal residence time in the rabbit eye was observed[84].

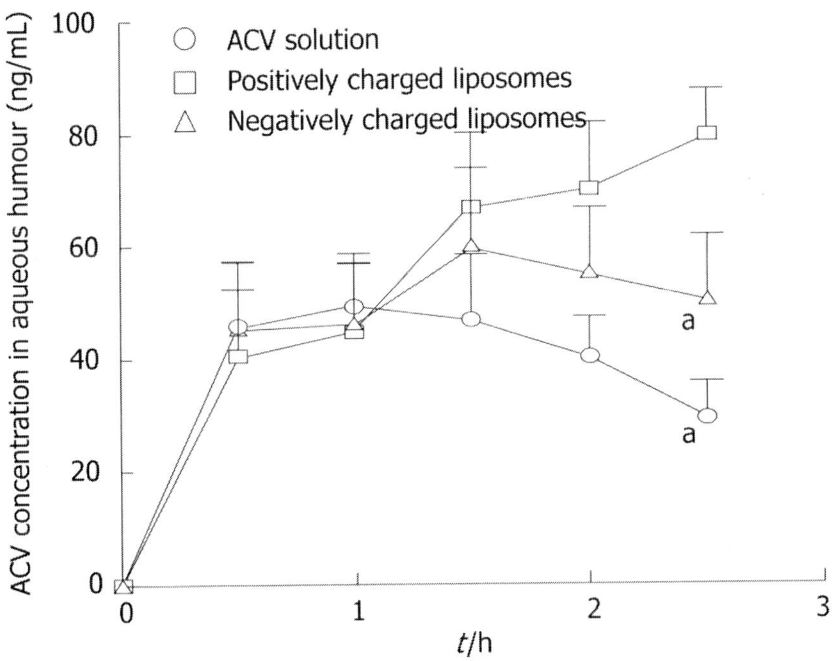

Figure 7. Acyclovir concentrations in aqueous humor after topical administration of Acyclovir solution and Acyclovir-containing liposomes

For posterior segment delivery, liposomes development is more focused toward improving half-life of drug by lessening clearance from vitreous humor, protecting labile molecules such as peptides and oligonucleotides from degradation and providing sustained drug release[5,85,86]. For instance, the vitreal half-life of fluconazole in rabbit eye was increased from 3.08 to 23.40 h after formulating into liposomes[86]. In another study, tacrolimus loaded liposomes were developed for the treatment of uveoretinitis. Following single intravitreous administration, tacrolimus vitreous level above 50 ng/mL was sustained for 14 d. The tacrolimus liposomal formulation demonstrated more effectiveness in suppressing uveoretinitis relative to drug alone and there was also reduced toxicity to inner retinal cells[87].

Several liposomal formulations for ocular drug delivery are being exploited, few are in pre-clinical and clinical study stage and few are commercially available. Visudyne® and Tears again® are the examples of commercially available liposomal formulations for the treatment of ocular diseases. Visudyne® (QLT Ophthalmics, Inc., Menlo Park, CA, United States) is a liposomal formulation containing photosensitizer, verteporfin. It is used in photodynamic therapy for subfoveal choroidal neovascularization in age related macular degeneration, presumed ocular histoplasmosis and pathological myopia[88]. Tears again® (Optima Pharmaceutical GmbH, Germany) is a phospholipid liposomes spray approved for the treatment of the Dry Eye syndrome. In clinical studies, this liposomal spray demonstrated significant advantages when compared with triglyceride-containing eye gel and a balanced salt solution[89,90].

Dendrimers

Dendrimers are characterized as nanosized, highly branched, star shaped polymeric systems. These branched polymeric systems are available in different molecular weights with terminal end amine, hydroxyl or carboxyl functional group. The terminal functional group may be utilized to conjugate targeting moieties[91]. Dendrimers are being employed as carrier systems in drug delivery. Selection of molecular weight, size, surface charge, molecular geometry and functional group are critical to deliver drugs. The highly branched structure of dendrimers allows incorporation of wide range of drugs, hydrophobic as well as hydrophilic. In ocular drug delivery, few promising results were reported with these branched polymeric systems[4,92,93].

Poly (amidoamine) (PAMAM) dendrimers are widely employed in ocular drug delivery[92]. Vandamme *et al*[94] demonstrated application of PAMAM dendrimers as ophthalmic vehicles for delivery of pilocarpine nitrate and tropicamide, for miotic and mydriatic activity. In this study, mean ocular residence time for fluorescein in saline and in PAMAM solutions were studied in rabbit eye. Fluorescein in 0.2% w/v Carbopol solution was used as reference bioadhesive polymer. The mean ocular residence time was significantly higher in case of PAMAM solutions and 0.2% w/v Carbopol solution compared to saline. Therefore, the use of dendrimers could be another option for increasing ocular residence time and therapy enhancing ocular bioavailability and achieving better therapeutic outcomes. For instance, PAMAM dendrimers when co-administrated with pilocarpine nitrate and tropicamide, showed higher miotic and mydriatic activity in albino rabbits[94].

In order to avoid scar tissue formation after glaucoma filtration surgery, conjugates of modified PAMAM dendrimers with glucosamine (DG) and glucosamine 6-sulfate (DGS) were synthesized to exert immunomodulatory and anti-angiogenic activities, respectively. The subconjunctival administration of these modified conjugates in rabbit model of glaucoma filtration surgery have shown significant inhibition of pro-inflammatory and pro-angiogenic responses and consequently reduced scar tissue formation. The results obtained from the experiment indicated that the ocular administration of DG and DGS might be effective and safe in clinical practice in avoiding scar tissue formation post glaucoma filtration surgery[95].

In-situ gelling systems

In-situ hydrogels refer to the polymeric solutions which undergo sol-gel phase transition to form viscoelastic gel in response to environmental stimuli. Gelation can be elicited by changes in temperature, pH and ions or can also be induced by UV irradiation. For ocular delivery, research studies have been more focused toward development of thermosensitive gels which respond to changes in temperature[96]. Several thermogelling polymers have been reported for ocular delivery which includes poloxamers, multiblock copolymers made of polycaprolactone, polyethylene glycol, poly (lactide), poly (glycolide), poly (*N*-isopropylacrylamide) and chitosan. These thermosensitive polymers form temperature dependent micellar aggregates which gellify after a further temperature increment due to aggregation or packing[96,97]. For drug delivery, these polymers are mixed with drugs in the solution state and solution can be administered which forms an *in situ* gel depot at physiological temperature. These thermosensitive gels demonstrated promising results for enhancing ocular bioavailability for both anterior and posterior segment. Gao *et al*[98] have evaluated suitability of thermosensitive gel made of triblock polymer PLGA-PEG-PLGA (poly-(DL-lactic acid co-glycolic acid)-polyethylene glycol-poly-(DL-lactic acid co-glycolic acid) as a ocular delivery carrier for dexamethasone acetate (DXA). It was formulated as either 0.1% w/v DXA solution or 0.1%, w/v DXA in 20% PLGA-PEG-PLGA *in situ* gel forming solution and administered topically in rabbit eye. Following topical administration, the C_{max} of DXA in the anterior chamber was significantly higher for the PLGA-PEG-PLGA solution

(125.2 μg/mL) relative to the eye drop (17.6 ± 2.18 ng/mL) along with higher AUC values. The increment in both C_{max} and AUC was approximately 7.00 and 7.98 fold for PLGA-PEG-PLGA *in situ*gel compared to the solution eye drops. These results suggest potentiality of PLGA-PEG-PLGA thermosensitive gel forming solution in enhancing ocular bioavailability.

Rieke *et al*[99] have reported applicability of ReGel™ (biodegradable and thermosensitive triblock copolymer consisting of PLGA and PEG, in providing sustained release of a large molecule ovalbumin to the choroid and retina following subconjunctival administration in the rat eye. The ovalbumin concentrations were maintained at measurable levels in the sclera, choroid, and retina of rats over a period of 14 d. These results suggest feasibility of thermosetting gel system in providing sustained delivery of macromolecules to the posterior segment ocular tissues such as choroid and retina. Cross linked poly (*N*-isopropylacrylamide) (PNIPAAm)-poly (ethylene glycol) diacrylate hydrogels were also synthesized for sustained release of macromolecules such as bovine serum albumin (BSA) and immunoglobulin G (IgG)[100]. The gel system has provided nearly 3 wk of sustained BSA release under *in vitro* condition. The results of research studies clearly signify the advantages of thermosensitive gels in providing sustained drug release, prolong contact time of drugs with the cornea, less frequency of applications, reduced side effects and higher ocular bioavailability over aqueous drops. In conclusion, the thermosensitive gels may be a viable option for the delivery of drugs for treating chronic ocular diseases.

Contact lens

Contact lenses are thin, and curved shape plastic disks which are designed to cover the cornea[101]. After application, contact lens adheres to the film of tears over the cornea due to the surface tension. Drug loaded contact lens have been developed for ocular delivery of numerous drugs such as β-blockers, antihistamines and antimicrobials. It is postulated that in presence of contact lens, drug molecules have longer residence time in the post-lens tear film which ultimately led to higher drug flux through cornea with less drug inflow into the nasolacrimal duct. Usually, drug is loaded into contact lens by soaking them in drug solutions. These soaked contact lenses demonstrated higher efficiency in delivering drug compared to conventional eye drops. Kim *et al*[102] observed much higher bioavailability of dexamethasone (DX) from poly (hydroxyethyl methacrylate) (PHEMA) contact lenses in comparison to eye drops. Indeed, efficient than topical drops, these soaked contact lenses suffers from disadvantages of inadequate drug loading and short term drug release. To overcome these obstacles, particle-laden contact lenses and molecularly imprinted contact lenses have been developed. In particle-laden contact lenses, drug is first entrapped in vesicles such as liposomes, nanoparticles or microemulsion and then these vesicles are dispersed in the contact lens material. Gulsen *et al*[103,104]developed particle-laden contact lenses for ocular delivery of lidocaine. In two different studies, they have prepared particle-laden contact lenses by dispersing lidocaine loaded microemulsion drops or liposome in poly-2 hydroxyethyl methacrylate (p-HEMA) hydrogels. Results of both the studies demonstrated the extended release of lidocaine over a period of 8 d. Indeed, particles-laden contact lenses look promising for extended ocular drug delivery; it needs to be stored in drug saturated solutions to avoid drug loss during storage. The designing of stimuli responsive such as pH or temperature sensitive "smart" particles which can release drug only in the eye could eliminate this problem. The imprinted contact lenses have also showed benefit in terms of both drug loading and drug release[105]. It has been demonstrated that soft contact lenses fabricated by the molecular imprinting method have 1.6 times higher timolol loading capacity than the contact lenses prepared by a conventional method and also provided sustained timolol delivery[106]. In another study, ketotifen fumarate loaded imprinted lenses have revealed higher tear fluid bioavailability compared to drug soaked lenses or ketotifen fumarate marketed eye drops. The relative bioavailability for the imprinted lenses was 3 times greater than that of non-imprinted lenses. The AUC value of ketotifen fumarate for imprinted lenses, non-imprinted lenses and eye drops were 4365 ± 1070 μg/h per milliliter, 493 ± 180 μg/h per milliliter, 46.6 ± 24.5 μg/h per milliliter, respectively[107]. The results clearly demonstrate more effectiveness of imprinted lenses over non-imprinted lenses and eye drops.

Implants

Intraocular implants are specifically designed to provide localized controlled drug release over a extended period. These devices help in circumventing multiple intraocular injections and associated complications[108,109]. Usually for drug delivery to posterior ocular tissues, implants are placed intravitreally by making incision through minor surgery at pars plana which is located posterior to the lens and anterior to the retina. Though implantation is invasive procedure, these devices are gaining interest due to their associated advantages such as sustained drug release, local drug release to diseased ocular tissues in therapeutic levels, reduced side effects and ability to circumvent blood retina barrier[109,110]. Several implantable devices have been developed for ocular drug delivery especially for the treatment of chronic vitreoretinal diseases.

Ocular implants are available as biodegradable and non-biodegradable drug releasing devices. Non-biodegradable implants offer long-lasting release by achieving near zero order release kinetics[110]. Polymers such as polyvinyl alcohol (PVA), ethylene vinyl acetate (EVA), and polysulfone capillary fiber (PCF) are being employed for fabricating non-biodegradable implants[108]. Vitrasert® and Retisert® are the examples of marketed non-biodegradable implants.

Vitrasert® (Bausch and Lomb Inc., Rochester, NY, United States) is a controlled-release intraocular implant of ganciclovir approved by Food and Drug Administration (FDA) for the treatment of acquired immune deficiency syndrome-associated cytomegalovirus retinitis. It is composed of a ganciclovir tablet of 4.5 mg surrounded by PVA/EVA that slowly release the drug over an extended period of 5–8 mo. The device provides long term sustained release without systemic toxicity at reduced cost[108,110,111]. Retisert®(Bauschnd Lomb Inc., Rochester, NY, United States) is approved by FDA for the treatment of chronic uveitis which affects the posterior segment of the eye. It is the first marketed silicone laminated PVA implant. It provides sustained release of fluocinolone acetonide up to 3 years. The implant had effectively controlled inflammation, reduced uveitis recurrences and improved vision acuity. The associated side effects are cataracts and elevated IOP[110–113]. Long term drug release may be achieved with these non-biodegradable implants but are associated with certain short comes. These devices have to be surgery implanted and removed after drug depletion, which makes the treatment expensive and patient non-compliance. Also, adverse events such as endophthalmitis, pseudoendophthalmitis, vitreous haze and hemorrhage, cataract development and retinal detachment limit their applications.

Another category of ocular implant includes biodegradable implants. These implants are gaining much attention and are being studied at large due to their biocompatible property and sustained drug release properties. Because of biodegradable nature, these implants are not required to be surgically removed which signify a distinctive advantage over the non-biodegradable implants. Polylactic acid (PLA), polyglycolic acid (PGA), PLGA, and polycaprolactones are the most commonly used polymers for the fabrication of biodegradable implants[108]. Examples of biodegradable implants for ocular delivery include Surodex™ and Ozurdex® which are designed for the sustained delivery of dexamethasone for the treatment of intraocular inflammation and macular edema (ME), respectively[110]. Surodex™ (Allergan, Inc., Irvine, CA, United States) composes PLGA and hydroxypropyl methylcellulose enclosing dexamethasone. The implant is inserted in the anterior chamber of eye to control postoperative inflammation in cataract patients. It provides sustained dexamethasone release for a period of 7–10 d with improved anti-inflammatory effect comparable to topical steroid administration[110].

Ozurdex® (Allergan Inc., Irvine, CA, United States) is another biocompatible and biodegradable intravitreal implant. It was approved by FDA in June 2009 for the treatment of macular edema. It employs Allergan's NO-VADUR® technology for delivering dexamethasone. The NOVADUR® system contains a PLGA polymer matrix which degrades slowly to lactic acid and glycolic acid allowing prolonged release of dexamethasone up to 6 mo. Randomized clinical trials have demonstrated its potency in reducing vision loss and improving vision acuity in eyes with macular edema associated with branch retinal vein occlusion (BRVO) or central retinal vein occlusion (CRVO). Also, clinical studies with Ozurdex® for treatment of diabetic retinopathy, and Irvine-Gass syndrome proved it as a promising treatment and drug delivery candidate[110].

Microneedles

Microneedle based technique is an emerging and minimally invasive mode of drug delivery to posterior ocular tissues. This technique may provide efficient treatment strategy for vision threatening posterior ocular diseases such as age related macular degeneration, diabetic retinopathy and posterior uveitis. This new microneedle based administration strategy may reduce the risk and complications associated with intravitreal injections such as retinal detachment, hemorrhage, cataract, endophthalmitis and pseudoendophthalmitis. Moreover, this strategy may help to circumvent blood retinal barrier and deliver therapeutic drug levels to retina/choroid. Microneedles are custom designed to penetrate only hundreds of microns into sclera, so that damage to deeper ocular tissues may be avoided. These needles help to deposit drug or carrier system into sclera or into the narrow space present between sclera and choroid called "suprachoroidal space" (SCS). Puncturing of sclera and depositing drug solution or carrier systems in sclera or SCS may facilitate diffusion of drug into deeper ocular tissues, choroid and neural retina[114]. For intraocular delivery of drugs Jason et al. investigated the application of microneedles surface coated with drugs[115]. Cadaver eyes were used to evaluate the role and scleral penetration of microneedle and intrascleral dissolution of microneedle surface coated drug (sulforhodamine). Results demonstrated that surface coated drug was rapidly dissolved in scleral tissue indicating high scleral sulforhodamine deposition within microneedle hole. In another study, Jiang *et al*[116] made attempts to evaluate the performance of microneedles to infuse drug solutions, nanoparticles and microparticles into scleral tissues. By use of microneedles, authors were able to infuse approximately 10–35 μL of fluid in to tissues. Nanoparticles suspensions and microparticles were also delivered into sclera by microneedles however; microparticles were delivered only in the presence of collagenase spreading enzymes and hyaluronidase. Study demonstrated that hollow microneedles may be employed for scleral infusion of drug or micro/nanoparticles with minimal invasive route.

Further, in another study Patel *et al*[117] made attempts to deliver drug solution, nanoparticles and microparticles in the SCS of rabbit, pig, and cadaver eyes with microneedles. Authors hypothesized that microneedle based minimally invasive strategy may help to deliver high level of both drug and nanocarriers to retinal tissues from SCS. Parameters for suprachoroidal delivery with microneedles such as microneedle length, pressure, and particle size were studied and optimized. Results demonstrated the strategy to be safe, minimally invasive and may sustain drug release. But, the study did not provide any evidence of drug reaching the inner retinal tissues from SCS. Same group made further attempts to study *in vivo*pharmacokinetics of SCS deposited solution/suspension post microneedle infusion. Results demonstrated that microneedle may provide a safe, reliable and targeted approach to chorio-retinal tissues[117].

CONCLUSION

Drug delivery to targeted ocular tissues has been a major challenge to ocular scientist, for decades. Administration of drug solutions as topical drop with conventional formulations was associated with certain drawbacks which initiated the introduction of different carrier systems for ocular delivery. Tremendous efforts are being put into ocular research toward the development of safe and patient compliant novel drug delivery strategies. Currently, researchers are thriving hard to improve *in vivo* performance of conventional formulations. On the other hand, advent of nanotechnology, new techniques, devices and their applications in drug delivery is developing immense interest to ocular scientists. Drug molecules are being encapsulated into nanosized carrier systems or devices and are being delivered by invasive/non-invasive or minimally invasive mode of drug administration. Several nanotechnology based carrier systems are being developed and studied at large such as nanoparticles, liposomes, nanomicelles, nanosuspensions and dendrimers. Few of these are commercially manufactured at large scale and are applied clinically. Nanotechnology is benefiting the patient body by minimizing the drug induced toxicities and vision loss. Also, these nanocarriers/devices sustain drug release; improve specificity, when targeting moieties are used, and help to reduce the dosing frequency. However, there is still need of developing a carrier system which could reach targeted ocular tissue, including back of the eye tissues, post non-invasive mode of drug administration. With the current pace of ocular research and efforts being made and put in, it is expected to result in a topical drop formulation that retains high precorneal residence time, avoids non-specific drug tissue accumulation and deliver therapeutic drug levels into targeted ocular tissue (both anterior and posterior). In near future, this delivery system may replace invasive mode of drug administration to back of the eye such as periocular and intravitreal injection.

FOOTNOTES

Author contributions: Patel A and Cholkar K contributed equally to this paper; all the authors wrote the manuscript.

REFERENCES

1. Bourlais CL, Acar L, Zia H, Sado PA, Needham T, Leverge R. Ophthalmic drug delivery systems--recent advances. Prog Retin Eye Res. 1998;17:33–58.

2. Gulsen D, Chauhan A. Ophthalmic drug delivery through contact lenses. Invest Ophthalmol Vis Sci.2004;45:2342–2347. \

3. Gaudana R, Ananthula HK, Parenky A, Mitra AK. Ocular drug delivery. AAPS J. 2010;12:348–360.

4. Gaudana R, Jwala J, Boddu SH, Mitra AK. Recent perspectives in ocular drug delivery. Pharm Res.2009;26:1197–1216.

5. Bochot A, Fattal E. Liposomes for intravitreal drug delivery: a state of the art. J Control Release.2012;161:628–634.

6. Kim SH, Lutz RJ, Wang NS, Robinson MR. Transport barriers in transscleral drug delivery for retinal diseases. Ophthalmic Res. 2007;39:244–254.

7. Lee SJ, He W, Robinson SB, Robinson MR, Csaky KG, Kim H. Evaluation of clearance mechanisms with transscleral drug delivery. Invest Ophthalmol Vis Sci. 2010;51:5205–5212.

8. Schoenwald RD. Ocular drug delivery. Pharmacokinetic considerations. Clin Pharmacokinet.1990;18:255–269.

9. Vaka SR, Sammeta SM, Day LB, Murthy SN. Transcorneal iontophoresis for delivery of ciprofloxacin hydrochloride. Curr Eye Res. 2008;33:661–667.

10. Tirucherai GS, Dias C, Mitra AK. Corneal permeation of ganciclovir: mechanism of ganciclovir permeation enhancement by acyl ester prodrug design. J Ocul Pharmacol Ther. 2002;18:535–548.

11. Gunda S, Hariharan S, Mitra AK. Corneal absorption and anterior chamber pharmacokinetics of dipeptide monoester prodrugs of ganciclovir (GCV): in vivo comparative evaluation of these prodrugs with Val-GCV and GCV in rabbits. J Ocul Pharmacol Ther. 2006;22:465–476.

12. Gallarate M, Chirio D, Bussano R, Peira E, Battaglia L, Baratta F, Trotta M. Development of O/W nanoemulsions for ophthalmic administration of timolol. Int J Pharm. 2013;440:126–134.

13. Tirucherai GS, Mitra AK. Effect of hydroxypropyl beta cyclodextrin complexation on aqueous solubility, stability, and corneal permeation of acyl ester prodrugs of ganciclovir. AAPS PharmSciTech. 2003;4:E45.

14. Vulovic N, Primorac M, Stupar M, Brown MW, Ford JL. Some studies on the preservation of indometacin suspensions intended for ophthalmic use. Pharmazie. 1990;45:678–679.

15. Meseguer G, Buri P, Plazonnet B, Rozier A, Gurny R. Gamma scintigraphic comparison of eyedrops containing pilocarpine in healthy volunteers. J Ocul Pharmacol Ther. 1996;12 :481–488.

16. Gebhardt BM, Varnell ED, Kaufman HE. Cyclosporine in collagen particles: corneal penetration and suppression of allograft rejection. J Ocul Pharmacol Ther. 1995;11:509–517.

17. Saettone MF, Chetoni P, Cerbai R, Mazzanti G, Braghiroli L. Evaluation of ocular permeation enhancers: in vitro effects on corneal transport of four beta-blockers, and in vitro/in vivo toxic activity. Int J Pharm. 1996;142:103–113.

18. van der Bijl P, van Eyk AD, Meyer D. Effects of three penetration enhancers on transcorneal permeation of cyclosporine. Cornea. 2001;20:505–508.

19. Burgalassi S, Chetoni P, Monti D, Saettone MF. Cytotoxicity of potential ocular permeation enhancers evaluated on rabbit and human corneal epithelial cell lines. Toxicol Lett. 2001;122:1–8.

20. Keister JC, Cooper ER, Missel PJ, Lang JC, Hager DF. Limits on optimizing ocular drug delivery. J Pharm Sci. 1991;80:50–53.

21. Hornof MD, Bernkop-Schnürch A. In vitro evaluation of the permeation enhancing effect of polycarbophil-cysteine conjugates on the cornea of rabbits. J Pharm Sci. 2002;91:2588–2592.

22. Kurz D, Ciulla TA. Novel approaches for retinal drug delivery. Ophthalmol Clin North Am.2002;15:405–410.

23. Cholkar K, Patel SP, Vadlapudi AD, Mitra AK. Novel strategies for anterior segment ocular drug delivery. J Ocul Pharmacol Ther. 2013;29:106–123.

24. Mannermaa E, Vellonen KS, Urtti A. Drug transport in corneal epithelium and blood-retina barrier: emerging role of transporters in ocular pharmacokinetics. Adv Drug Deliv Rev. 2006;58:1136–1163.

25. Shen J, Gan L, Zhu C, Zhang X, Dong Y, Jiang M, Zhu J, Gan Y. Novel NSAIDs ophthalmic formulation: flurbiprofen axetil emulsion with low irritancy and improved anti-inflammation effect. Int J Pharm. 2011;412:115–122.

26. Vandamme TF. Microemulsions as ocular drug delivery systems: recent developments and future challenges. Prog Retin Eye Res. 2002;21:15–34.

27. Liang H, Brignole-Baudouin F, Rabinovich-Guilatt L, Mao Z, Riancho L, Faure MO, Warnet JM, Lambert G, Baudouin C. Reduction of quaternary ammonium-induced ocular surface toxicity by emulsions: an in vivo study in rabbits. Mol Vis. 2008;14:204–216.

28. Tajika T, Isowaki A, Sakaki H. Ocular distribution of difluprednate ophthalmic emulsion 0. 05% in rabbits. J Ocul Pharmacol Ther. 2011;27:43–49.

29. Liu Y, Lin X, Tang X. Lipid emulsions as a potential delivery system for ocular use of azithromycin.Drug Dev Ind Pharm. 2009;35:887–896.

30. Karasawa F, Ehata T, Okuda T, Satoh T. Propofol injection pain is not alleviated by pretreatment with flurbiprofen axetil, a prodrug of a nonsteroidal antiinflammatory drug. J Anesth. 2000;14:135–137.

31. Yamaguchi M, Ueda K, Isowaki A, Ohtori A, Takeuchi H, Ohguro N, Tojo K. Mucoadhesive properties of chitosan-coated ophthalmic lipid emulsion containing indomethacin in tear fluid. Biol Pharm Bull.2009;32:1266–1271.

32. Lang J, Roehrs R, Jani R. Remington: The Science and Practice of Pharmacy. 21. Philadelphia: Lippincott Williams & Wilkins; 2009. Ophthalmic preparations; p. 856.

33. Scoper SV, Kabat AG, Owen GR, Stroman DW, Kabra BP, Faulkner R, Kulshreshtha AK, Rusk C, Bell B, Jamison T, Bernal-Perez LF, Brooks AC,

Nguyen VA. Ocular distribution, bactericidal activity and settling characteristics of TobraDex ST ophthalmic suspension compared with TobraDex ophthalmic suspension. Adv Ther. 2008;25:77–88.

34. Kinoshita S, Awamura S, Oshiden K, Nakamichi N, Suzuki H, Yokoi N. Rebamipide (OPC-12759) in the treatment of dry eye: a randomized, double-masked, multicenter, placebo-controlled phase II study.Ophthalmology. 2012;119:2471–2478.

35. Sasaki H, Yamamura K, Mukai T, Nishida K, Nakamura J, Nakashima M, Ichikawa M. Enhancement of ocular drug penetration. Crit Rev Ther Drug Carrier Syst. 1999;16:85–146.

36. Fukuda M, Hanazome I, Sasaki K. The intraocular dynamics of vancomycin hydrochloride ophthalmic ointment (TN-011) in rabbits. J Infect Chemother. 2003;9:93–96. Eguchi H, Shiota H, Oguro S, Kasama T. The inhibitory effect of vancomycin ointment on the manifestation of MRSA keratitis in rabbits. J Infect Chemother. 2009;15:279–283.

37. Gray C. Systemic toxicity with topical ophthalmic medications in children. Paediatric and Perinatal Drug Therapy. 2006;7:23–29.

38. Ishibashi T, Yokoi N, Kinoshita S. Comparison of the short-term effects on the human corneal surface of topical timolol maleate with and without benzalkonium chloride. J Glaucoma. 2003;12:486–490.

39. Whitson JT, Ochsner KI, Moster MR, Sullivan EK, Andrew RM, Silver LH, Wells DT, James JE, Bosworth CF, Dickerson JE, Landry TA, Bergamini MV. The safety and intraocular pressure-lowering efficacy of brimonidine tartrate 0. 15% preserved with polyquaternium-1. Ophthalmology. 2006;113:1333–1339.

40. Ayaki M, Yaguchi S, Iwasawa A, Koide R. Cytotoxicity of ophthalmic solutions with and without preservatives to human corneal endothelial cells, epithelial cells and conjunctival epithelial cells. Clin Experiment Ophthalmol. 2008;36:553–559.

41. Ayaki M, Iwasawa A, Yaguchi S, Koide R. Preserved and unpreserved 12 anti-allergic ophthalmic solutions and ocular surface toxicity: in vitro assessment in four cultured corneal and conjunctival epithelial cell lines. Biocontrol Sci. 2010;15:143–148.

42. Cholkar K, Patel A, Vadlapudi DA, Mitra AK. Novel Nanomicellar Formulation Approaches for Anterior and Posterior Segment Ocular Drug Delivery. Recent Patents on Nanomedicine. 2012;2:82–95.

43. Civiale C, Licciardi M, Cavallaro G, Giammona G, Mazzone MG. Polyhydroxyethylaspartamide-based micelles for ocular drug delivery. Int J Pharm. 2009;378:177–186.

44. Liaw J, Chang SF, Hsiao FC. In vivo gene delivery into ocular tissues by eye drops of poly(ethylene oxide)-poly(propylene oxide)-poly(ethylene oxide) (PEO-PPO-PEO) polymeric micelles. Gene Ther.2001;8:999–1004.

45. Tong YC, Chang SF, Liu CY, Kao WW, Huang CH, Liaw J. Eye drop delivery of nanopolymeric micelle formulated genes with cornea-specific promoters. J Gene Med. 2007;9:956–966.

46. Ideta R, Yanagi Y, Tamaki Y, Tasaka F, Harada A, Kataoka K. Effective accumulation of polyion complex micelle to experimental choroidal neovascularization in rats. FEBS Lett. 2004;557:21–25.

47. Ideta R, Tasaka F, Jang WD, Nishiyama N, Zhang GD, Harada A, Yanagi Y, Tamaki Y, Aida T, Kataoka K. Nanotechnology-based photodynamic therapy for neovascular disease using a supramolecular nanocarrier loaded with a dendritic photosensitizer. Nano Lett. 2005;5:2426–2431.

48. Parveen S, Mitra M, Krishnakumar S, Sahoo SK. Enhanced antiproliferative activity of carboplatin-loaded chitosan-alginate nanoparticles in a retinoblastoma cell line. Acta Biomater. 2010;6:3120–3131.

49. Nagarwal RC, Kumar R, Pandit JK. Chitosan coated sodium alginate-chitosan nanoparticles loaded with 5-FU for ocular delivery: in vitro characterization and in vivo study in rabbit eye. Eur J Pharm Sci.2012;47:678–685.

50. Gupta H, Aqil M, Khar RK, Ali A, Bhatnagar A, Mittal G. Sparfloxacin-loaded PLGA nanoparticles for sustained ocular drug delivery. Nanomedicine. 2010;6:324–333.

51. Singh KH, Shinde UA. Development and Evaluation of Novel Polymeric Nanoparticles of Brimonidine Tartrate. Curr Drug Deliv. 2010 May 24;

52. Gupta H, Aqil M, Khar RK, Ali A, Bhatnagar A, Mittal G. Biodegradable levofloxacin nanoparticles for sustained ocular drug delivery. J Drug Target. 2011;19:409–417.

53. Agnihotri SM, Vavia PR. Diclofenac-loaded biopolymeric nanosuspensions for ophthalmic application.Nanomedicine. 2009;5:90–95.

54. Nair KL, Vidyanand S, James J, Kumar GSV. Pilocarpine-loaded poly(DL-lactic-co-glycolic acid) nanoparticles as potential candidates for controlled drug delivery with enhanced ocular pharmacological response. J Appl Polym Sci. 2012;124:2030–2036.

55. Ibrahim HK, El-Leithy IS, Makky AA. Mucoadhesive nanoparticles as carrier systems for prolonged ocular delivery of gatifloxacin/prednisolone bitherapy. Mol Pharm. 2010;7:576–585.

56. Pignatello R, Ricupero N, Bucolo C, Maugeri F, Maltese A, Puglisi G. Preparation and characterization of eudragit retard nanosuspensions for the ocular delivery of cloricromene. AAPS PharmSciTech.2006;7:E27.

57. Bhagav P, Upadhyay H, Chandran S. Brimonidine tartrate-eudragit long-acting nanoparticles: formulation, optimization, in vitro and in vivo evaluation. AAPS PharmSciTech. 2011;12:1087–1101.

58. Bu HZ, Gukasyan HJ, Goulet L, Lou XJ, Xiang C, Koudriakova T. Ocular disposition, pharmacokinetics, efficacy and safety of nanoparticle-formulated ophthalmic drugs. Curr Drug Metab.2007;8:91–107.

59. Bhatta RS, Chandasana H, Chhonker YS, Rathi C, Kumar D, Mitra K, Shukla PK. Mucoadhesive nanoparticles for prolonged ocular delivery of natamycin: In vitro and pharmacokinetics studies. Int J Pharm.2012;432:105–112.

60. Musumeci T, Bucolo C, Carbone C, Pignatello R, Drago F, Puglisi G. Polymeric nanoparticles augment the ocular hypotensive effect of melatonin in rabbits. Int J Pharm. 2013;440:135–140.

61. Amrite AC, Kompella UB. Size-dependent disposition of nanoparticles and microparticles following subconjunctival administration. J Pharm Pharmacol. 2005;57:1555–1563.

62. Amrite AC, Edelhauser HF, Singh SR, Kompella UB. Effect of circulation on the disposition and ocular tissue distribution of 20 nm nanoparticles after periocular administration. Mol Vis. 2008;14:150–160.

63. Zhang L, Li Y, Zhang C, Wang Y, Song C. Pharmacokinetics and tolerance study of intravitreal injection of dexamethasone-loaded nanoparticles in rabbits. Int J Nanomedicine. 2009;4:175–183.

64. Bourges JL, Gautier SE, Delie F, Bejjani RA, Jeanny JC, Gurny R, BenEzra D, Behar-Cohen FF. Ocular drug delivery targeting the retina and

retinal pigment epithelium using polylactide nanoparticles.Invest Ophthalmol Vis Sci. 2003;44:3562–3569.

65. Koo H, Moon H, Han H, Na JH, Huh MS, Park JH, Woo SJ, Park KH, Kwon IC, Kim K, Kim H. The movement of self-assembled amphiphilic polymeric nanoparticles in the vitreous and retina after intravitreal injection. Biomaterials. 2012;33 :3485–3493.

66. Kim H, Robinson SB, Csaky KG. Investigating the movement of intravitreal human serum albumin nanoparticles in the vitreous and retina. Pharm Res. 2009;26:329–337.

67. Patravale VB, Date AA, Kulkarni RM. Nanosuspensions: a promising drug delivery strategy. J Pharm Pharmacol. 2004;56 :827–840.

68. Kassem MA, Abdel Rahman AA, Ghorab MM, Ahmed MB, Khalil RM. Nanosuspension as an ophthalmic delivery system for certain glucocorticoid drugs. Int J Pharm. 2007;340:126–133.

69. Ali HS, York P, Ali AM, Blagden N. Hydrocortisone nano-suspensions for ophthalmic delivery: A comparative study between microfluidic nanoprecipitation and wet milling. J Control Release.2011;149:175–181.

70. Kaur IP, Garg A, Singla AK, Aggarwal D. Vesicular systems in ocular drug delivery: an overview. Int J Pharm. 2004;269:1–14. j.ijpharm.2003.09.016.

71. Hathout RM, Mansour S, Mortada ND, Guinedi AS. Liposomes as an ocular delivery system for acetazolamide: in vitro and in vivo studies. AAPS PharmSciTech. 2007;8:1.

72. Abdelbary G. Ocular ciprofloxacin hydrochloride mucoadhesive chitosan-coated liposomes. Pharm Dev Technol. 2011;16 :44–56.

73. Zhang J, Guan P, Wang T, Chang D, Jiang T, Wang S. Freeze-dried liposomes as potential carriers for ocular administration of cytochrome c against selenite cataract formation. J Pharm Pharmacol. 2009;61:1171–1178.

74. Lajavardi L, Bochot A, Camelo S, Goldenberg B, Naud MC, Behar-Cohen F, Fattal E, de Kozak Y. Downregulation of endotoxin-induced uveitis by intravitreal injection of vaso-active intestinal Peptide encapsulated in liposomes. Invest Ophthalmol Vis Sci. 2007;48:3230–3238.

75. Sasaki H, Karasawa K, Hironaka K, Tahara K, Tozuka Y, Takeuchi H. Retinal drug delivery using eyedrop preparations of poly-l-lysine-modified liposomes. Eur J Pharm Biopharm. 2013;83:364–369.

76. Abrishami M, Zarei-Ghanavati S, Soroush D, Rouhbakhsh M, Jaafari MR, Malaekeh-Nikouei B. Preparation, characterization, and in vivo evaluation of nanoliposomes-encapsulated bevacizumab (avastin) for intravitreal administration. Retina. 2009;29:699–703.

77. Inokuchi Y, Hironaka K, Fujisawa T, Tozuka Y, Tsuruma K, Shimazawa M, Takeuchi H, Hara H. Physicochemical properties affecting retinal drug/coumarin-6 delivery from nanocarrier systems via eyedrop administration. Invest Ophthalmol Vis Sci. 2010;51:3162–3170.

78. Habib FS, Fouad EA, Abdel-Rhaman MS, Fathalla D. Liposomes as an ocular delivery system of fluconazole: in-vitro studies. Acta Ophthalmol. 2010;88:901–904.

79. Shimazaki H, Hironaka K, Fujisawa T, Tsuruma K, Tozuka Y, Shimazawa M, Takeuchi H, Hara H. Edaravone-loaded liposome eyedrops protect against light-induced retinal damage in mice. Invest Ophthalmol Vis Sci. 2011;52:7289–7297.

80. Fujisawa T, Miyai H, Hironaka K, Tsukamoto T, Tahara K, Tozuka Y, Ito M, Takeuchi H. Liposomal diclofenac eye drop formulations targeting the retina: formulation stability improvement using surface modification of liposomes. Int J Pharm. 2012;436:564–567.

81. Natarajan JV, Chattopadhyay S, Ang M, Darwitan A, Foo S, Zhen M, Koo M, Wong TT, Venkatraman SS. Sustained release of an anti-glaucoma drug: demonstration of efficacy of a liposomal formulation in the rabbit eye. PLoS One. 2011;6:e24513.

82. Law SL, Huang KJ, Chiang CH. Acyclovir-containing liposomes for potential ocular delivery. Corneal penetration and absorption. J Control Release. 2000;63:135–140.

83. Zhang J, Wang S. Topical use of coenzyme Q10-loaded liposomes coated with trimethyl chitosan: tolerance, precorneal retention and anti-cataract effect. Int J Pharm. 2009;372:66–75.

84. Alghadyan AA, Peyman GA, Khoobehi B, Milner S, Liu KR. Liposome-bound cyclosporine: clearance after intravitreal injection. Int Ophthalmol. 1988;12:109–112.

85. Gupta SK, Velpandian T, Dhingra N, Jaiswal J. Intravitreal pharmacokinetics of plain and liposome-entrapped fluconazole in rabbit eyes. J Ocul Pharmacol Ther. 2000;16:511–518.

86. Zhang R, He R, Qian J, Guo J, Xue K, Yuan YF. Treatment of experimental autoimmune uveoretinitis with intravitreal injection of tacrolimus (FK506) encapsulated in liposomes. Invest Ophthalmol Vis Sci.2010;51:3575–3582.

87. Keam SJ, Scott LJ, Curran MP. Verteporfin : a review of its use in the management of subfoveal choroidal neovascularisation. Drugs. 2003;63:2521–2554.

88. Dausch D, Lee S, Dausch S, Kim JC, Schwert G, Michelson W. Comparative study of treatment of the dry eye syndrome due to disturbances of the tear film lipid layer with lipid-containing tear substitutes. Klin Monbl Augenheilkd. 2006;223:974–983.

89. Lee S, Dausch S, Maierhofer G, Dausch D. A new therapy concept for the treatment of dry eye--the usefulness of phospholipid liposomes. Klin Monbl Augenheilkd. 2004;221:825–836.

90. Fischer M, Vögtle F. Dendrimers: From Design to Application-A Progress Report. Angew Chem Int Ed.1999;38:884–905.

91. Abdelkader H, Alany RG. Controlled and continuous release ocular drug delivery systems: pros and cons. Curr Drug Deliv. 2012;9:421–430.

92. Spataro G, Malecaze F, Turrin CO, Soler V, Duhayon C, Elena PP, Majoral JP, Caminade AM. Designing dendrimers for ocular drug delivery. Eur J Med Chem. 2010;45:326–334.

93. Vandamme TF, Brobeck L. Poly(amidoamine) dendrimers as ophthalmic vehicles for ocular delivery of pilocarpine nitrate and tropicamide. J Control Release. 2005;102:23–38.

94. Shaunak S, Thomas S, Gianasi E, Godwin A, Jones E, Teo I, Mireskandari K, Luthert P, Duncan R, Patterson S, Khaw P, Brocchini S. Polyvalent dendrimer glucosamine conjugates prevent scar tissue formation. Nat Biotechnol. 2004;22:977–984.

95. Rajoria G, Gupta A. In-Situ Gelling System: A Novel Approach for Ocular Drug Delivery. AJPTR.2012;2:24–53. Available from: URL: http://www.scopemed.org/?jft=42&ft=42-1348137747.

96. Bonacucina G, Cespi M, Mencarelli G, Giorgioni G, Palmieri GF. Thermosensitive Self-Assembling Block Copolymers as Drug Delivery Systems. Polymers. 2011;3:779–811.

97. Gao Y, Sun Y, Ren F, Gao S. PLGA-PEG-PLGA hydrogel for ocular drug delivery of dexamethasone acetate. Drug Dev Ind Pharm. 2010;36:1131–1138.

98. Rieke ER, Amaral J, Becerra SP, Lutz RJ. Sustained sub-conjunctival protein delivery using a thermosetting gel delivery system. J Ocul Pharmacol Ther. 2010;26:55–64.

99. Kang Derwent JJ, Mieler WF. Thermoresponsive hydrogels as a new ocular drug delivery platform to the posterior segment of the eye. Trans Am Ophthalmol Soc. 2008;106:206–213.

100. Gupta H, Aqil M. Contact lenses in ocular therapeutics. Drug Discov Today. 2012;17:522–527.

101. Kim J, Chauhan A. Dexamethasone transport and ocular delivery from poly(hydroxyethyl methacrylate) gels. Int J Pharm. 2008;353:205–222.

102. Gulsen D, Li CC, Chauhan A. Dispersion of DMPC liposomes in contact lenses for ophthalmic drug delivery. Curr Eye Res. 2005;30:1071–1080.

103. Gulsen D, Chauhan A. Dispersion of microemulsion drops in HEMA hydrogel: a potential ophthalmic drug delivery vehicle. Int J Pharm. 2005;292:95–117.

104. White CJ, Byrne ME. Molecularly imprinted therapeutic contact lenses. Expert Opin Drug Deliv.2010;7:765–780.

105. Hiratani H, Fujiwara A, Tamiya Y, Mizutani Y, Alvarez-Lorenzo C. Ocular release of timolol from molecularly imprinted soft contact lenses. Biomaterials. 2005;26:1293–1298.

106. Tieppo A, White CJ, Paine AC, Voyles ML, McBride MK, Byrne ME. Sustained in vivo release from imprinted therapeutic contact lenses. J Control Release. 2012;157:391–397.

107. Bourges JL, Bloquel C, Thomas A, Froussart F, Bochot A, Azan F, Gurny R, BenEzra D, Behar-Cohen F. Intraocular implants for extended drug delivery: therapeutic applications. Adv Drug Deliv Rev.2006;58:1182–1202.

108. Del Amo EM, Urtti A. Current and future ophthalmic drug delivery systems. A shift to the posterior segment. Drug Discov Today. 2008;13:135–143.

109. Lee SS, Hughes P, Ross AD, Robinson MR. Biodegradable implants for sustained drug release in the eye. Pharm Res. 2010;27:2043–2053.

110. Choonara YE, Pillay V, Danckwerts MP, Carmichael TR, du Toit LC. A review of implantable intravitreal drug delivery technologies for the treatment of posterior segment eye diseases. J Pharm Sci.2010;99:2219–2239.

111. Mohammad DA, Sweet BV, Elner SG. Retisert: is the new advance in treatment of uveitis a good one?Ann Pharmacother. 2007;41:449–454.

112. Jaffe GJ, McCallum RM, Branchaud B, Skalak C, Butuner Z, Ashton P. Long-term follow-up results of a pilot trial of a fluocinolone acetonide implant to treat posterior uveitis. Ophthalmology. 2005;112:1192–1198.

113. Donnelly RF, Raj Singh TR, Woolfson AD. Microneedle-based drug delivery systems: microfabrication, drug delivery, and safety. Drug Deliv. 2010;17:187–207.

114. Jiang J, Gill HS, Ghate D, McCarey BE, Patel SR, Edelhauser HF, Prausnitz MR. Coated microneedles for drug delivery to the eye. Invest Ophthalmol Vis Sci. 2007;48:4038–4043.

115. Jiang J, Moore JS, Edelhauser HF, Prausnitz MR. Intra-scleral drug delivery to the eye using hollow microneedles. Pharm Res. 2009;26:395–403.

116. Patel SR, Lin AS, Edelhauser HF, Prausnitz MR. Suprachoroidal drug delivery to the back of the eye using hollow microneedles. Pharm Res. 2011;28:166–176.

Chapter 3

Cell-Mediated Drugs Delivery

Elena V. Batrakova,[1,2,*] **Howard E. Gendelman,**[1,2,3,4] **and Alexander V. Kabanov**[1,2,4,5]

[1]*Center for Drug Delivery and Nanomedicine, University of Nebraska Medical Center, Omaha, Nebraska, USA*

[2]*Department of Pharmaceutical Sciences, University of Nebraska Medical Center, Omaha, Nebraska, USA*

[3]*Center for Neurodegenerative Disorders, University of Nebraska Medical Center, Omaha, Nebraska, USA*

[4]*Department of Pharmacology and Experimental Neuroscience, University of Nebraska Medical Center, Omaha, Nebraska, USA*

[5]*Department of Chemistry, M.V. Lomonosov Moscow State University, Moscow, Russia.*

ABSTRACT

Introduction

Drug targeting to sites of tissue injury, tumor or infection with limited toxicity is the goal for successful pharmaceutics. Immunocytes (including mononuclear phagocytes (dendritic cells, monocytes and macrophages), neutrophils, and lymphocytes) are highly mobile; they can migrate across impermeable barriers and release their drug cargo at sites of infection or tissue injury. Thus immune cells can be exploited as trojan horses for drug delivery.

Areas Covered In This Review

This paper reviews how immunocytes laden with drugs can cross the blood brain or blood tumor barriers, to facilitate treatments for infectious diseases, injury, cancer, or inflammatory diseases. The promises and perils of cell-mediated drug delivery are reviewed, with examples of how immunocytes can be harnessed to improve therapeutic end points.

Expert Opinion

Using cells as delivery vehicles enables targeted drug transport, and prolonged circulation times, along with reductions in cell and tissue toxicities. Such systems for drug carriage and targeted release represent a novel disease combating strategy being applied to a spectrum of human disorders. The design of nanocarriers for cell-mediated drug delivery may differ from those used for conventional drug delivery systems; nevertheless, engaging different defense mechanisms into drug delivery may open new perspectives for the active delivery of drugs.

Keywords: *cell-carriers*; *drug delivery*; *immunocytes*; *nanoparticles*; *targeted drug transport*

INTRODUCTION

The development of targeted drug delivery is amongst the most important goals of pharmaceutical research. Its realization can lead to improved therapeutic efficacy, reductions in drug dosing intervals, and decreased toxicities. However, this is not a simple task as drug homing to pathologically relevant disease sites has only recently been investigated. Obstacles are substantial and include sustained time-based plasma concentrations and local blood flow.

1. Promise and Perils of Cell-Mediated Drug Delivery

Using immunocytes, mononuclear phagocytes (MP; monocytes, macrophages, and dendritic cells) lymphocytes, and neutrophils and stem cells as drug carrier systems offers several advantages over common drug administration regimens. These include targeted drug transport to disease sites; prolonged drug half-lives; time-controlled drug release; and diminished drug immunogenicity and cytoxicity profiles. Immunocytes and stem cells exhibit an intrinsic homing property enabling them to migrate to sites of injury, inflammation, and tumor. In addition, they can act as Trojan horses carrying concealed drug cargoes while migrating across impermeable barriers (for example, the blood brain or blood tumor barriers) to sites of disease.

Despite such advantages, there is as yet limited success for several reasons. *First*, drug loading in cell carriers is low. *Second*, there are limitations due to the ability of immunocytes and stem cells to efficiently disintegrate and clear entrapped therapeutic agents. *Third*, the loaded drug should not be prematurely released, but unloaded, in continuous action, upon the cell's arrival to the site of action or disease. *Fourth*, the cell carriers should migrate to the disease site in substantial quantities. This should not be compromised during drug loading. *Finally*, all used formulations must be safe for both the cell carrier and the organism.

Many of these limitations can be addressed by incorporation of drugs into protective polymeric nanocarriers (liposomes and lipid nanoparticles [1-4], micelles [5-8], nanogels [9, 10], nanospheres and nanocapsules [11, 12], solid nanoparticles and nanosuspensions [13-16], block ionomer complexes [17, 18] or nanofibers and nanotubes[19, 20]) that preserve drugs inside subcellular organelles (Figure 1). The ideal drug carrier for cell-mediated delivery should have optimal size, shape and surface characteristics; multivalent attachment; controlled drug release; and biocompatibility. All these characteristics are essential for translating cell-mediated drug delivery systems for human use.

Drug loading into cell-carriers

Nanocarriers commonly have a core-shell structure. Their central part (such as aqueous pool of nanospheres, hydrophobic core of micelles, or polyelectrolyte core of nanogels) permits drug entrapment. It is surrounded with a polymer shell (or lipids in liposomes and lipid nanoparticles), which defines the nanocarrier dispersion stability, circulation time and cell interactions [1-8, 10, 11, 13, 14, 17, 18]. Both the core and shell are crucial for successful cell-mediated delivery. For example, the surface coating affects the ability of particles to internalize into the cells and therefore, affects the loading of cell-carriers. In general, charged nanocarriers are rapidly taken up by MP and other immunocytes or stem cells compared to neutral nanoparticles [1, 21-25]. The recognition occurs by receptors located on the cell plasma membranes, in particular, mannose, complement, and Fc receptors (MR, CR and FcR) [26-28]. In

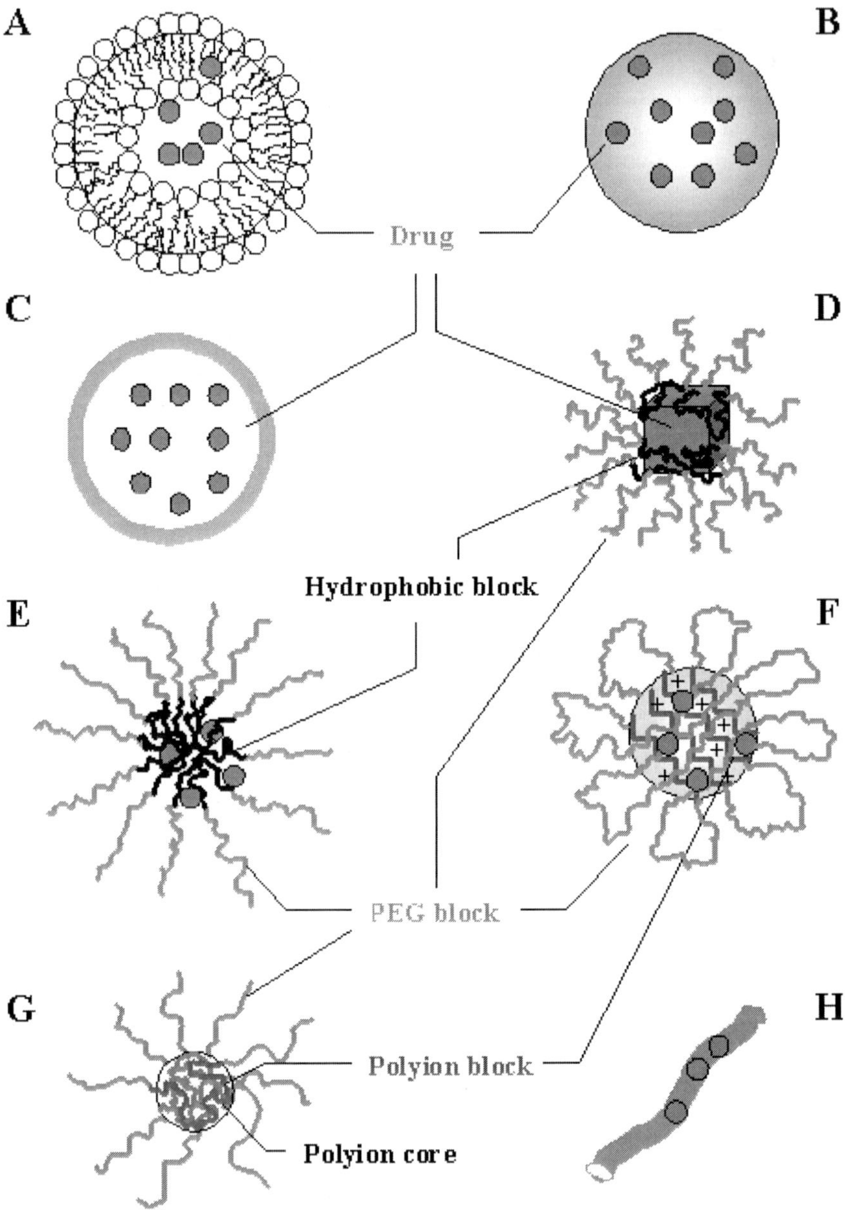

Figure 1. Types of nanocarriers for drug delivery. **A**: liposomes; B: polymer and lipid nanoparticles; **C**: nanospheres and nanocapsules; **D**: nanosuspensions; **E**: polymer micelles; **F**: nanogels; **G**: block ionomer complexes; **H**: nanofibers and nanotubes. Used with permission **...**

general, MR recognize mannans, as well as integrins (for example CD11b/CD18), CR interact with particles after nonspecific complement opsonization, and FcR recognize particles after specific antibody opsonization [28]. Furthermore, positively charged nanoparticles accumulate in MP to a greater extent than negatively charged particles. Thus, positively charged nanoparticles prepared by high pressure homogenization with antiretroviral drugs, indinavir (IDV), ritonavir (RTV), and efavirenz (EFV) accumulate in MP at about two-fold greater than the same size negatively charged carriers [14]. Similar effect was reported for nanoparticles ("nanozymes") comprising of redox enzyme, catalase, and variant synthetic polyelectrolyte block copolymers [29]. Loading capacity of nanozymes comprising of positively-charged mono-polymers (polyethyleneimine-(PEI) and poly-L-lysine- (PL) based) is greater than the nanozyme prepared with negatively-charged block poly-L-glutamic acid (PGLU)-based copolymer, This likely occurs due to the greater absorption and internalization of positively-charged nanoparticles to negatively-charged outside plasma membrane of cell-carriers.

An electrostatically neutral and hydrophilic poly(ethylene glycol) (PEG) is perhaps the most common shell-forming polymer currently used in injectable nanocarriers [30-32]. Particularly, commercial PEGylated liposome-encapsulated Dox, Doxil, is approved for use in the treatment of recurrent ovarian cancer, AIDS-related Kaposi's sarcoma [33] and metastatic breast cancer [34]. However, PEG is by far not the only polymer that should be considered in the context of the cell-mediated delivery, as it can limit cellular uptake as was demonstrated for microspheres bearing high-density surface PEG chains that were resistant to phagocytosis[35]. Similar results were obtained with nanoparticles comprised of catalase and positively charged polymers based on PEI [29]. Thus, loading capacity of nanozymes without PEG corona in MP was significantly greater than those with PEG. To this end, modification of nanocarrier shell with specific vector moieties can provide for targeted cell delivery of nanocarriers [1, 26, 36-38]. In particular, modification of PEG shell of a poly(amidoamine) nanoparticles (PAMAM) with streptavidin facilitates targeting to biotinylated T-cell markers such as anti-CD3 or peptide/ major histocompatibility (MHC) complexes [39] (Figure 2).

Size and shape of nanocarriers are also of importance for cell uptake, although the phagosome may have different sizes depending of the size of the particles, which can range from as little as few hundred nanometers [35, 40, 41] to several of microns [42]. For example, murine bone marrow-derived macrophages accumulate IgG-opsonized latex beads greater than 20 μm in diameter [43]. Furthermore, particles with the size about 1μm were accumulated at greater extent (2.56 fold) than smaller drug carriers at 500 μm [14].

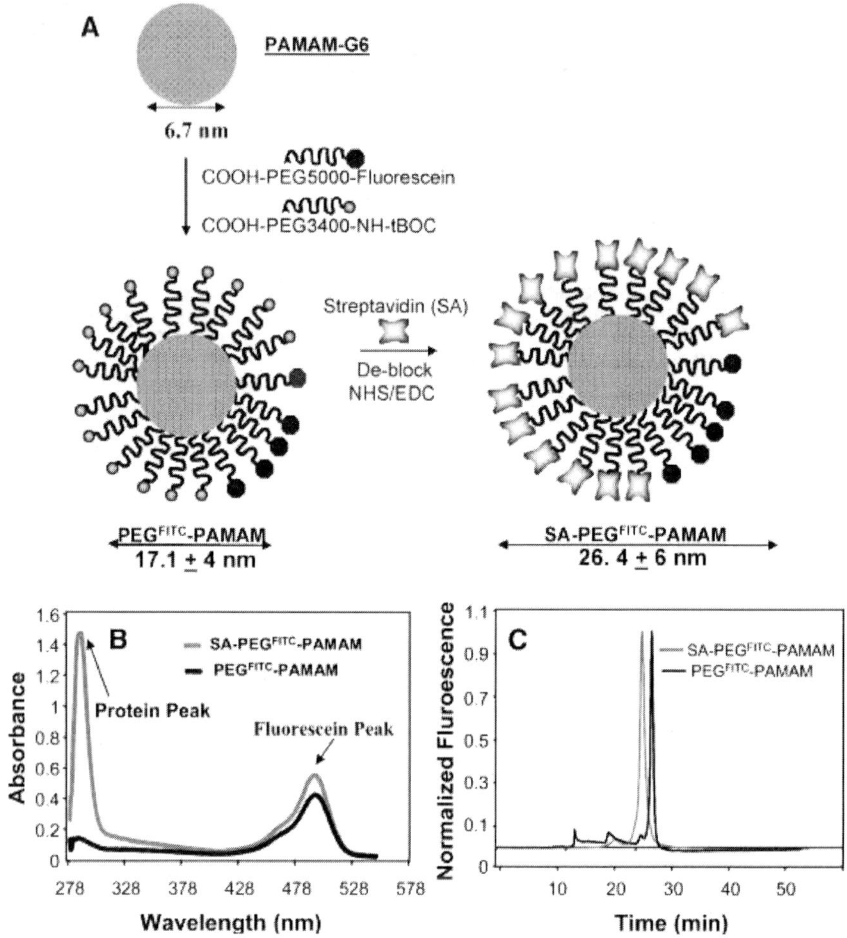

Figure 2. Schematic representation of SA-PEG^{FITC}/PAMAM construction

A recent study by Champion et al. reported striking effects of shape of particles on phagocytosis in alveolar rat macrophages [44]. Polystyrene-based particles of more than twenty shapes including spheres, rectangles, rods, worms, oblate ellipses, elliptical disks and UFO-like particles were manufactured. The local particle shape at the point where it was attached to the cell played the crucial role in phagocytosis. For example, a macrophage attached to a sharper side of the ellipse would internalize the particle in a few minutes. In contrast, a macrophage attached to a dull side would not internalize the same ellipse for hours. Although, particle size played a reduced role in the initiation of the phagocytosis it could of course affect its completion especially when the particle volume exceeded that of a cell.

Drug preservation inside cells

The stability of the drug loaded into cell carriers among other factors can depend on the intracellular trafficking of the nanocarriers. In general, drug-loaded nanocarriers need to avoid lysosomes to reduce drug disintegration inside cells [14]. In this regard, it is noteworthy that cationic and anionic nanoparticles show divergent fates inside MP [38]. For example, phagocytosis of cationic polyamine-coated nanoparticles leads to the diminished phagosomal acidification when compared to anionic protein-coated particles. Such cationic nanoparticles protect the incorporated drug against lisosomal degradation [29,38]. Furthermore, loading of "nanozymes" containing positively-charged block copolymers (PEI-PEG and PL-PEG) protected the enzyme in macrophages. Increasing the amount of the block copolymer in the nanozyme formulation improved the stability of the enzyme. In contrast, catalase loaded in a polyion complex with a negatively charged block copolymer (PGLU-PEG) was degraded in macrophages to an even greater extent than catalase loaded alone. Protection of the enzymatic activity inside carrier cells may be, in part, due to a "proton sponge" effect of block copolymers [45], when an excess of amino groups on the surface of the nanoparticles buffers acidification of the cell's endocytic compartments. This serves to inhibit protease activity and decrease drug degradation [38]. In other words, a positively charged block-copolymer prevents phagosome-lysosomal clearance functions and as a result, enzyme degradation.

In extreme case, intracellular drug degradation can be avoided by attaching the drug to the surface of cell-carriers. This "back pack" approach would still provide targeted transport and increased blood circulation along with the preserved drug activity. Thus, attaching avidin-coated nanoparticles to the biotinylated plasma membrane can be achieved through the avidin-biotin complex [46]. Another possibility is red blood cells (RBCs) with attached drugs on their surface [47]. Specifically, glycoprotein A covalently conjugated to the surface of the RBCs may provide extended half-life, controlled volume of distribution, and multivalent therapeutic interactions. However, general limitations of the "back pack" approach may include decreased loading of cell carriers, impeded drug release at the disease site, as well as increased immunogenicity and toxicity.

Drug release from cell-carriers

Mechanism of nanoparticle unloading at the site of action remains an active area of investigation. Controlled release of drugs from the cell-carriers modulates the rate of drug appearance, dose and duration of exposure at the diseased sites. To this end, utilizing cellular responses to various conditions could provide desired triggered release. Obviously, targeting of cell-carriers and their prolonged residence at the disease site should provide opportunities for drug unloading. In addition, MP, in particular, are known to produce and store various compounds in intracellular vesicles, and liberate them *via*exocytosis at the site of the disease. A similar mechanism can trigger release of drug-loaded nanocarriers, when macrophages serve as drug delivery vehicles [48]. Furthermore, the drug release can be also triggered by the increase of intracellular concentration of Ca^{2+} [49]. Finally, mild hypothermia was shown to facilitate controlled release of drug-loaded liposomes from macrophages in anti-cancer therapy [50].

Overall the structure and composition of protective nanocontainers play a crucial role in the effectiveness of formulations for cell-based drug delivery systems. For example, our recent studies indicate that structure of block copolymer used for catalase nanoformulation, nanozyme, affect its cytotoxicity, loading and release capacities, as well as preservation of catalase enzymatic activity inside cell-carriers [29]. Thus, nanozymes containing a negatively-charged block copolymer (PGLU-PEG/catalase) demonstrated low toxicity, high loading capacity and effective release from macrophages. However, the polymer provided limited protection of the enzyme against cell-associated protease degradation. In contrast, nanozymes based on positively charged block copolymers, especially the PLs, showed increased cytotoxicity and low loading and release rates, but were protective of the catalase. Increasing the amount of positively-charged block-copolymer in nanozyme leaded to protection of catalase enzymatic activity but substantially reduced loadings and release. Importantly, nanozymes with PEG corona show water stability, limited cytoxicity and efficient catalase protection. Nonetheless, these formulations also demonstrated decreased loading capacity and release rates. In addition, nanozymes based on mono-polymer (without PEG corona) have higher loading and release levels, but did not protect catalase inside the cell-carriers. Taken together, the most optimal nanozyme formulation was the one based on positively charged block copolymers (PEI-PEG/catalase and PL-PEG/catalase) that demonstrate the most efficient protection of catalase enzymatic activity along with relatively high loading and release rates with limited cytotoxicity (Figure 3).

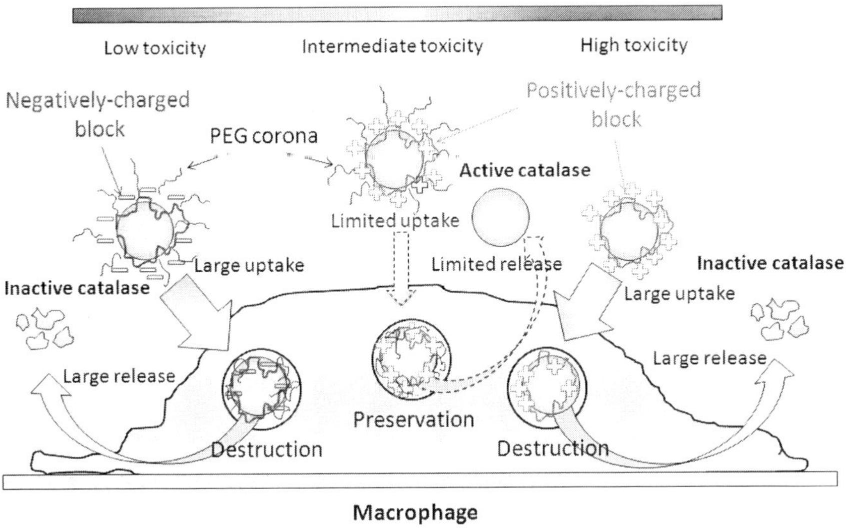

Figure 3. Optimization of nanocarriers structure for catalase delivery in macrophages

Homing of drug-loaded cell-carriers at disease sites

The numbers of cell-carriers that can reach the disease site is especially crucial in the case of CNS disorders when drug-loaded cells need to penetrate the blood brain barrier (BBB) to mediate therapeutic effect. Many neurological diseases, such as Alzheimer's and Parkinson's diseases (AD and PD), Prion disease, meningitis, encephalitis and HIV-associated neurocognitive disorders (HAND), have in common an inflammatory component [51]. The process of inflammation is characterized by extensive MP recruitment. Notably, MP migrate toward the inflammation site *via* the processes known as diapedesis and chemataxis [52]. Such cells efficiently cross the BBB due to their margination and extravasation properties causing barrier breakdown as a consequence of brain inflammation[53-57]. Many reports in the literature indicate that blood borne monocytes traffic primarily between adjacent endothelial cells, *i.e.* paracellularly through the junctional complexes[58, 59]. Therefore, these cells can be loaded with a required drug and administered intravenously to reach the brain. For example in an experimental model of PD, considerable levels of catalase loaded into BMM were reported in 1-methyl-4-phenyl-1,2,3,6-tetrahydropyridine (MPTP)-intoxicated brains[48]. Approximately 2.1% of the injected catalase dose was delivered to the brain during MPTP-associated inflammation. Furthermore, several studies confirmed the migration of inflammatory-response cells, in particular, bone-marrow-derived mesenchymal stromal cells toward injury sites such as infarcted myocardium[60, 61], spinal cord injury[62], and cerebral ischemia[63].

Neural stem cells (NSCs) were also suggested as drug delivery vehicles for gene therapy in the CNS [64-66]. Indeed, these cells are highly migratory and migrate to areas of brain pathology including ischemic and neoplastic brain lesions that are commonly present in AD, PD, brain cancer, stroke, and multiple sclerosis. How the mobility of stem cells are directed is not understood, although NSCs express a wide variety of receptors that should enable them to respond to many chemotactic signals present in brain pathologies (Figure 4) [66]. In particular, activated microglia induced NSCs migration to the brain. Delivery of various neurotropic factors is in the focus of cell-mediated strategies for neurological diseases treatment.

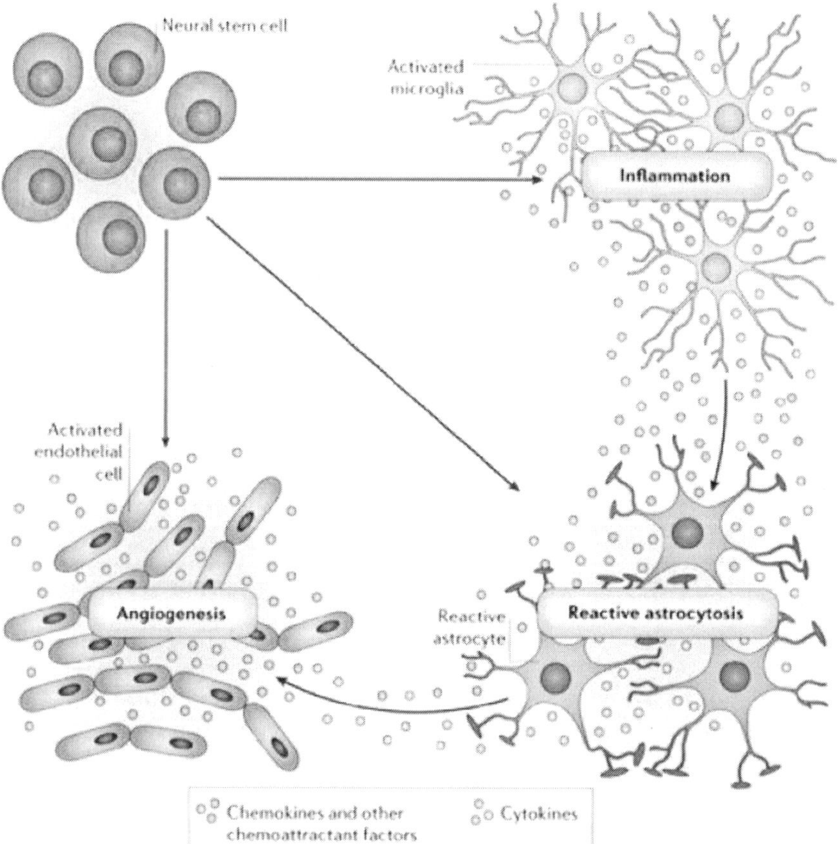

Figure 4. Using neural stem cells for drug delivery to CNS

In regards to cell-mediated drug delivery of antineoplastic agents, MP are known to accumulate in large numbers in vascular, hypoxic sites in cancer tumors, for example, in breast and prostate carcinomas [67, 68]. Hypoxia is widespread in malignant human tumors due to their poorly organized vasculature. The cytokines released by tumor cells in response to hypoxia and other physiological stresses usually attract macrophages and monocytes, which should facilitate anticancer drug delivery using these cells as vehicles, avoiding indiscriminate drug distribution and decreasing severe toxicity.

Finally, targeting to disease sites can be improved by drug-incorporated magnetic nanoparticles loaded into the cell-carriers and followed by application of local magnetic fields [37]. Thus, albino rats with brain inflammation (induced by intrastratial microinjection of human recombinant IL-1β) received intravenous injections of RGD-coated magnetic liposomes and non-magnetic liposomes as a control. RGD peptide (*i.e.*small peptide domane Arg-Gly-Asp) was used for selective binding by monocytes and neutrophils that express integrin receptors on their surface. Magnetic liposomes demonstrated about a ten-fold increase in brain levels compared with non-magnetic controlled carriers when local magnetic field was applied. In addition, magnetic neutrophils prepared *in vitro* target lungs under magnetic guidance following intravenous injection [69]. Moreover, drug loaded magnetic liposomes can be targeted for selective and preferential presentation to blood monocytes/neutrophils that result in both drug and magnetic incorporation into these cells and can be guided to target tissue sites.

Safety of cell-mediated drug delivery systems

An obvious concern for inflammatory-response cell based drug delivery relates to possible cytotoxic effects of cell-carriers. MP attracted to the site of pathology by cytokines release reactive oxygen species (ROS) that cause cell damage. Moreover, a number of therapeutic strategies for CNS neurodegenerative disorders are based on the prevention of monocyte-macrophage infiltration [70]. Therefore, precluding cytotoxicities for cell-based drug formulations is a requirement for formulation developments. Furthermore, genetically-modified and immortalized cell-carriers show atypical characteristics, such as higher degree of multipotency, which may increase the probability of tumor formation. Nevertheless, studies so far report no cytotoxic effects after macrophage-mediated drug delivery in the brain of healthy C57BL/6 mice adoptively transferred with macrophages carrying nanoformulated catalase [71]. In addition, propagation and expansion of immortalized and genetically-modified cell-carriers should allow improve quality controls [66].

Cell-mediated drug delivery systems in clinical settings

Finally, developing such methods in a clinical setting may be successful when cell-carriers are harvested from peripheral blood by apheresis, then loaded with particles and re-infused into the patient. An alternative approach may be harvesting stem cells from bone marrow, propagation them in culture to obtain monocytes, and then loading the cells with a drug and adoptively transferring them. This procedure will allow for expansion of the cell population, although this would require a more invasive procedure. To this end, immortalization of cell-carriers can allow propagation of cells with definable properties so that clonal populations with particular traits can be established. Furthermore, drug loading into cell-carriers can be achieved directly in the patient's peripheral blood, when nanoformulated drugs would be injected and selectively taken up by circulating cell carriers. Targeting of such nanocarriers could be achieved by coating the nanocarrier surface with the receptor-specific moieties (e.g., folate, gelatin, fibronectin, A-protein, mannose, or RGD peptide), which is recognized by specific MP surface receptors. Overall, chronic diseases are more suitable for this type of drug delivery than pathologies that require acute intervention. Based on such considerations, we now present some successful investigations using living cells as drug delivery vehicles for therapeutics (Table 1).

2. Neurodegenerative disorders and cell-based carriage of therapeutic nanoparticles

AD and PD

The progressive impairment of short-term memory and emotional disturbances that typify AD results from synaptic dysfunction and neuronal death in the hippocampus and linked regions in the cerebral cortex and limbic system. Because of the increasing numbers of affected people, there is an urgent need to develop strategies able to interfere with disease progression and protect neurons. In this respect, delivery of neurotrophic factors (NTFs) including nerve growth factor (NGF) [65], brain-derived neurotropic factor (BDNF) [64], and choline acetyltransferase (ChAT) [72] is of urgent need. To overcome limited drug transport across the BBB, direct brain implantation of cells engineered to produce NTFs was developed [73]. In particular, ciliary neurotrophic factor delivered in cells implanted in the brain prevents Aβ oligomer-induced neuronal damage and neurobehavioral impairments in mouse models of AD. In another study NSCs overexpressing ChAT cDNA targeted acetylcholine deficits [74]. The ability cell-carriers, placed by cortical transplants, to improve cognition was tested after induced cholinergic lesions in rats. Significant improvements were recorded in Water maze acquisition as well as in the retention and spatial probe trials [74]. A robust enhancement of hippocampal synaptic density, mediated by BDNF and delivered by NSCs was seen [75]. NSCs ameliorated complex behavioral deficits associated with AD pathology by

BDNF. In these experiments the intracranial transplantation of the cell-carriers were utilized.

Table 1. Examples of cell-mediated drug delivery systems

Type of cell-carrier	Therapeutic agent	Rote of administration[1]	References
Alziemer's disease			
Neural stem cells (NSCs)	ciliary neurotrophic factor (NTFs)	*i.c.*	73
NSCs	nerve growth factor (NGF)	*i.c.*	65
NSCs	brain-derived neurotropic factor (BDNF)	*i.c.*	64, 75
NSCs	choline acetyltrausferase (ChAT)	*i.c.*	72, 74
Parkinson's disease			
Monocytes, macrophages	catalase	*i.v.*	29, 48, 71
Fetal dopamine neurons	dopamine	*i.c.*	76
NSCs	glial cell-line derived neurotropic factor (GDNF)	*i.c.*	77, 80
NSCs	vascular endothelial growth factor (VGEF)	*i.c.*	78, 79
HIV-1 infection and neuroAIDS			
Monocytes, macrophages	atazanavir, ritonavir, indinavir	*i.v.*	14, 81-83
Cancer			
T-cells	doxorubicin	*In vitro*	94
Mesenchymal stem cells (MSCs)	tumor necrosis factor-related apoptosis-inducing ligand (TRAIL)	*ipsilateral*	93
MSCs	interferon-β	*i.v.*	86, 90, 95
MSCs	oncolytic adenovirus	*i.v.*	91, 92
MSCs	antineoplastic agent	*In vitro, in 3D collagen gels*	46
Erythrocytes	5-fluorouracil	*i.v.*	96
Monocytes and macrophages	Au-containing nanoparticles	*In vitro, in spheroids*	85, 89
Introperitoneal macrophages	5-fluorouracil	*i.p.*	50
Bacterially-derived mini-cells	small interfering RNA (siRNA), cytotoxic drugs	*i.v., i.t.*	97, 98
Bacterially-derived mini-cells	doxorubicin	*i.v., i.t.*	98
Macrophages	imaging agents, quantum dots	*i.v.*	103
Macrophages	*N*-(2-hydroxypropyl)methacrylamide (HPMA) copolymer-gadolinium (Gd) chelates	*i.v.*	104
Hybridoma cell lines 2A and OX86	anti-CD137 and anti-OX40 mAb	*s.c.*	102
Inflammation			
Macrophages	betamethasone phosphate	*in situ*	25
Erythrocytes	glycoprotein A		25
Sertoli cells	anti-inflammatory agent, curcumin	*i.v.*	105
Microbial Infections			
Erythrocytes	antifungal agent, amphotericin B	*in viro*	106
Macrophages	chloroquine	*i.v.*	107
Epilepsy			
Human MSCs and human embryonic stem cells	adenosine	*i.c. in* cell-encapsulation device	108, 109

[1] *i.c.*, intracranial; *i.v.*, intravenous; *i.p.*, intraperitoneal; *i.t.*, intratumoral; *s.c.*, subcutaneous injections.

Due to specific nigrostriatal degeneration, PD was one of the first targets for cell therapy transplants. Fetal dopamine neurons were inoculated into the putamen, where the cells worked as dopamine pumps as seen during systemic administration of L-DOPA [76]. It was suggested that the transplanted cells were gene therapy vehicles for dopamine delivery rather than replacement neurons [76]. Delivery of neurotropic factors such as glial cell-line derived neurotropic factor (GDNF) or vascular endothelial growth factor (VGEF) in neural stem cell transplants were tested in PD mouse models [77-79]. GDNF was delivered in bone marrow-derived macrophages to the effected brain [80]. The cells were transduced *ex vivo* with lentivirus expressing a GDNF gene driven by a synthetic macrophage-specific promoter and then transplanted into recipient mice. Eight weeks after transplantation, the mice were injected with MPTP for seven days to induce nigrostriatal neurodegeneration. Macrophage GDNF treatments dramatically ameliorated MPTP-induced degeneration of tyrosine hydroxylase (TH)-positive neurons seen in the substantia nigra and TH(+) terminals in the striatum. This resulted in axon regeneration and reversed hypoactivity in the open field test.

Development of novel CNS drug delivery using macrophages carriers for delivery of the antioxidant enzyme, catalase, in MPTP mice was reported by our laboratories [48, 71]. In this system, nanoformulated catalase (nanozyme) was obtained by coupling the enzyme to a cationic block copolymer, PEI-PEG, leading to a polyion complex micelle. Bone marrow-derived macrophages (BMM) carried significant amounts of catalase then slowly released the active enzyme over several days [48]. The enzyme released upon stimulation of nanozyme-loaded cell-carriers decomposed microglial hydrogen peroxide produced upon nitrated alpha-synuclein (N-α-syn) or tumor necrosis factor alpha (TNF-α) induced activation *in vitro*. Subsequent studies examined relationships between the composition of catalase nanozyme, their physicochemical characteristics (morphology, size, and ξ-potential), and cell loadings, release, and enzymatic activities for macrophage carriage [29].

Significant amount of catalase was detected in brains of mice after transfer of BMM loaded with nanoformulated catalase following MPTP intoxication. It was demonstrated that such nanozyme-loaded BMM injected into MPTP-intoxicated mice reduce neuroinflammation and attenuate nigrostriatal degeneration [71]. In particular, MPTP intoxication decreased the number of TH-positive nigral dopaminergic neurons (32 % survival) compared to saline-treated controls (Figure 5). In contrast, the number of surviving dopaminergic neurons in MPTP-intoxicated mice treated with nanozyme-loaded BMM was greater than the total number of neurons in MPTP-treated mice (62.4 % survival). Furthermore, treatment with nanozyme alone (without cell carriers) also produced some neuroprotection effect, although with fewer neurons (41.3

% survival) compared to the mice treated with cell/nanozyme formulation. Finally, treatment with empty monocytes did not preserve neurons in MPTP-intoxicated animals (31.1 % survival). This signified a neuroprotective effect of nanozyme-loaded monocytes in MPTP-induced neurodegeneration.

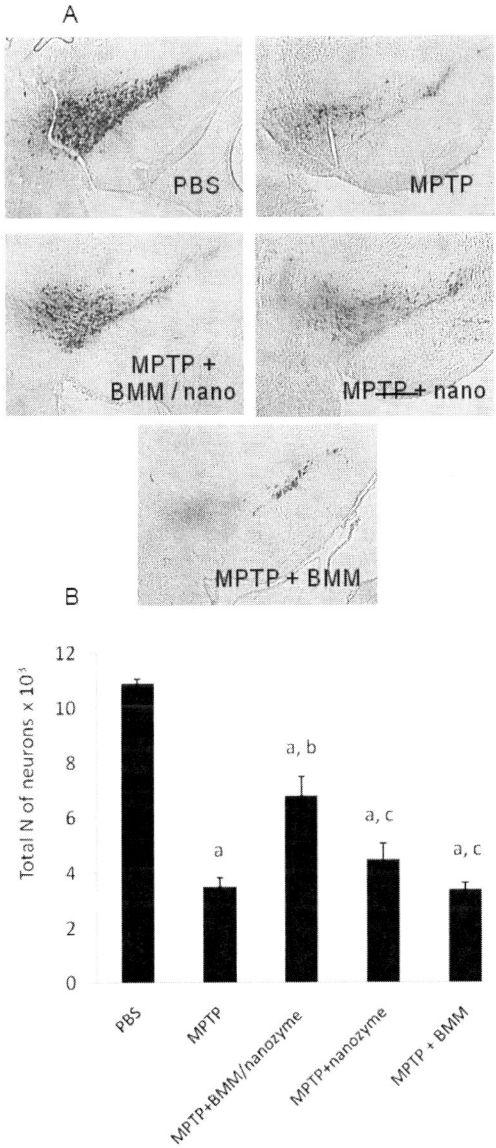

Figure 5. Neuroprotective effect of nanozyme-loaded BMM in an MPTP model of PD

The possible means by which BMM-mediated therapeutic effects of the nanozymes were suggested (Figure 6) : 1) nanozyme-loaded BMM cross the BBB and release catalase at the site of inflammation (particularly, in the substantia nigra pars compacta, *SNpc*); 2) a "depot" was established such that nanozyme is slowly released from BMM to the blood stream and bypasses the BBB independently of cell-carriers; and 3) catalase nanozyme released from BMM in the liver and spleen suppresses peripheral leukocyte activation that results in significant protection of SNpc neurons against MPTP-induced neurodegeneration [71]. Overall, as few therapeutic modalities exist which affect progression of PD and aimed at neuroregenerative therapies, the cell-mediated delivery of catalase and GDNF may represent efficacious strategies that attenuate neuroinflammation and provide neuroprotection for disease.

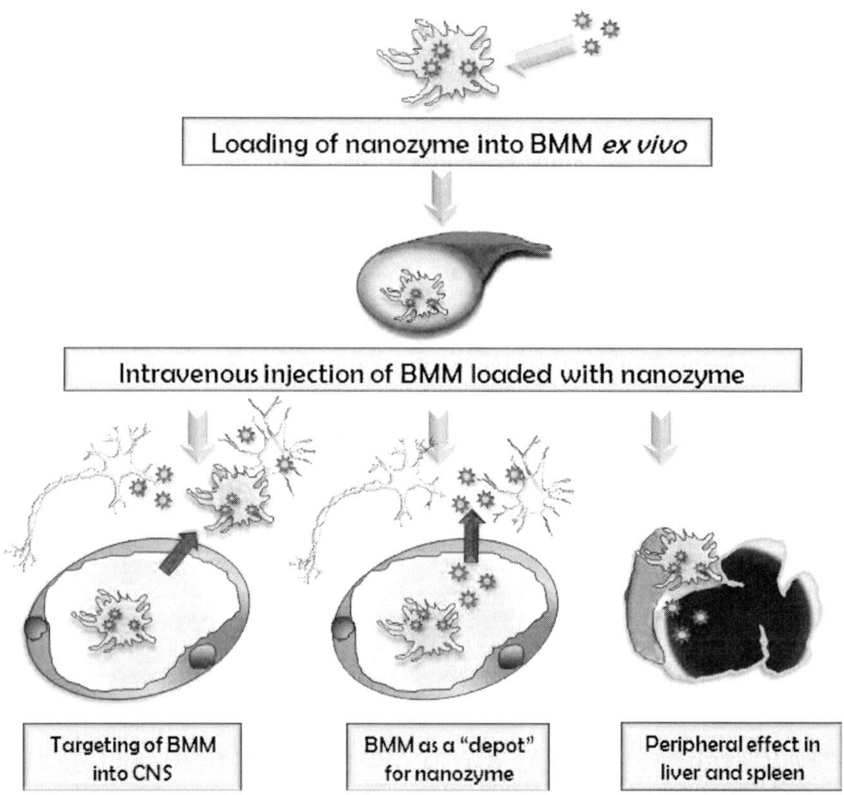

Figure 6. A pictorial scheme for cell-based nanoformulated drug delivery

HIV-1 infections and neuroAIDS

NeuroAIDS is a clinical disorder that encompasses neurologic disorders seen primarily as a consequence of damage to the central nervous system by HIV. The clinical syndromes include cognitive, motor and behavioral disorders of varying severity. Such syndromes affect 30 to 40% of adults and children with AIDS and, despite the advent of potent combination antiretroviral therapy (cART), incidence rates remain constant although disease severity is less. MP carriage has been utilized as a delivery system for antiretroviral therapy of HIV in mice [14, 57, 81-83]. To protect the drug against degradation, as well as the cell-carriers against the drug, it was incorporated into nanosized drug crystals (termed nano antiretroviral therapy or nanoART, Figure 7). NanoART were made by high-pressure homogenization of crystalline drug with various poloxamer and/or phospholipid surfactants, or by wet-milling homogenization, which produced nanoformulations with high loading capacities (over 80%) and relatively small particle sizes (200 to 350 nm). Scanning electron microscopy analysis revealed particles of various sizes and morphologies, which were either polygonal, rod, cuboidal, or spherical in shape with smooth-surfaces.

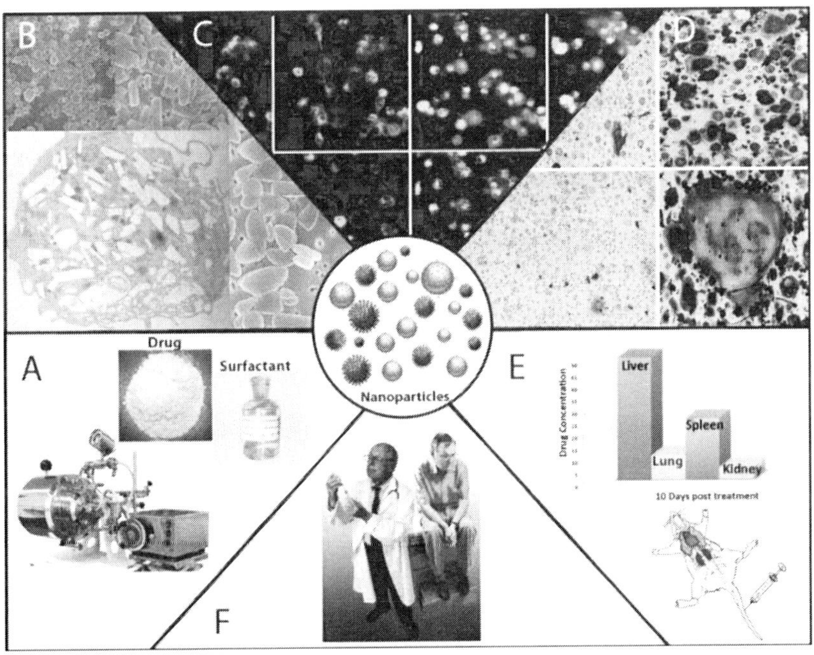

Figure 7. NanoART

(A) NanoART of antiretroviral agents were prepared by high-pressure homogenization using an AvestinC-5 homogenizer; (**B**) scanning-electron microscopy demonstrated different nanoparticle morphology; (**C**) uptake of drug-loaded nanoART (red) in monocyte-derived macrophages (green); (**D**) cells treated with nanoART showed complete or near complete suppression of HIV1 p24 antigen production; (**E**) biodistribution of nanoART in mice following intravenious injections; (**F**) \patient treatment with nanoART in clinical settings. Used with permission 111.

Simultaneous uptake of nanoformulated cART into the cell-carriers was shown [83]. It was also demonstrated that size, charge, coating, and shape of the nanoparticles are crucial for efficient MP uptake, drug release, and cell migration [14]. Coating of nanoART greatly affected drug accumulation in cell-carriers. Positively-charged nanoART are taken up better than negatively charged ones. To determine whether any of the wet-milled nanoformulations could induce long-term antiretroviral effects, macrophages were pretreated with individual nanoART preparations and then challenged with $HIV\text{-}1_{ADA}$ up to 15 days after drug treatment. Nanoparticles loaded with atazanavir, ritonavir and indinavir were released from macrophages and demonstrated >85, 80 and 40% inhibition, respectively, of progeny virion production and HIV-1p24 antigen through challenge day 15. All formulations of EFV showed almost complete suppression of viral infection[83]. These results supported the continued development of macrophage-mediated nanoART carriage for treating HIV-1 disease. In addition, delivery of antisense oligodeoxynucleotides, ribozymes, incorporated into liposomes to HIV-1-infected cells was reported in human monocyte-derived macrophages [84]. Thus, in addition to nanoART a ribozyme complementary to HIV-1 5'-long terminal repeat delivered in pH-sensitive liposomes inhibited virus production by 90%, while the free ribozyme caused limited viral inhibitions.

3. CANCER

Cell-mediated delivery is also a potential therapeutic and diagnostic strategy for cancer [46, 50, 85-93]. The advantage herein is that cytotoxic activity of the cell-carriers can be used in tandem with site-specific delivery of antineoplastic agents. In addition, antineoplastic agents are safely packaged into cell-carriers to reduce secondary cytotoxicities. However, adverse effects of the cytotoxic drugs on the cell-carriers themselves need be considered. For example, when nanoparticles coated with cytotoxic antibiotic, doxorubicin, were loaded into T-cells, the release of drug inside the cell-carrier caused cell damage [94]. As a result over 60% T cells loaded with the nanoparticles were dead over 15 hours after the loading.

To overcome these limitations, mesenchymal stem cells (MSCs) were genetically engineered to produce antitumor proteins [86, 90-93, 95]. MSCs possess a set of several unique properties, which make them ideally suited for both cellular therapies in regenerative medicine and as vehicles for gene and drug delivery to treat cancer. These properties include: 1) relative ease of isolation; 2) the ability to differentiate into a wide variety of functional cell types of both mesenchymal and non-mesenchymal origin; 3) the ability to be extensively expanded in culture without loss of differentiative capacity; 4) the hypoimmunogenicity and even ability to produce immunosuppression upon transplantation; 5) the pronounced anti-inflammatory properties; and last but not the least, 6) the ability to home to damaged tissues, tumors, and metastases following *in vivo*administration. Thus, MSCs of human origins were genetically modified to express TNF related apoptosis-inducing ligand (TRAIL) at the glioma tumor site. First, the retention of tumor tropic ability of hMSC S-TRAIL cells by *in vitro* and *in vivo* migration assays was clearly demonstrated [93]. Next, for the *in vivo*assessment of therapeutic efficacy, hMSCs were injected ipsilateral to an established intracranial glioma tumor in a mouse xenograft model. Genetically engineered hMSC S-TRAIL cells were effective in inhibiting intracranial U87 glioma tumor growth (81.6%) *in vivo* and resulted in significantly longer animal survival. Immunohistochemical studies demonstrated a significant, 8-fold greater tumor cell apoptosis in the hMSC S-TRAIL-treated group compared to controls. Overall, the study demonstrated the therapeutic efficacy of hMSC S-TRAIL cells and confirmed that hMSCs can serve as a powerful cell-based delivery vehicle for the site-specific release of therapeutic proteins. Furthermore, reduction of tumor growth by genetically-modified hMSCs expressing interferon-β was reported in models of metastatic breast cancer, melanoma [86, 90], and gliomas [95]. The hMSC were also used as carriers for conditionally replicating adenoviruses in a breast cancer metastasis model [91] and in a model of intracranial malignant glioma [92].

Attachment of nanoparticles with cytotoxic agent to the cell surface allowed cell carriers to remain intact. Thus, NeutrAvidin-coated nanoparticles were anchored on biotinylated plasma membrane of hMSC where they remained attached for up to two days [46]. The hMSC with such nanoparticulate patches retained their inherent tumoritropic properties in a tumor model with a 3D-extracellular matrix. These results provide a novel strategy to actively deliver nanostructures and therapeutics to tumors utilizing stem cells as carriers.

Besides MSCs, erythrocytes, MP, and bacterially-derived minicells were also evaluated as carriers for delivery of antineoplastic agents. In particular, erythrocytes were used as carriers for anticancer agent, 5-fluorouracil (5-FU), to treat malignant ascites [96]. Intravenous injections of the cell-carriers loaded with 5-FU resulted in significant regression of the quantity of malignant ascites and the increase of survival time in mice. Notably, MP can serve as "Trojan Horses" for nanoparticle transport into tumor regions [85, 89]. Thus, infiltration of monocytes and active delivery of Au-containing nanoparticles to the center of tumor spheroids (which contain hypoxic centers) was reported [89]. Following the delivery the tumor spheroids were destroyed by irradiation with near-infrared light, which actuated the delivered Au-containing nanoparticles. In another study, macrophages acted as a cellular vehicle for 5-FU encapsulated in oligomannose-coated liposomes (OMLs) [50]. The successful decrease of tumor growth by co-administration of OML-encased 5-FU and OML-encased magnetic nanoparticles, followed by treatment with an alternating magnetic field was reported in a mice. These OMLs were injected into the peritoneal cavity, and then gradually accumulated in the omentum and the other lymphoid tissues within 24 hours. Treatment of macrophages at 39 °C for 30 min. led to the release of 5-FU from the macrophages, suggesting that controlled release from macrophages could be achieved by mild hyperthermia. The encased magnetic nanoparticles, which are known to convert electromagnetic energy to heat in the OMLs allowed achieving *in vivo* hyperthermia. Finally, MP were utilized for delivery of therapeutic DNA constructs into tumors [85].

Bacterially-derived minicells with diameters of 400 nm were suggested for targeted delivery of different chemotherapeutics and small interfering RNA (siRNA) duplexes that cause stabilization and regression of tumor size in human cancer xenografts in mice [97, 98]. The cell-carriers were specifically targeted to tumor cell-surface proteins with biospecific antibodies (BsAb). About 80% efficiency of packing minicells with siRNA was reported. Thus, minicells carrying at least 12,000 siRNA molecules and 100 copies of short hairpin RNA (shRNA)-encoding plasmid can specifically and sequentially deliver to tumor xenografts these therapeutics and compromise drug resistance by knocking down a drug resistance protein. This cell-mediated drug delivery system enables to use several thousand-fold less drug than needed for conventional systemic administration of cancer therapies. Evaluation of uptake and intracellular kinetics of minicell-delivered siRNA indicates that after endocytosis, the minicells traverse the well-established early and late endosomal pathways, terminating into acidified organelles, the lysosomes, where mini-cells are degraded and release their cargo[98]. In addition, several methods of packaging in minicells of a range of anticancer chemotherapeutic drugs, such as doxorubicin, paclitaxel, irinotecan, 5-FLU, cisplatin, carboplatin monastrol, and vinblastine despite their disparate structure, charge, hydrophobicity, and solubility were developed [99-101].

Antibody-producing hybridoma cells were used for delivery of anti-CD137 and anti-OX40 mAb [102]. Microcapsules containing viable cells that secrete antibody were implanted by injection into the subcutaneous tissue of Balb/C mice bearing CT26 colon carcinomas. CD137 and OX40 are members of the TNF receptor family that provide costimulatory effect on T lymphocytes. The treatment resulted in complete tumor eradication in an elevated fraction of cases and strong tumor-specific cytotoxic T lymphocyte responses with either anti-CD137 or anti-OX40 producing hybridomas.

Besides cytotoxic agents, delivery of imaging agents is of importance for cancer therapy. MP-mediated delivery of quantum dots to brain tumors was explored for this purpose[103]. Thus, the surgical management of brain tumors requires the precise localization of tumor tissues within normal brain parenchyma in order to achieve accurate diagnostic biopsy and complete surgical resection. To this end, quantum dots are optical semiconductor nanocrystals that exhibit stable, bright fluorescence. It was demonstrated that the intravenous injection of quantum dots is accompanied by reticuloendothelial system and macrophage sequestration. Macrophages infiltrate brain tumors and phagocytize intravenously injected quantum dots, optically labeling the tumors.

Another class of imaging agents, *N*-(2-hydroxypropyl)methacrylamide (HPMA) copolymer-gadolinium (Gd) chelates, were used for enhanced magnetic resonance imaging (MRI) of macrophage-mediated malignancies[104]. To increase the nanoparticles uptake in macrophages, copolymers were vectorized with mannosamine, which was shown to increase mannose receptor mediated uptake in human MP. It was suggested that intravenously administered GD-nanoparticles will be accumulated in macrophages and then transported to the pathological sites (e.g., solid tumor). Overall, the tumor recruitment of inflammatory-response cells may be exploited for cell-mediated delivery of antineoplastic and imaging agents.

4. LUNG INJURY

Drug delivery to small airways, terminal bronchioles and alveoli is complicated due to the methodological limitations in targeting the deep lung with high efficiency drug distribution to the site of pathology. To overcome these limitations, chitosan nanoparticles with entrapped curcumin (an anti-inflammatory compound) was loaded into Testis-derived Sertoli cells and then injected intravenously into mice with bronchial or alveolar injuries[105]. By 24-hours post-injection, most of the curcumin load (~90%) delivered in the injected Sertoli cells was present and distributed throughout the lungs including the perialveloar sac area in the lower lungs. These results identify a novel and efficient method for targeted delivery of drugs for treatment of bronchitis and alveolitis. In addition, lungs are suitable for both local and systemic drug delivery, therefore this route may be used broadly for other diseases.

5. MICROBIAL INFECTIONS

The efficiency of antifungal agents has been diminished by their severe side effects and poor pharmacokinetics. To this end, erythrocytes were suggested as delivery vehicles for antifungal agent, amphotericin B (AmB). In order to avoid toxic effects and achieve efficient drug loading, AmB was encapsulated in nanosuspension (AmB-NS) by high-pressure homogenization [106]. AmB-NS was loaded into RBC by using hypotonic hemolysis, leading to intracellular AmB amounts of 3.81 +/- 0.47 pg RBC^{-1} and an entrapment efficacy of 15-18%. Hence, RBC served as primary carriers. Then, the uptake of RBC loaded with AmB-NS by leucocytes was studied by flow cytometry. More than 98% of the phagocyte population (granulocytes and peripheral monocytes accumulated AmB-NS after four hours of incubation, and then showed a slow AmB release over ten days without any alteration in cell viability. This results in an immediate, permanent inhibition of intra- and extracellular fungal activity. AmBNS-RBC-leukocyte-mediated delivery of AmB was efficient in amounts 1000 times lower than the toxic dose.

Another antifungal agent, chloroquine, loaded into phosphatidylserine-containing negatively charged liposomes was used in MP-mediated delivery against *C. neoformans* infection in the mouse brain [107]. Administration of chloroquine-loaded liposomes accumulated inside macrophage phagolysosomes resulted in a remarkable reduction in fungal load in the brain even in low doses compared to free drug in high doses thus increasing antifungal activity of macrophages. Therefore, this drug-delivery method is effective for the transport of water-insoluble substances, such as AmB, and this warrants consideration for further testing.

6. EPILEPSY

Relatively straightforward method of engineered adenosine-releasing cells was developed for treatment of epilepsy [108]. Lack of adenosine, a modulator of neuronal activity with anticonvulsant and neuroprotective properties, was found to contribute to ictogenesis. Therefore, focal reconstitution of adenosine within an epileptogenic brain region constitutes a rational therapeutic approach, whereas systemic augmentation of adenosine is precluded by side effects. To this end, human mesenchymal stem cells and human embryonic stem cells were embedded in a cell-encapsulation device to release adenosine and reduce acute injury and seizures, as well as chronic seizures [108]. Human embryonic stem cells (hESCs) have a high proliferative capacity and can be subjected to specific cellular differentiation pathways. hESCs, differentiated *in vitro* into neuroepithelial cells and grafted into the mouse brain, displayed intrahippocampal location and neuronal morphology. Overall, this therapeutic approach demonstrated antiepileptic and neuroprotective properties when grafted into the mouse hippocampus. The therapeutic potential of this approach suggests the feasibility to engineer autologous adenosine-releasing stem cells derived from a patient [109].

7. EXPERT OPINION

The concept of using cells as drug delivery vehicles is under development. Challenges include sufficient loading capacity and triggered drug release, preservation of therapeutic agents against degradation inside the cells, protection of cell-carriers from drug cytotoxic effects, and efficient homing of cells to the disease site. In addition, harvesting of cells sufficient quantities or their expansion is also important issue.

Two different approaches are utilized in cell-mediated drug delivery. *First*, cell-carriers (predominantly MP) are genetically modified to produce therapeutically active molecules. *Second*, host cells are loaded with a drug usually incorporated into protective container, and then used as "Trojan horses" to deliver the drug to the disease side (Figure 8). Most of cell-based delivery systems suggest the unloading the therapeutic agent at the disease site. However, a few utilized another approach, when the targeted cells accumulate the carrier along with the drug (mini-cells with anticancer agents and red blood cells with antifungal agents).

Regarding the first approach, the ideal cell delivery vehicle would be stable in tissue culture and capable of sustained, preferably regulated expression of therapeutic molecules. The cells should have appropriate and predictable differentiation pattern and survive long time *in vivo* after transplantation. Obviously, for both approaches, the cell carriers should demonstrate responsiveness to the chemotactic signals produced by the type of pathology that they are used to treat. Regarding the second approach, loading cell-carriers with drug in the blood stream by injecting nanoparticlular drug targeted to these cells is the most attractive strategy. To this end, it is important to note that these cells are moving targets. Nanosystems that bind to these targets are potentially a powerful approach for imaging their movement and delivering a therapeutic drug dose to the disease site. Clearly, a multifunctional molecular scaffold that is aimed at delivery of drug specifically to the cell-carrier, but in addition could be engineered for imaging purposes would offer substantial benefits over current approaches.

Cell-based Drug Delivery Systems

disease	Genetically-modified cell-carriers	Drug-loaded cell-carriers
PD, AD	NSCs or BMM producing neurotrophic factors	BMM loaded with redox enzymes
HIV		Monocytes and BMM with antiretroviral agents
cancer	• MSCs producing antitumor proteins • Antibody-producing hybridoma cells	• T-cells, RBC, and BMM loaded with chemotherapeutics • BMM loaded with Au- or magnetic nanoparticles • MSCs with nanoparticles attached to the surface • Mini-cells loaded with chemotherapeutics or siRNA • BMM loaded with imaging agents
Lung injury		Sertoli cells loaded with anti-inflammatory agents
Microbial infections		RBC loaded with antifungal agents
Epilepsy	MSCs producing adenosine	

Figure 8. Cell-based Drug Delivery Systems.

Such strategy is opposite to the common approach when drug-loaded nanoparticles are designed to minimize their entrapment in reticulo-endothelail system and avoid drug decomposition and clearance. Consequently, design of nanocarriers for cell-mediated drug delivery (size, shape, surface charge, etc.) may differ from those used for conventional drug delivery systems. Nevertheless, engaging different defense mechanisms into drug delivery may open a new perspective of active delivery of drugs for treatment of various devastating diseases.

ACKNOWLEDGMENTS

This work was supported by the United States National Institutes of Health grants 1RO1 NS057748 (to EVB); 2R01 NS034239, 2R37 NS36126, P01 NS31492, P20RR 15635, P01 MH64570, 5 P01 DA026146, 1P01 DA028555 and P01 NS43985 (to HEG); RO1 NS051334, 2RO1 CA89225, 1 R01 CA116591, UO1 CA151806, 1P20 RR021937, the Unites States Department of Defense grants W81XWH-09-1-0386 and DoD USAMRMC 06108004, and the Government of Russian Federation grants 02.740.11.523 and 11.G4.31.0004 (to AVK).

FOOTNOTES

Article Highlights

Using cell-mediated drug delivery systems offers several advantages including: a) targeted drug transport to disease sites; b) prolonged drug half-lives; c) time-controlled release of loaded drugs; and d) diminished drug immunogenicity and cytoxicity profiles.

The goals need to be achieved for successful cell-mediated drug delivery include: a) high drug loading into cell-carriers; b) efficient preservation of entrapped therapeutic agents against disintegration and clearance in the host cells; c) drug triggered release at the site of action; d) efficient homing of cell-carriers to a disease site; e) safety of cell-based drug formulations for the whole organism.

Two different approaches are utilized in cell-mediated drug delivery: genetically modified cell-carriers producing therapeutically active molecules; and drug loaded cell-carriers used as "Trojan horses" to deliver the drug to the disease side.

Living cells for drug carriage and release represent a novel disease combating strategy that can be applied to a spectrum of human infectious, cancerous, and degenerative disorders.

REFERENCES

1. Fujiwara M, Baldeschwieler JD, Grubbs RH. Receptor-mediated endocytosis of poly(acrylic acid)-conjugated liposomes by macrophages. Biochim Biophys Acta. 1996 Jan 12;1278(1):59–67.

2. Torchilin VP. Drug targeting. Eur J Pharm Sci. 2000 Oct;11(Suppl 2):S81–91.

3. Mora M, Sagrista ML, Trombetta D, et al. Design and characterization of liposomes containing long-chain N-acylPEs for brain delivery: penetration of liposomes incorporating GM1 into the rat brain. Pharm Res.2002 Oct;19(10):1430–8.

4. Aoki H, Kakinuma K, Morita K, et al. Therapeutic efficacy of targeting chemotherapy using local hyperthermia and thermosensitive liposome: evaluation of drug distribution in a rat glioma model. Int J Hyperthermia. 2004 Sep;20(6):595–605.

5. Kabanov AV, Chekhonin VP, Alakhov V, et al. The neuroleptic activity of haloperidol increases after its solubilization in surfactant micelles. Micelles as microcontainers for drug targeting. FEBS Lett. 1989 Dec 4;258(2):343–5.

6. Kabanov AV, Vinogradov SV, Suzdaltseva YG, et al. Water-soluble block polycations as carriers for oligonucleotide delivery. Bioconjug Chem. 1995 Nov-Dec;6(6):639–43.

7. Kabanov A, Alakhov V. Pluronic block copolymers in drug delivery: from micellar nanocontainers to biological response modifiers. Crit Rev Ther Drug Carrier Syst. 2002;19(1):1–72.

8. Kwon GS. Polymeric micelles for delivery of poorly water-soluble compounds. Crit Rev Ther Drug Carrier Syst. 2003;20(5):357–403.

9. Vinogradov S, Batrakova E, Kabanov A. Poly(ethylene glycol)-polyethyleneimine NanoGel (TM) particles: novel drug delivery systems for antisense oligonucleotides. Colloids and Surfaces B-Biointerfaces.1999 Nov;16(1-4):291–304.

10. Vinogradov SV, Batrakova EV, Kabanov AV. Nanogels for oligonucleotide delivery to the brain.Bioconjug Chem. 2004 Jan-Feb;15(1):50–60.

11. Gref R, Minamitake Y, Peracchia M, et al. Biodegradable long-circulating polymeric nanospheres.Science. 1994 Mar 18;263(5153):1600–3.

12. Hyuk IS, Jeong U, Xia Y. Polymer hollow particles with controllable holes in their surfaces. Nat Mater.2005 Sep;4(9):671–5.

13. Calvo P, Gouritin B, Chacun H, et al. Long-circulating PEGylated polycyanoacrylate nanoparticles as new drug carrier for brain delivery. Pharm Res. 2001 Aug;18(8):1157–66.

14. 14** Nowacek AS, Miller RL, McMillan J, et al. NanoART synthesis, characterization, uptake, release and toxicology for human monocyte-macrophage drug delivery. Nanomed. 2009 Dec;4(8):903–17.

15. [This study described characterization of nanoformulated antiretroviral drugs for HIV-1 infection and its application for human cell-mediated drug delivery.]

16. Muller RH, Jacobs C, Kayser O. Nanosuspensions as particulate drug formulations in therapy. Rationale for development and what we can expect for the future. Adv Drug Deliv Rev. 2001 Mar 23;47(1):3–19.

17. Friedrich I, Reichl S, Muller-Goymann CC. Drug release and permeation studies of nanosuspensions based on solidified reverse micellar solutions (SRMS). Int J Pharm. 2005 Nov 23;305(1-2):167–75.

18. Harada A, Kataoka K. Chain length recognition: core-shell supramolecular assembly from oppositely charged block copolymers. Science. 1999 Jan 1;283(5398):65–7.

19. Jaturanpinyo M, Harada A, Yuan X, et al. Preparation of bionanoreactor based on core-shell structured polyion complex micelles entrapping trypsin in the core cross-linked with glutaraldehyde. Bioconjug Chem.2004 Mar-Apr;15(2):344–8.

20. Bull SR, Guler MO, Bras RE, et al. Self-assembled peptide amphiphile nanofibers conjugated to MRI contrast agents. Nano Lett. 2005 Jan;5(1):1–4.

21. Guler MO, Pokorski JK, Appella DH, et al. Enhanced oligonucleotide binding to self-assembled nanofibers. Bioconjug Chem. 2005 May-Jun;16(3):501–3.

22. Juliano R, Stamp D. The effect of particle size and charge on the clearance rates of liposomes and liposome encapsulated drugs. Biochem Biophys Res Commun. 1975;63:651–8.

23. Lee KD, Hong K, Papahadjopoulos D. Recognition of liposomes by cells: in vitro binding and endocytosis mediated by specific lipid headgroups and surface charge density. Biochim Biophys Acta. 1992 Jan 31;1103(2):185–97.

24. Nishikawa K, Arai H, Inoue K. Scavenger receptor-mediated uptake and metabolism of lipid vesicles containing acidic phospholipids by mouse peritoneal macrophages. J Biol Chem. 1990 Mar 25;265(9):5226–31.

25. Miller CR, Bondurant B, McLean SD, et al. Liposome-cell interactions in vitro: effect of liposome surface charge on the binding and endocytosis of conventional and sterically stabilized liposomes.Biochemistry. 1998 Sep 15;37(37):12875–83.

26. Ishihara T, Izumo N, Higaki M, et al. Role of zinc in formulation of PLGA/PLA nanoparticles encapsulating betamethasone phosphate and its release profile. J Control Release. 2005 Jun 20;105(1-2):68–76.

27. Tempone AG, Perez D, Rath S, et al. Targeting Leishmania (L.) chagasi amastigotes through macrophage scavenger receptors: the use of drugs entrapped in liposomes containing phosphatidylserine. J Antimicrob Chemother. 2004 Jul;54(1):60–8.

28. Krieger M. Molecular flypaper and atherosclerosis: structure of the macrophage scavenger receptor.Trends Biochem Sci. 1992 Apr;17(4):141–6.

29. Aderem A, Underhill DM. Mechanisms of phagocytosis in macrophages. Annu Rev Immunol.1999;17:593–623.

30. Zhao Y, Haney MJ, Klyachko NL, et al. Polyelectrolyte complex optimization for macrophage delivery of redox enzyme nanoparticles. Nanomedicine (Lond) 2011 Jan;6(1):25–42.

31. Harada A, Kataoka K. Pronounced activity of enzymes through the incorporation into the core of polyion complex micelles made from charged block copolymers. J Control Release. 2001 May 14;72(1-3):85–91.

32. Harada A, Kataoka K. Switching by pulse electric field of the elevated enzymatic reaction in the core of polyion complex micelles. J Am Chem Soc. 2003 Dec 17;125(50):15306–7.

33. Knop K, Hoogenboom R, Fischer D, et al. Poly(ethylene glycol) in drug delivery: pros and cons as well as potential alternatives. Angew Chem Int Ed Engl. Aug 23;49(36):6288–308.

34. Gabizon A, Shmeeda H, Barenholz Y. Pharmacokinetics of pegylated liposomal Doxorubicin: review of animal and human studies. Clin Pharmacokinet. 2003;42(5):419–36.

35. Papaldo P, Fabi A, Ferretti G, et al. A phase II study on metastatic breast cancer patients treated with weekly vinorelbine with or without trastuzumab according to HER2 expression: changing the natural history of HER2-positive disease. Ann Oncol. 2006 Apr;17(4):630–6.

36. Gbadamosi JK, Hunter AC, Moghimi SM. PEGylation of microspheres generates a heterogeneous population of particles with differential surface characteristics and biological performance. FEBS Lett. 2002 Dec 18;532(3):338–44.

37. Daleke DL, Hong K, Papahadjopoulos D. Endocytosis of liposomes by macrophages: binding, acidification and leakage of liposomes monitored by a new fluorescence assay. Biochim Biophys Acta. 1990 May 24;1024(2):352–66.

38. Jain S, Mishra V, Singh P, et al. RGD-anchored magnetic liposomes for monocytes/neutrophils-mediated brain targeting. Int J Pharm. 2003 Aug 11;261(1-2):43–55.

39. Thiele L, Merkle HP, Walter E. Phagocytosis and phagosomal fate of surface-modified microparticles in dendritic cells and macrophages. Pharm Res. 2003 Feb;20(2):221–8.

40. Fahmy TM, Schneck JP, Saltzman WM. A nanoscopic multivalent antigen-presenting carrier for sensitive detection and drug delivery to T cells. Nanomedicine. 2007 Mar;3(1):75–85.

41. Tabata Y, Ikada Y. Effect of the size and surface charge of polymer microspheres on their phagocytosis by macrophage. Biomaterials. 1988 Jul;9(4):356–62.

42. Mizushima Y, Hamano T, Yokoyama K. Tissue distribution and anti-inflammatory activity of corticosteroids incorporated in lipid emulsion. Ann Rheum Dis. 1982 Jun;41(3):263–7. [PMC free article]

43. Hillaireau H, Couvreur P. Nanocarriers' entry into the cell: relevance to drug delivery. Cell Mol Life Sci.2009 Sep;66(17):2873–96.

44. Cannon GJ, Swanson JA. The macrophage capacity for phagocytosis. J Cell Sci. 1992 Apr;101(Pt 4):907–13.

45. Champion JA, Katare YK, Mitragotri S. Making polymeric micro- and nanoparticles of complex shapes.Proc Natl Acad Sci U S A. 2007 Jul 17;104(29):11901–4.

46. Behr J-P. The Proton Sponge: a Trick to Enter Cells the Viruses Did Not Exploit CHIMIA. International Journal for Chemistry. 1997;51(1):34–6.

47. Cheng H, Kastrup CJ, Ramanathan R, et al. Nanoparticulate cellular patches for cell-mediated tumoritropic delivery. ACS Nano. 2010 Feb 23;4(2):625–31.

48. Krantz A. Red cell-mediated therapy: opportunities and challenges. Blood Cells Mol Dis. 1997;23(1):58–68.

49. 48* Batrakova EV, Li S, Reynolds AD, et al. A macrophage-nanozyme delivery system for Parkinson's disease. Bioconjug Chem. 2007 Sep-Oct;18(5):1498–506.

50. [This manuscript reported development and characterization of cell-mediated delivery system of antioxidant enzyme, catalase. Loading, release, and protection of enzymatic activity of catalase in bone marrow-derived monocytes (BMM) are evaluated in in vitro model of PD.]

51. 49. Sollner T, Bennett MK, Whiteheart SW, et al. A protein assembly-disassembly pathway in vitro that may correspond to sequential steps of synaptic vesicle docking, activation, and fusion. Cell. 1993 Nov 5;75(3):409–18.

52. 50. Ikehara Y, Niwa T, Biao L, et al. A carbohydrate recognition-based drug delivery and controlled release system using intraperitoneal macrophages as a cellular vehicle. Cancer Res. 2006 Sep 1;66(17):8740–8.

53. 51. Perry VH, Bell MD, Brown HC, et al. Inflammation in the nervous system. Curr Opin Neurobiol. 1995 Oct;5(5):636–41.

54. 52. Kuby J. Immunology. Freeman, WH. and Co.; New York: 1994.

55. 53. Anthony DC, Bolton SJ, Fearn S, et al. Age-related effects of interleukin-1 beta on polymorphonuclear neutrophil-dependent increases in blood-brain barrier permeability in rats. Brain. 1997 Mar;120(Pt 3):435–44.

56. 54. Anthony DC, Blond D, Dempster R, et al. Chemokine targets in acute brain injury and disease. Prog Brain Res. 2001;132:507–24.

57. 55. Blamire AM, Anthony DC, Rajagopalan B, et al. Interleukin-1beta - induced changes in blood-brain barrier permeability, apparent diffusion coefficient, and cerebral blood volume in the rat brain: a magnetic resonance study. J Neurosci. 2000 Nov 1;20(21):8153–9.

58. 56. Persidsky Y, Ghorpade A, Rasmussen J, et al. Microglial and astrocyte chemokines regulate monocyte migration through the blood-brain barrier in human immunodeficiency virus-1 encephalitis. Am J Pathol.1999 Nov;155(5):1599–611.

59. 57. Dou H, Destache CJ, Morehead JR, et al. Development of a macrophage-based nanoparticle platform for antiretroviral drug delivery. Blood. 2006 Oct 15;108(8):2827–35.

60. 58. Pawlowski NA, Kaplan G, Abraham E, et al. The selective binding and transmigration of monocytes through the junctional complexes of human endothelium. J Exp Med. 1988 Nov 1;168(5):1865–82.

61. 59. Lossinsky AS, Shivers RR. Structural pathways for macromolecular and cellular transport across the blood-brain barrier during inflammatory conditions. Review. Histol Histopathol. 2004 Apr;19(2):535–64.

62. 60. Orlic D, Kajstura J, Chimenti S, et al. Bone marrow cells regenerate infarcted myocardium. Nature. 2001 Apr 5;410(6829):701–5.

63. 61. Orlic D, Kajstura J, Chimenti S, et al. Mobilized bone marrow cells repair the infarcted heart, improving function and survival. Proc Natl Acad Sci U S A. 2001 Aug 28;98(18):10344–9. [PMC free article]

64. 62. Hofstetter CP, Schwarz EJ, Hess D, et al. Marrow stromal cells form guiding strands in the injured spinal cord and promote recovery. Proc Natl Acad Sci U S A. 2002 Feb 19;99(4):2199–204. [PMC free article]

65. 63. Mahmood A, Lu D, Lu M, et al. Treatment of traumatic brain injury in adult rats with intravenous administration of human bone marrow stromal cells. Neurosurgery. 2003 Sep;53(3):697–702. discussion -3.

66. 64. Martinez-Serrano A, Hantzopoulos PA, Bjorklund A. Ex vivo gene transfer of brain-derived neurotrophic factor to the intact rat forebrain: neurotrophic effects on cholinergic neurons. Eur J Neurosci.1996 Apr;8(4):727–35.

67. 65. Martinez-Serrano A, Bjorklund A. Ex vivo nerve growth factor gene transfer to the basal forebrain in presymptomatic middle-aged rats prevents the development of cholinergic neuron atrophy and cognitive impairment during aging. Proc Natl Acad Sci U S A. 1998 Feb 17;95(4):1858–63. [PMC free article]

68. 66. Muller FJ, Snyder EY, Loring JF. Gene therapy: can neural stem cells deliver? Nat Rev Neurosci. 2006 Jan;7(1):75–84.

69. 67. Leek RD, Lewis CE, Whitehouse R, et al. Association of macrophage infiltration with angiogenesis and prognosis in invasive breast carcinoma. Cancer Res. 1996 Oct 15;56(20):4625–9.

70. 68. Lewis JS, Landers RJ, Underwood JC, et al. Expression of vascular endothelial growth factor by macrophages is up-regulated in poorly vascularized areas of breast carcinomas. J Pathol. 2000 Oct;192(2):150–8.

71. 69. Ranney DF, Huffaker HH. Magnetic microspheres for the targeted controlled release of drugs and diagnostic agents. Ann N Y Acad Sci. 1987;507:104–19.

72. Hendriks JJ, Teunissen CE, de Vries HE, et al. Macrophages and neurodegeneration. Brain Res Brain Res Rev. 2005 Apr;48(2):185–95.

73. 71** Brynskikh AM, Zhao Y, Mosley RL, et al. Macrophage delivery of therapeutic nanozymes in a murine model of Parkinson's disease. Nanomedicine (Lond) 2010 Apr;5(3):379–96.

74. [This study supports the feasibility of cell-mediated drug delivery to the brain by examining i) the nanozyme loading capacity for cell carriers; ii) the effect of nanozymes on cell viability and function; and iii) the neuroprotective activities of BMM-carried nanozyme against MPTP intoxication in in vivo model of PD.]

75. Low WC, Lewis PR, Bunch ST, et al. Function recovery following neural transplantation of embryonic septal nuclei in adult rats with septohippocampal lesions. Nature. 1982 Nov 18;300(5889):260–2.

76. Garcia P, Youssef I, Utvik JK, et al. Ciliary neurotrophic factor cell-based delivery prevents synaptic impairment and improves memory in mouse models of Alzheimer's disease. J Neurosci. 2010 Jun 2;30(22):7516–27.

77. Pizzo DP, Coufal NG, Lortie MJ, et al. Regulatable acetylcholine-producing fibroblasts enhance cognitive performance. Mol Ther. 2006 Jan;13(1):175–82.

78. Blurton-Jones M, Kitazawa M, Martinez-Coria H, et al. Neural stem cells improve cognition via BDNF in a transgenic model of Alzheimer disease. Proc Natl Acad Sci U S A. 2009 Aug 11;106(32):13594–9.

79. Zurn AD, Tseng J, Aebischer P. Treatment of Parkinson's disease. Symptomatic cell therapies: cells as biological minipumps. Eur Neurol. 1996;36(6):405–8.

80. Akerud P, Canals JM, Snyder EY, et al. Neuroprotection through delivery of glial cell line-derived neurotrophic factor by neural stem cells in a mouse model of Parkinson's disease. J Neurosci. 2001 Oct 15;21(20):8108–18.

81. Casper D, Engstrom SJ, Mirchandani GR, et al. Enhanced vascularization and survival of neural transplants with ex vivo angiogenic gene transfer. Cell Transplant. 2002;11(4):331–49.

82. Yasuhara T, Shingo T, Muraoka K, et al. Neurorescue effects of VEGF on a rat model of Parkinson's disease. Brain Res. 2005 Aug 16;1053(1-2):10–8.

83. Biju K, Zhou Q, Li G, et al. Macrophage-mediated GDNF delivery protects against dopaminergic neurodegeneration: a therapeutic strategy for Parkinson's disease. Mol Ther. 2010 Aug;18(8):1536–44.

84. Dou H, Morehead JR, Destache C, et al. Laboratory investigations for the morphologic, pharmacokinetic, and anti-retroviral properties of indinavir nanoparticles in human nomocyte-derived macrophages. Virology. 2007 Feb;358(1):148–58.

85. Dou H, Grotepas CB, McMillan JM, et al. Macrophage delivery of nanoformulated antiretroviral drug to the brain in a murine model of neuroAIDS. J Immunol. 2009 Jul 1;183(1):661–9. [PMC free article]

86. Nowacek AS, McMillan J, Miller R, et al. Nanoformulated Antiretroviral Drug Combinations Extend Drug Release and Antiretroviral Responses in HIV-1-Infected Macrophages: Implications for NeuroAIDS Therapeutics. J Neuroimmune Pharmacol. 2010 Mar 17;

87. Duzgunes N, Pretzer E, Simoes S, et al. Liposome-mediated delivery of antiviral agents to human immunodeficiency virus-infected cells. Mol Membr Biol. 1999 Jan-Mar;16(1):111–8.

88. Chokri M, Lopez M, Oleron C, et al. Production of human macrophages with potent antitumor properties (MAK) by culture of monocytes in the presence of GM-CSF and 1,25-dihydroxy vitamin D3. Anticancer Res. 1992 Nov-Dec;12(6B):2257–60.

89. Studeny M, Marini FC, Champlin RE, et al. Bone marrow-derived mesenchymal stem cells as vehicles for interferon-beta delivery into tumors. Cancer Res. 2002 Jul 1;62(13):3603–8.

90. Stagg J, Lejeune L, Paquin A, et al. Marrow stromal cells for interleukin-2 delivery in cancer immunotherapy. Hum Gene Ther. 2004 Jun;15(6):597–608.

91. Nakamura K, Ito Y, Kawano Y, Kurozumi K, et al. Antitumor effect of genetically engineered mesenchymal stem cells in a rat glioma model. Gene Ther. 2004 Jul;11(14):1155–64.

92. Choi MR, Stanton-Maxey KJ, Stanley JK, et al. A cellular Trojan Horse for delivery of therapeutic nanoparticles into tumors. Nano Lett. 2007 Dec;7(12):3759–65.

93. Studeny M, Marini FC, Dembinski JL, et al. Mesenchymal stem cells: potential precursors for tumor stroma and targeted-delivery vehicles for anticancer agents. J Natl Cancer Inst. 2004 Nov 3;96(21):1593–603.

94. Stoff-Khalili MA, Rivera AA, Mathis JM, et al. Mesenchymal stem cells as a vehicle for targeted delivery of CRAds to lung metastases of breast carcinoma. Breast Cancer Res Treat. 2007 Oct;105(2):157–67.

95. Sonabend AM, Ulasov IV, Tyler MA, et al. Mesenchymal stem cells effectively deliver an oncolytic adenovirus to intracranial glioma. Stem Cells. 2008 Mar;26(3):831–41.

96. Menon LG, Kelly K, Yang HW, et al. Human bone marrow-derived mesenchymal stromal cells expressing S-TRAIL as a cellular delivery vehicle for human glioma therapy. Stem Cells. 2009 Sep;27(9):2320–30.

97. Steinfeld U, Pauli C, Kaltz N, et al. T lymphocytes as potential therapeutic drug carrier for cancer treatment. Int J Pharm. 2006 Mar 27;311(1-2):229–36.

98. Nakamizo A, Marini F, Amano T, et al. Human bone marrow-derived mesenchymal stem cells in the treatment of gliomas. Cancer Res. 2005 Apr 15;65(8):3307–18.

99. Wang GP, Guan YS, Jin XR, et al. Development of novel 5-fluorouracil carrier erythrocyte with pharmacokinetics and potent antitumor activity in mice bearing malignant ascites. J Gastroenterol Hepatol.2010 May;25(5):985–90.

100. MacDiarmid JA, Mugridge NB, Weiss JC, et al. Bacterially derived 400 nm particles for encapsulation and cancer cell targeting of chemotherapeutics. Cancer Cell. 2007 May;11(5):431–45.

101. MacDiarmid JA, Amaro-Mugridge NB, Madrid-Weiss J, et al. Sequential treatment of drug-resistant tumors with targeted minicells containing siRNA or a cytotoxic drug. Nat Biotechnol. 2009 Jul;27(7):643–51.

102. Nikaido H. Multidrug efflux pumps of gram-negative bacteria. J Bacteriol. 1996 Oct;178(20):5853–9.

103. Nikaido H. Molecular basis of bacterial outer membrane permeability revisited. Microbiol Mol Biol Rev. 2003 Dec;67(4):593–656.

104. Poole K. Outer membranes and efflux: the path to multidrug resistance in Gram-negative bacteria. Curr Pharm Biotechnol. 2002 Jun;3(2):77–98.

105. Dubrot J, Portero A, Orive G, et al. Delivery of immunostimulatory monoclonal antibodies by encapsulated hybridoma cells. Cancer Immunol Immunother. 2010 Nov;59(11):1621–31.

106. Popescu MA, Toms SA. In vivo optical imaging using quantum dots for the management of brain tumors. Expert Rev Mol Diagn. 2006 Nov;6(6):879–90.

107. Zarabi B, Nan A, Zhuo J, et al. Macrophage targeted N-(2-hydroxypropyl)methacrylamide conjugates for magnetic resonance imaging. Mol Pharm. 2006 Sep-Oct;3(5):550–7.

108. Kumar A, El-Badri N, Glaum M, et al. Initial Observations of Cell Mediated Drug Delivery to the Deep Lung. Cell Transplant. 2010 Nov 5; Nov 5, (Epub ahead of print)

109. Staedtke V, Brahler M, Muller A, et al. In vitro inhibition of fungal activity by macrophage-mediated sequestration and release of encapsulated amphotericin B nanosupension in red blood cells. Small. 2010 Jan;6(1):96–103.

110. Khan MA, Jabeen R, Nasti TH, et al. Enhanced anticryptococcal activity of chloroquine in phosphatidylserine-containing liposomes in a murine model. J Antimicrob Chemother. 2005 Feb;55(2):223–8.

111. Boison D. Engineered adenosine-releasing cells for epilepsy therapy: human mesenchymal stem cells and human embryonic stem cells. Neurotherapeutics. 2009 Apr;6(2):278–83.

112. Boison D. Adenosine augmentation therapies (AATs) for epilepsy: prospect of cell and gene therapies.Epilepsy Res. 2009 Aug;85(2-3):131–41.

113. Kabanov AV, Batrakova EV. Polymer nanomaterials. In: Gendelman HE, Ikezu T, editors.Neuroimmune Pharmacology. Springer; Omaha: 2008. pp. 691–707.

114. Nowacek A, Gendelman HE. NanoART, neuroAIDS and CNS drug delivery. Nanomedicine (Lond)2009 Jul;4(5):557–74.

Chapter 4

Design Strategies and Applications of Circulating Cell-Mediated Drug Delivery Systems

Yixue Su, Zhiwei Xie, Gloria B. Kim, Cheng Dong,* and Jian Yang*

Department of Biomedical Engineering, Materials Research Institutes, the Huck Institutes of Life Sciences, The Pennsylvania State University, W340 Millennium Science Complex, University Park, PA 16802

ABSTRACT

Drug delivery systems, particularly nanomaterial-based drug delivery systems, possess a tremendous amount of potential to improve diagnostic and therapeutic effects of drugs. Controlled drug delivery targeted to a specific disease is designed to significantly improve the pharmaceutical effects of drugs and reduce their side effects. Unfortunately, only a few targeted drug delivery systems can achieve high targeting efficiency after intravenous injection, even with the development of numerous surface markers and targeting modalities. Thus, alternative drug and nanomedicine targeting approaches are desired. Circulating cells, such as erythrocytes, leukocytes, and stem cells, present innate disease sensing and homing properties. Hence, using living cells as drug delivery carriers has gained increasing interest in recent years. This review highlights the recent advances in the design of cell-mediated drug delivery systems and targeting mechanisms. The approaches of drug encapsulation/conjugation to cell-carriers, cell-mediated targeting mechanisms, and the methods of controlled drug release are elaborated here. Cell-based "live" targeting and delivery could be used to facilitate a more specific, robust, and smart payload distribution for the next-generation drug delivery systems.

Keywords: drug delivery; nanoparticles; immune cells; circulating cells; stem cells; targeting

INTRODUCTION

Drug delivery systems (DDS) have been extensively studied in past several decades as a powerful tool to achieve improved therapeutic efficacy and diagnostic effects by remarkably improving the pharmacokinetics and pharmacodynamics. Recently, a great deal of effort has been made to design novel nano-scale DDS with controlled therapeutic outcomes, targeted modalities towards specific diseases, and minimized side effects. However, traditional DDS, including injection formulations, particles, liposomes, and hydrogels, cannot meet the ever-growing requirements of modern medicine, such as truly targeted therapies and personalized medicine. Thus, cell-mediated drug delivery has emerged as a new frontier in medicine. In this review, we will analyze the current design strategies of circulating cell-based drug delivery systems in terms of 1) selecting cells, 2) loading therapeutics into live cells or conjugating therapeutics to the surface of live cells, 3) allowing cell-mediated DDS to target specific diseases, and 4) achieving controlled drug release. The purpose of this review is to summarize the existing designs in constructing circulating cell-mediated DDS and to provide our perspectives on future directions of "live" drug delivery.

Conventional DDS and their challenges

Current DDS typically employ vehicles to carry therapeutics in order to improve the drug solubility, reduce toxicity, prolong circulating time, limit bio-distribution, achieve specific targeting, control drug release, and diminish immunogenicity. Ideal drug delivery vehicles should be biocompatible, biodegradable, easy to modify, and targeted towards specific diseases. A wide range of vehicles have been employed as controlled DDS, e.g. liposomes[1], nanoparticles (NPs)[2], micelles[3], hydrogels[4-5], fibers[6-7], and films[8]. Particularly, nano- and micro-materials decorated with targeting ligands or molecules, such as peptides, antibodies, aptamers, and proteins, that are specific to the receptors expressed or overexpressed on aberrant cells and their surrounding microenvironments have been widely studied in targeted drug delivery[9-10]. Unfortunately, very few of these nanomedicines have shown clinical efficacy due to significant challenges associated with the trafficking and targeting *in vivo*.

Current targeting strategies can be categorized into passive and active targeting. Passive targeting is typically dependent on enhanced permeability and the retention (EPR) effect caused by leaky vasculature and poor lymphatic drainage. The EPR effect is a unique phenomenon, by which macromolecules and nanoparticles (NPs) escape from the blood flow and preferentially accumulate more in tumors rather than in normal tissues. It occurs frequently in solid tumors, in which blood vessels commonly have defects. Additionally, defective lymphatic drainage leads to the loss of the ability to clear infiltrating substances. These distinguishing anatomical and pathophysiological characteristics of solid tumors have been generally utilized in tumor targeting. However, passive targeting is largely chance-dependent. Therefore, active targeting, especially molecular targeting, has been the focus of drug delivery research in recent years. Until now, numerous surface markers have been found on abnormal cells or in their surrounding microenvironment. Integrins, folate, growth factors, and cytokines offer possibility for counter-ligand functionalized drug vehicles to recognize cells and tissue targets[11-12]. With the capability to bind to specific surface markers, circulating drug carriers may have a higher tendency to attach and accumulate at the sites of diseases such as cancers. In addition, a number of cell-penetrating molecules have also been discovered that can further improve the drug delivery efficiency[13]. Certainly, the combination of passive and active targeting for drug delivery may yield amplified results.

Despite tremendous efforts made towards discovering novel materials and biomolecule markers for targeted DDS, very few of them are truly specific after intravenous injection and the targeting still remains chance-dependent. Both active and passive targeting approaches require exogenous drug vehicles to disperse and voyage in circulation for a long time to pass through the leaky vasculature or detect the surface markers. However, the circulating environment, in which many drug vehicles cannot have a long enough circulating time to achieve targeted binding, is extremely complicated[14-15]. In addition, the human body has an innate defense mechanism for invasion. For example, the reticuloendothelial system (RES) rapidly recognizes foreign bodies and destroys them via a series of biological processes. The RES, also called the mononuclear phagocyte system, is comprised primarily of bone marrow progenitors, blood monocytes, and tissue macrophages[16]. Furthermore, the EPR effect is somehow heterogeneous in the tumor microenvironments and varies among patients[17]. For example, hypoxic regions of solid tumor generally do not even exhibit EPR effects due to poor angiogenesis[17-18]. Considering the complexity and sophistication of *in vivo* conditions, conventional passive and active targeting strategies still remain inadequate. Hence, developing novel DDS with truly specific targeting is a formidable challenge for modern medicine and nanotechnology.

Cell-mediated DDS

Recently, cell-mediated DDS have emerged as a promising strategy that is poised to address the above challenges. This novel strategy takes advantage of cell properties, such as long circulation time, abundant surface ligands, flexible morphology, cellular signaling, and metabolism, to offer a unique opportunity to maximize therapeutic outcomes as well as minimizing side effects. The increasing attention towards this field can be seen from the increasing numbers of publications according to Web of Science™ (Figure 1). Here, we review recent advances in the design of cell-mediated targeting and drug delivery systems. It is our hope that this review will synergize the current ongoing efforts and lead to future innovations in live cell-mediated DDS.

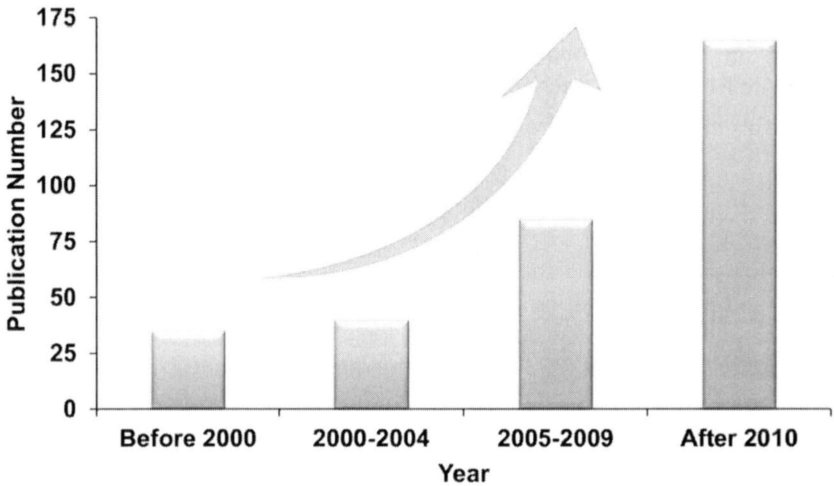

Figure 1. The numbers of publications searched with the keywords of "Cell Mediated" and "Drug Delivery". Source: Web of Science™.

CIRCULATING CELLS

Circulating cells can serve as ideal drug delivery carriers for a number of reasons. They are highly mobile and able to travel through blood flow without immunogenicity. Circulating cells are involved in various disease processes, including infection, inflammation, and cancers development, so they can offer multiple advantages for disease targeting. For instance, leukocytes have the innate ability to cross the blood brain barrier (BBB) to access tumor cells in the brain[19]. Inflammatory responses and wound healing of many diseases are involved in cell homing processes that spontaneously attract circulating cells to disease sites. Furthermore, using circulating cells as delivery vehicles is advantageous as it significantly reduces immune clearance and prolongs the biological half-time for drug delivery. Candidate cells to mediate drug delivery include erythrocytes, leukocytes, platelets, and stem cells, whose properties are summarized in Table 1.

Table 1. Properties of erythrocytes, leukocytes, and stem cells.

	Erythrocytes	Leukocytes	Stem Cells
Amount (/ml blood)	$4.2\text{-}6.0\times10^6$	$4\times10^3\text{-}11\times10^3$	Unknown
Diameter (μm)	~7	7-15	30-40
Location	Blood circulation	Circulation, tissues and organs	Circulation, tissues and organs
Life Span	~120 days	A few days; years for memory lymphocytes	Depending on cell types
Functions	Transport oxygen	Immune defense	Repair tissues and organs
Pros as DDS	Large encapsulating volume; large surface areas; longer life span; reversible deformation; RES targeting.	Capable of cross biological membrane; tumor-homing; immune response.	Potency; homing to injured cells, tissues and organs.
Cons as DDS	Rapid RES elimination; drug released off-site	Short life-span; hard to handle.	Hard to maintain potency *in vitro*.

Red blood cells

Erythrocytes, or red blood cells (RBCs), make up the largest population of blood cells (>99%). Approximately 2 million new erythrocytes are continuously produced per second in the human body. RBCs are non-nuclear biconcave discs that average ~7 μm in diameter and ~2.5 μm in thickness, and have a large internal capacity volume of 185-191 μm^3. Utilizing hemoglobin as an iron-containing protein, RBCs transport oxygen from the respiratory organs to the rest of body. In addition to oxygen, RBCs can carry a range of valuable payloads from therapeutics to imaging contrast agents.

RBC-based drug delivery has attracted increasing attention for many reasons. RBCs can be easily isolated, frozen, and stored for an extended period of at least ten years[20]. RBCs in the blood circulation have a life span of 120 days and thus may act as a reservoir for sustained drug release[21]. The biconcave shape and non-nuclear architecture allow RBCs to encapsulate a large amount of drugs. The membranes of RBCs have reversible deformability, making them capable for taking up payloads via physical methods. Additionally, RBCs are completely biodegradable without producing toxic byproducts, as the RES recognizes old and incompatible RBCs and rapidly removes them. The clearance pathways of RBCs have been widely employed in targeting the RES of the liver, spleen and bone marrow.

The use of RBCs as drug delivery vehicles, however, still encounters several challenges. RBC carriers can be rapidly eliminated by the RES after drug encapsulation or cellular modification due to potential morphological and functional alterations. The RES rapidly recognizes and eliminates modified RBCs before they reach disease sites, which makes non-RES targeting particularly challenging. Drugs may also be released from destructed cells and result in cytotoxicity. The abundance of RBCs could be a double-edged sword that prevents efficient delivery of therapeutics to disease sites. In addition, RBCs are generally not as selective as other circulating cells in terms of targeting disease sites and promoting the healing process.

Leukocytes

Leukocytes, or white blood cells (WBCs,) play significant roles in the immune system, by cleaning cellular debris and defending the body against infections and diseases. Normal human blood contains $4\text{-}11\times10^{9}$leukocytes per mL[22]. Leukocytes are found in five major types, neutrophils (40% - 75%), eosinophils (1% - 6%), basophils (less than 1%), monocytes (2%-10%), and lymphocytes (20%-45%). Although the lifespan of leukocytes (up to 20 days) is typically shorter than that of RBCs, their specialized functions make them attractive as drug delivery carriers because leukocytes are involved in various immune responses, cellular interactions, cell adhesion, and are capable of penetrating through biological barriers into non-vascular areas. Neutrophils, monocytes/macrophages, and lymphocytes are all potential candidates for delivering therapeutics to treat various diseases.

Neutrophils, also known as polymorphonuclear granulocytes (PMNs) that contain distinctive cytoplasmic granules, are the most abundant leukocyte found in the human body. They are the first cells that arrive at the sites of infection or inflammation, produce cytokines to recruit other cells, and are cleared after a few days[23]. Neutrophils also engulf invading microorganisms or foreign substances and consequently eliminate the invaders using digestive enzymes (e.g. lysozyme, hydrolytic enzymes and myeloperoxidase) or respiratory burst[24]. Unfortunately, neutrophils have the average lifespan of 5.4 days in circulation and only a few hours after their isolation from blood[25]. The short life span of neutrophils limits their applications in DDS, but their ability to immediately migrate and transport make them attractive as drug carrier candidates.

Monocytes are mononuclear leukocytes with kidney-shaped nuclei and clear cytoplasm. They are produced from stem cell precursors in the bone marrow. Monocytes circulate in the bloodstream and migrate to tissues, particularly the liver, lymph nodes, and lungs. Meanwhile, monocytes continuously migrate to and accumulate at disease sites in association with infection or inflammation[26-27]. Once leaving the blood flow, monocytes differentiate into macrophages in response to various stimulations. Otherwise, they return to the bone marrow without activation. Macrophages play versatile roles in inflammation, cell recruitment, secretion of cytokines and growth factors, and bacteria/cellular debris clearance. Emerging studies also indicate that macrophages are the major players in disease microenvironments and disease progression, such as in cancer invasion[28-29]. Additionally, monocytes/macrophages present phagocytic capability that enables the spontaneous encapsulation of drugs/particles.

Lymphocytes are identified by their large nucleus surrounded by a thin layer of cytoplasm. The average diameter of lymphocytes is 7-15 μm. Lymphocytes are primarily found in the circulation and central lymphoid organs such as the spleen, tonsils, and lymph nodes[30]. T cells and B cells are the major types of lymphocytes and are responsible for the adaptive immune system. T cells mature in the thymus and play a critical role in cell-mediated immunity and can be broadly divided into helper T cells, cytotoxic T cells, and regulatory T cells[31]. When an antigen appears, antigen-presenting cells (APCs) recognize and present the antigen to T cells. Then, helper T cells secrete various cytokines, which stimulate toxic T cells to directly kill abnormal cells. Regulatory T cells are also activated to suppress immune response in order to maintain immunological tolerance. B-cells are produced in the bone marrow and involved in humoral immunity. B cells make antibodies against antigens and can be characterized by the presence of immunoglobulin on their surface[32]. Meanwhile, B cells can differentiate into memory B cells, which respond rapidly when exposed to the same antigen. Thus, both lymphocytes present multiple functions in human immunity and are involved in numerous diseases: detecting antigens, infiltrating disease sites, and attacking abnormal cells. Clearly, lymphocytes could serve as a potential platform to deliver drugs specifically.

Overall, leukocytes have a rapid response and intrinsic homing properties with respect to infections, inflammations, and tumors. Such sensitive detections and biological barriers infiltration abilities give rise to opportunities for leukocytes-mediated drug delivery. However, vulnerable leukocytes are difficult to harvest and handle with relative short life spans, which hinder the manipulation processes for loading drugs and NPs.

Stem cells

Stem cells are self-renewable with a potential to differentiate into various cell types and therefore essential in tissue repair and regeneration. They are broadly classified into two categories according to their source and plasticity: embryonic stem cells and adult stem cells. Embryonic stem cells originate from the inner cell mass of a blastocyst in an early-stage embryo, which are pluripotent that can generate all cell types in the body. Adult stem cells are tissue-specific stem cells that can differentiate into limited specialized cells[33]. Stem cell therapy has been significantly explored in tissue engineering, regenerative medicine, and even translated into clinical trials[34-35]. Utilizing stem cells as drug carriers can be particularly beneficial since it may add another dimension into existing stem cell therapy. Current stem cell therapies including those for neurological disease, heart disease, and cancer can all be improved by incorporating therapeutics into stem cells. Interestingly, stem cells are inherently tumor homing, which is mediated by inflammatory factors, tumor-derived factors, and tumor-specific receptors[36-37]. Generally, stem cells take 2-4 days to migrate to tumor sites[38]. Thus, stem cells also have natural living targeting capabilities that response to diseases and tissue regeneration. Additionally, stem cells are relatively easy to harvest, culture *in vitro*, and can be induced to differentiate into specialized cells under certain conditions. These properties make stem cells promising candidates for targeted drug delivery.

ENCAPSULATION/CONJUGATION OF DRUGS/PARTICLES TO CIRCULATING CELLS

In order to obtain cell-mediated drug delivery, drugs or drug-loaded particles should be able to attach onto or internalized into cells to form drug-cell or particle-cell complexes. The basic requirements for drugs and drug-loaded particles in cell-mediated delivery are no or low toxicity to cell carriers, particle degradability, and controlled release of drugs. For drug/particle-cell complexes, strong binding between payloads and cell carriers, long circulating time, and low immunogenicity are desired. The use of cell carriers should not only transport drugs in high specificity to target tissues, but also regulate pharmacokinetics and pharmacodynamics via altering the properties of biomaterials that are used to encapsulate therapeutics. To date, drug vehicles can be added to cells by biological pathways, physical approaches or chemical modifications (Figure 2).

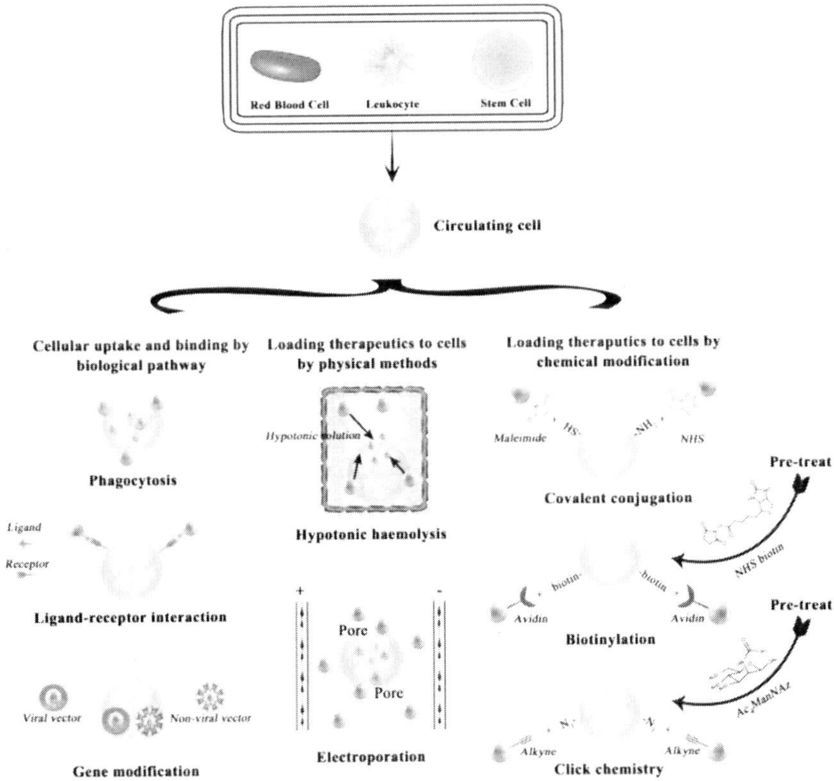

Figure 2. Strategies of encapsulation and conjugation of drugs/particles onto/into circulating cells, including red blood cells (RBCs), leukocytes, and stem cells, by biological, physical and chemical means.

Cellular uptake and binding by biological pathways

The most straightforward way to load drugs with circulating cells is using unmodified cells, which possesses natural biological pathways that can be utilized. Biological pathways represent a series of actions that regulate cell behavior and fate at the molecular level. Various cells are able to load drugs via different pathways. In these cases, their morphology and functional integrity can be largely preserved. In this section, we will review the pathways that facilitate the encapsulation of drugs or genes within cells and/or conjugating nanoparticles onto the surface of cells. Loading therapeutics via physical and chemical modifications of cells will be reviewed in later two sections, respectively.

Endocytosis

Endocytosis is a fundamental cellular process, by which eukaryotic cells engulf fluid, molecules, and even other cells[39]. Most substances, especially large and polar molecules, cannot directly pass through the hydrophobic cell membranes. Endocytosis provides a pathway to internalize these substances via plasma membrane deformation. During this process, plasma membranes invaginate and form intracellular vesicles around the substances to be internalized, followed by the membrane fusion. Endocytosis is dominated by several parameters such as sizes, shapes, charges, and mechanical properties of the foreign substances[40]. It offers an opportunity to regulate the encapsulation of drug vehicles by an endocytosis-dependent pathway.

Pinocytosis is a mode of endocytosis that can non-specifically "drink" the surrounding liquid and molecules. Eukaryotic cells continuously conduct pinocytosis and are prone to engulf relatively small molecules. This plasma membrane homeostasis process has been utilized encapsulation bare drugs into erythrocytes in early attempts. Drugs, such as hydrocortisone, primaquine, vinblastine, and chlorpromazine, can directly induce erythrocyte membrane internalization by pinocytosis[41-43]. The formation of vacuoles is age-related[44]. Pinocytosis is adenosine triphosphate (ATP)-dependent and influenced by the drug concentration, temperature, pH, and the balance between the concentrations of magnesium and calcium in the cell membrane[45-46]. These drug-loading methods are based on simple absorption; however, they tend to induce potential cytotoxicity.

In contrast to pinocytosis, phagocytosis sporadically occurs in specialized cells (e.g. neutrophils, macrophages, lymphocytes, and dendritic cells). Stimulated by immune responses, these white blood cells can internalize particles in nano or micro sizes via an actin-dependent mechanism associated with receptor reorganizations such as Fc receptors, complements, and mannoses[47]. The sizes of drug vehicles, including NPs, micelles, and liposomes, commonly fall in this range. Then, the engulfed particles fuse with lysosomes that are later broken down by abundant digestive enzymes in lysosomes. Since phagocytosis is a type of immune defense procedures against foreign invaders, including drugs or drug-encapsulated particles, it gains tremendous interest in cell-mediated DDS design. Drug-laden particles are susceptible to rapid RES clearance *in vivo*, which means immune cells are the first cells to sense and catch particles during circulation. Instead of designing particles to overcome rapid RES elimination, phagocytic cells can be rationally utilized as delivery carriers. Further, phagocytosis can be significantly enhanced by opsonization, by which particles are coated with proteins and/or antibodies to enable the rapid reorganization of particles by the mononuclear phagocytic system[48]. For example, immunoglobulin and albumin were coated on superparamagnetic iron oxide nanoparticles (SPIO NPs) to facilitate the monocyte-macrophage uptake[49]. Compared with pristine SPIO NPs, a larger amount of opsonized-SPIO NPs accumulated in monocytes nuclear periphery[49].

Numerous nanomaterials, including inorganic NPs, polymeric NPs, and liposomes, have been successfully loaded into phagocytic cells and delivered to target various diseases (Figure 3A)[50-54]. One important understanding in the application of phagocytic cell carriers is that the material geometry plays a significant role in regulating cell internalization. Specifically, particle size is crucial in this process, since phagocytosis takes place at the nano/microscale level[55-56]. IgG-opsonized and non-opsonized polystyrene (PS) microspheres with an average diameter of 2-3 μm exhibited the highest internalization in a rat alveolar macrophages phagocytosis model[55-56]. Material geometry also affects phagocytosis. Champion *et al.* studied phagocytosis of alveolar macrophages and cultured the cells with PS particles with various sizes and shapes[57]. Results suggested that macrophage internalization largely depends on the local particle shape at point of initial contact. The local particle shape is defined as the angel (Ω) between the normal of initial cell/particle contact point and a vector from the initial contact point to the center point of the particle contour. Thus, Ω represents the local curvature. There is a critical point, $\Omega = 45°$, where the internalization cannot be observed. Phagocytosis does not occur any more where is $\Omega > 45°$[57]. The relationship between local particle shape and phagocytosis leads to so-called shape engineering, which is important in cell-mediated drug delivery. Additionally, macrophage phagocytosis is regulated by the mechanical properties of the substrates, as macrophages show a strong preference to engulf rigid objects[58]. For example, macrophages internalized about 6 times IgG opsonized polyacrylamide micro-beads with a higher crosslinking rate (stiff beads) than those soft beads with 4 times lower crosslinking rate.[58]. The effect of surface charge on phagocytosis, however, raises a controversy. Ari *et al.* coated drugs with poloxamer and/or phospholipid surfactants to form NPs with various surface charges. They found that positively charged NPs accumulate more in macrophages than negatively charged NPs[51]. But liposomes with negative charges on surfaces were uptaken more by macrophages through the scavenger receptor recognition[59-60]. These paradoxical observations might be due to various phagocytic pathways adopted in the cellular uptake of different materials.

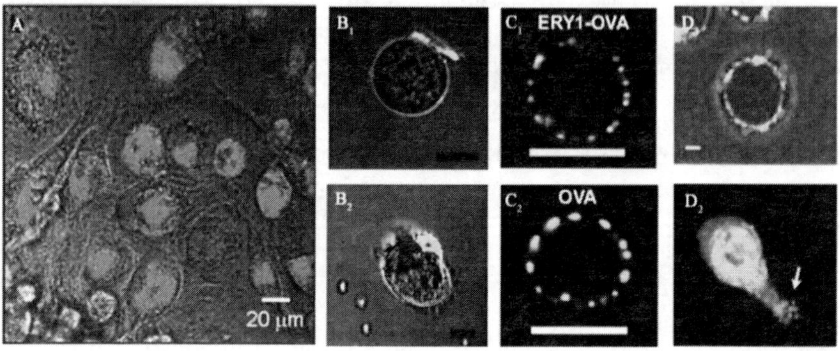

Figure 3. Examples of drugs/NPs internalized into or conjugated onto circulating cells via innate cellular uptake/binding. (A): Endocytosis pathway. Fluorescence microscopy images of phagocytosis of gold silica nanoshells (red) by macrophages[54]. Reprinted with **...**

Certainly, it is also important to prevent drug release within phagocytic cells before reaching the target disease site, since it may compromise the therapeutic effects and also damage cell carriers. In some cases, it is necessary to inhibit phagocytosis since cells may rapidly degrade the internalized particles. Thus, the design of drug-loaded particles is critical to achieve ideal particle location. Generally, it is considered that surface anchored particles would be ideal. For example, worm-like particles with very high aspect ratios (>20) were designed to partially inhibit macrophage phagocytosis in order to protect the drug during the circulating and targeting processes[61].

Therefore, endocytosis is a powerful and tricky pathway to load drugs and/or NPs. However, due to its complexity involving diverse receptors and a series of fusion and fission events, precisely controlling cell internalization is still a significant challenge. New techniques to fine-tune particle properties and endocytosis processes are still needed to improve endocytic cell-mediated nanoparticle delivery.

Ligand-receptor interactions

Ligand-receptor interactions are one of the most basic ways for cells to communicate with each other and the environment. Cellular receptors are generally embedded within plasma membrane surfaces, which contain specific sites for ligand binding and *vice versa*. Ligand-receptor interactions have been utilized by currently active nanoparticle targeting strategies due to its selectivity and specificity. Thus, this approach also can be used to load particles onto/into circulating cells. Considering that circulating cells are relatively rich in receptors on the surface, a variety of drug carriers have been linked to them via ligand-receptor interactions.

As an example, CD44 markers are involved in cell-adhesion and extensively expressed on a large number of circulating cells, e.g. leukocytes and stem cells. CD44 is also a receptor for hyaluronic acid (HA), which is a glycosaminoglycan that can be found in the extracellular matrix. HA is a promising drug delivery material that is known for its biocompatibility and intrinsic binding capability to CD44. HA coated polyelectrolyte multilayer (PEM) cell patches were developed to attach on the surfaces of B and T lymphocyte by binding to the CD44 on surfaces (Figure 3B)[62-63]. The cell-PEM-HA complexes retained strong retention that can be controlled by varying the size of the patches and the ratio of cells to patches[64]. The binding efficiency between PEM-HA and B cells increased at lower pH and higher NaCl concentrations [63]. This phenomenon may result from the change of HA conformation that exposed more binding sites to CD44 receptor[63]. Compared to conventional spherical particles of equal volume, the low curvature particles significantly decreased cell uptake. Interestingly, flat PEM-based disks with ~6 μm average diameters that exhibited low phagocytosis were able to tightly attach on cell surfaces[65]. This so called cellular backpack is an alternative approach for cell-mediated drug delivery by attaching drug-loaded particles on cell surfaces, maximizing protection for therapeutic payloads and cell integrity. The cellular backpack achieved controlled release of drugs since the drugs were encapsulated in biodegradable poly(lactic-co-glycolic acid)[65]. These approaches are promising in preventing drug release off-site, avoiding RES elimination, and enhancing the retention time of NPs in blood flow. Most recently, self-assembled thermo-sensitive poly(ethylene glycol) methacrylate-HA nanogels were found rapidly phagocytized by macrophages and accumulated in the liver and spleen 13 minutes after intravenous injection in mice. This study also suggested that HA-CD44 binding was effective for attaching nanocarriers on macrophages[66]. These macrophage/nanogel complexes have shown potential in drug delivery for macrophage-associated diseases.

The cell-particle encapsulation/conjugation also can be achieved by antibody–antigen interactions. Silica nanorattles, namely hollow silica spheres with moving cores, were bioconjugated with monoclonal antibody to recognize membrane proteins of CD73 and CD90 on mesenchymal stem cells (MSCs)[67]. After antibody–antigen recognition, antibody-modified nanorattles were internalized within MSCs, suggesting that multiple pathways were involved in the process[67]. Each MSC can retain up to 1,500 nanorattles for at least 48 hours, which is sufficient for MSCs to migrate to tumor sites[67].

Additionally, some receptors only exist in particular cell types, granting opportunities for cells taking up particles in the circulation instead of assembling the cell-drug complex *in vitro*. For instance, complement receptor type 1 (CR1) is primarily expressed on RBCs. Monoclonal antibody (mAb) against CR1 (Anti-CR1) was conjugated on a tissue-type plasminogen (tPA) activator in order to couple with RBC carriers[68]. Anti-CR1/tPA can rapidly bind to RBCs to prolong circulation time[68]. Glycophorin A is another erythrocyte specific surface marker. ERY1 is a 12 amino acid peptide that is able to specifically bind to the RBC surface via glycophorin A[69]. Antigen ovalbumin conjugated with ERY1 spontaneously recognized RBCs within 30 minutes and remained tightly bind for at least 72 hours after intravenous injection in mice (Figure 3C)[70]. However, to the best of our knowledge, no drug-loaded particle has been conjugated to circulating cells through these specific interactions.

Thus, ligand-receptor binding is a powerful method to attach particles to circulating cells. Advances in cellular engineering with discoveries of new surface markers and binding ligands are beneficial for ligand-based drug carrier design. Improving the binding efficiency on particular cell types is still remaining as a challenge. It is also desired to accomplish *in vivo* specific binding within the blood circulation.

Gene modification and delivery

Great efforts have been made to insert therapeutic plasmid DNA, mRNA, and siRNA into living cells to build novel gene or drug delivery systems. Transduction and transfection, both of which are powerful ways to obtain a large number of functional cells *in vitro* and *in vivo*, can introduce foreign genes, including therapeutic genes, into host cells[72-73].

Cell-mediated viral delivery to tumors has been well studied. Oncolytic virus (OV) is an anti-cancer agent that can selectively induce cell death in cancer cells, while leaving normal cells healthy[74]. Transduction can effectively introduce OVs into circulating cells based on a virus-mediated DNA transfer mechanism. Furthermore, OVs can be transported with cell carriers to tumors, followed by local replication and amplification to suppress tumor growth. A variety of OVs, such as adenovirus, herpes simplex virus, poxvirus, vesicular stomatitis virus, measles virus, Newcastle disease virus, influenza virus, and reovirus, has been combined with different cell-based delivery systems. For example, mesenchymal progenitor cells (MPCs) have served as vehicles to deliver oncolytic adenoviruses[75]. To enhance transduction efficiency, oncolytic adenoviruses were first genetically modified with cellular integrin-binding motifs[76]. Transfused MPCs maintained their intrinsic capacity of sensing tumors and further improved the therapeutic effects in an ovarian carcinoma model[75]. Additionally, human activated T lymphocytes infected with oncolytic measles viruses were able to transfer viruses to tumor cells via the heterologous cell fusion, and subsequently deliver viruses to myeloma[77].

Virus therapy is widely studied, but its clinical safety is still a concern due to rapid proliferation of viruses, off-target cytotoxicity, and body antiviral immunity. Nonviral transfection is an alternative strategy to introduce therapeutic nucleic acids into cellular carriers. Hu *et al.* developed nonviral vehicles, based on low molecular weight polyethylenimine-co-β-cyclodextrins (β-CD), to transfect tumor necrosis factor (TNF)-related apoptosis-inducing ligands (TRAILs) into circulating MSCs[78]. MSC-TRAILs efficiently killed tumor cells in a lung metastasis model. In order to achieve non-invasive drug delivery and MRI contrast capabilities, light sensitive, gold-coated SPIO core-shell magnetic NPs (SPIO@AuNPs) were transfected into cell adipose-derived MSCs[79]. Adipose-derived -MSCs retained high cell viability, osteogenic differentiation, and disease homing ability after transfection. Interestingly, SPIO@AuNPs stimulated cell growth at a certain concentration (<10 μg/mL) and did not induce pro-inflammation, which is typically promoted by other SPIO NPs[80].

With the help of genetic modifications, cell carriers are capable of constantly expressing desired therapeutic or imaging molecules. Since transduction/transfection permanently alters cell properties, it inevitably gives rise to safety concerns and even ethical issues. Researchers should take them into account when applying genetic modifications.

Loading therapeutics to cells by physical methods

RBCs possess various reversible transformation capabilities that allow for drug loading simply with external physical treatments such as hypotonic haemolysis and electroporation (Figure 2). These approaches are relatively easier to handle *in vitro* and eliminates the use of chemical reagents to reduce possible alterations of RBCs. RBCs can recover to their normal morphology and maintain the role of transporting therapeutics and particles to targeted sites.

Hypotonic haemolysis

Hypotonic haemolysis has been widely applied in RBC encapsulation. RBCs behave like osmometers that shrink in hypertonic conditions and swell in hypotonic conditions. Pores open on cell membranes, when the critical haemolytic volume is reached. The increased cell membrane permeability in hypotonic solution enables the entrapment of drugs, biomacromolecules, and NPs in RBCs by a passive mechanism[81-82]. The principle of reversible swelling is based on the fact that the absence of a superfluous membrane allows cells to accommodate additional volume by changing the shape from biconcave to spherical. Several hypotonic haemolysis techniques have been generated, such as hypotonic dialysis, hypotonic dilution, and hypotonic pre-swelling[83-84]. Hypotonic dialysis is predominantly applied in encapsulating enzymes, proteins, and contrast agents due to its relative ease of use, ability to preserve cell characteristics and high encapsulation rate. In the process, erythrocytes with a hematocrit value of 70-80% are prepared in a dialysis tube and immersed in a hypotonic buffer for a few hours under gentle stirring. The techniques have been applied in encapsulating drugs, enzymes, and proteins since 1970s[85-90]. Peptide nucleic acids (PNAs), as therapeutic agents, were loaded into RBCs via hypotonic dialysis and achieved 82% encapsulation after 18 hours incubation[88]. PNAs-laden RBCs underwent opsonization with $ZnCl_2$ and bis-sulfosuccynimidil-suberate treatment, and then specifically targeted macrophages. PNAs were effectively delivered and resulted in reduced production of nitric oxide and inhibited protein expression of the enzyme nitric oxide synthase[88].

Electroporation

Electroporation is a physical method that uses an electrical pulse to create temporary pores on cell surfaces[91]. Cells are typically suspended in a conductive solution and exposed to a high-intensity external electrical field. The disturbance of the phospholipid bilayer induces temporary dysfunctions of the cell membranes, resulting in the encapsulation of exogenous molecules in cells. During the process, cells keep their morphology when engulfed molecules are larger than the size of electropores. However, cells may swell and the cell membranes may rupture if the substances are smaller than the size of electropores[92]. Various compounds such as drugs, enzymes, and genes have been encapsulated within RBCs accomplished with sustained release[93-95]. The drawbacks of electroporation are obvious. The electrical force causes irreversible deterioration and the recovery rate is commonly low[95]. Thus, it is necessary to optimize the applied voltage to keep cell integrity in electroporation.

Loading/conjugating therapeutics into/onto cells by chemical modifications

Compared to biological and physical methods to load drugs into cells, chemical modification may fix drugs and drug vehicles onto/into circulating cells and avoid the metabolic degradation of materials and drugs within the carrier cells. The surface of circulating cells is a highly heterogeneous structure composed of different proteins, lipids, and carbohydrates with various functional groups for chemical modification opportunities as mentioned previously[96-97]. Furthermore, chemical modification is not solely dependent on the naturally available cell surface molecules. The complexity of the cell membrane offers alternatives to introduce additional functional groups for a number of conjugating reactions. Compared to pristine cells, modified cells enable convenient conjugations with therapeutics and particles via versatile chemistry tools (Figure 2).

Covalent conjugation onto surface markers

A number of functional groups have been discovered on different cell surfaces that allow bioconjugation with molecules or particles. Among the available functional groups, primary amine and thiol groups are mostly used for covalent reactions due to the ease of labeling and low toxicity of the reactions. N-hydroxysuccinimide (NHS) ester-based crosslinkers are most popular for primary amines modification via carbodiimide reaction. Rossi *et al.* grafted hyperbranched polyglycerol (HPG) onto the surface of RBCs to generate immunocamouflaged cells[98]. HPG is a hemocompatible polymer with abundant hydroxyl groups that are ideal for surface modification. HPG was functionalized with succinimidyl succinate (SS) groups by initially reacting with succinic anhydride and then with NHS[98]. SS-HPG was eventually grafted with primary amine groups presented on the RBC surfaces. Also, the thiol groups in cysteine containing membrane proteins are available for coupling via maleimide-thiol conjugation. Maleimide-functionalized liposome-coated NPs were conjugated on the surface of lymphocytes and MSCs[99]. The attachment of NPs did not activate lymphocytes or alter stem cell tumor-homing ability[99]. Stephan *et al.* also found maleimide-functionalized NPs rapidly relocated to T-cell/tumor cell contact zone following antigen recognition (Figure 3D)[71].

Covalent coupling typically exerts stronger binding than ligand-receptor interactions to avoid drug vehicle detachment during cell migration. Versatile conjugation chemistry enables exclusively covalent reactions with amino or thiol groups that are naturally expressed on cells. These covalent bindings help anchoring drug carriers on the surfaces of cells, which have advantages over engulfing inside cells in terms of cellular safety and drug protection.

Biotinylation

Biotin is a water-soluble vitamin that has a high affinity for avidin and streptavidin proteins[100]. Biotinylated cells can be conjugated with various proteins, enzymes, and NPs without compromising their viability and biological activities. NHS ester-activated biotins are widely-accepted biotinylation reagents that can introduce biotins on living cell membranes that contain primary amino groups[101-102]. NHS esters form stable amide bonds with primary amino groups in buffers within a few hours at room temperature, which can be tolerated by most cells. NeutrAvidin-coated NPs were anchored on the biotinylated membranes of human mesenchymal stem cells (hMSCs) for up to 2 days, while a certain amount of NPs were observed within the cytoplasm of hMSCs possibly resulting from the membrane recognition of NPs[103]. hMSCs maintained the tumor-homing capability after nanoparticle attachment[103]. RBCs were also coupled with a thrombolytics agent on their surface via avidin-biotin bridges, serving as Trojan horses for prophylactic fibrinolysis[104-105]. With increasing attention to cell surface biotinylation, additional biotinylation reagents have been developed to target different functional groups such as sulfhydryls, carboxyls and carbohydrates. This diversity provides multiple choices in terms of surface modification of circulating cells to bind and load drugs or NPs.

Click chemistry

"Click" chemistry represents a rapid, selective, and high-yielding bio-orthogonal reaction that is also capable of immobilizing materials on cell surfaces[106]. Numerous click chemistry strategies have been recently developed for living cells, including copper(I)-catalyzed azide-alkyne (CuAAC), strain-promoted azide–alkyne cycloaddition (SPAAC), thiol-ene, Diels–Alder, and Pseudo-Click reactions. Unlike many traditional synthetic routes, click reactions take place in water solution at physiological pH (6-8) and temperature (37°C) without toxic by-products, which are ideal platforms for loading drug vehicles onto circulating cells[107-108]. To improve the binding efficiency, abiotic functional groups are initially introduced onto cells allowing counter groups on drugs and particles to undergo click chemistry. To date, CuAAC, SPAAC, and thiol-ene reactions have been applied to click living cells[109]

CuAAC and SPAAC are [3+2] azide-alkyne cycloaddition in the context of Staudinger Ligation[110]. Azide groups, which are naturally absent on mammalian cell surfaces, need to be introduced to enable click reactions between cells and therapeutics via biosynthesis. Cell glycosylation, which is also called carbohydrate reaction, provides accessibility to incorporate azido sugars onto membranes through sialic acid. Generally, azido sugars can be metabolically inserted into the cell membranes after 24-48 hours of incubation. The azide-functionalized cells subsequently undergo reaction with drugs or NPs containing activated alkyne groups. The reaction, however, is thermodynamically favorable at high temperature or pressure, which is not applicable in living systems. Thus, CuAAC utilizes Cu^+ catalyst to accelerate the reaction within 30 minutes at physiological conditions[111]. Unfortunately, the cytotoxic Cu^+ catalyst limits the*in vitro* and *in vivo* applications. Importantly, studies have indicated that alkyne activation can be achieved by introducing ring strain, namely SPAAC, which is a Cu-free click chemistry for selective biomolecule labeling in living organisms based on different glycosylation degrees between abnormal and normal cells[107, 112]. Xu*et al.* immobilized highly branched polyamidoamine dendrimers on azido-modified macrophage surfaces by SPAAC copper-free click chemistry[113]. Cell motility and its common stress-activated signaling pathways, including AKT, p65, and p38, have no significant alteration during the cell-nanoparticle hybridization process. An alternative approach to employ click chemistry is the thiol-ene reaction. Briefly, methacryloyl groups can be incorporated onto cell surfaces through a glycosylation pathway and then clicked with a thiol-terminal polymer. The thiol-ene reaction was explored on clicking cancer cells[114]. Despite few developments on clicking circulating cells, click chemistry has shown potential in delivering therapeutics and imaging agents. A shortcoming of linking cell and therapeutics by click chemistry is that relatively complicated surface chemistry is required.

TARGETING STRATEGIES OF CELL-MEDIATED DRUG DELIVERY

Once drug vehicles are assembled with circulating cells either *in vitro* or *in vivo*, the complexes can travel and diffuse within the blood circulation and further navigate to the desired targets (Table 2). By taking advantages of carrier cells' intrinsic capabilities, controlled drug/particle delivery to specific disease sites can be achieved (Figure 4.). In this section, we will review the existing circulating cell-based targeting strategies for various objectives.

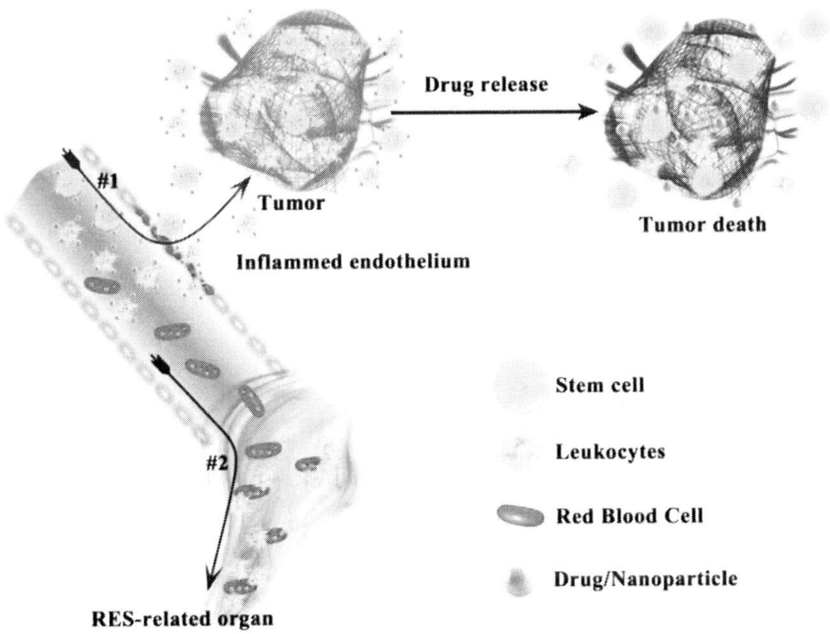

Figure 4. Illustration of circulating cell-mediated targeting and drug delivery pathways. #1: living targeting to disease sites via leukocytes and stem cells. #2: RES targeting via RBCs.

Table 2. Summary of current drug vehicle encapsulation/conjugation methods for different cell-mediated drug delivery and their targeted applications.

Pathway		Mechanism	Cell Type	Drug/ Particles	Target	Ref.
Unmodified Cells	Endocytosis	Endocytosis	Monocyte, macrophage	(IgG/ albumin coated) SPIO NPs,	Lymphoid tissues	49
			Monocyte, macrophage	Au nanoshells	BBB and hypoxia	53, 54
			Macrophage	Liposomal doxorubicin	Lung cancer	50
			Macrophage	Catalyse-polyethyleneimine-poly(ethylene glycol)	Parkinson's disease	52
			Macrophage	Indinavir NPs	HIV-1	115
		Anti-endocytosis	Macrophage	Polyelectrolyte multilayer-based disks	Macrophage associated diseases	65
	Ligand-receptor binding	Hyaluronic acid (HA)-CD44	B and T lymphocyte	Polyelectrolyte multilayer	ICAM-coated surface	62-64
			Macrophage	poly(ethylene glycol) methacrylate-HA nanogel	Liver and spleen macrophage-associated diseases	66
		Antibody-CD73 and CD90	MSC	Doxorubicin-silica nanorattles-antibody	Glioma tumor	67
		Antibody-complement receptor type 1	RBC	Tissue-type plasminogen activator-antibody	Thromboprophylaxis	68
		Peptide (ERY1)-glycophorin A	RBC	Antigen ovalbumin-ERY1	T cells	70
	Covalent conjugation	NHS-NH_3	RBC	Hyperbranched polyglycerols	RES-related organs	98
		Maleimide-thiol	T cell, MSC	Liposome-coated NPs	Prostate cancer, lymphoma	71, 99
Physical Modification	Hypotonic haemolysis	Membrane flexibility	RBC	Peptide nucleic acids	Macrophages	88
	Electroporation	Temporary dysfunctions of membrane semi-permeability	RBC	Alcohol dehydrogenase and aldehyde dehydrogenase	RES-related organs	116

Pathway		Mechanism	Cell Type	Drug/ Particles	Target	Ref.
Chemical Modification	Biotinylation	Avidin-biotin bridges	hMSC	NeutrAvidin-coated NPs	Liver tumor	103
			RBC	Thrombolytics agent	Thrombosis	104, 105
	Click Chemistry	Strain-promoted azide–alkyne cycloaddition	Macrophage	Polyamidoamine dendrimers		98
	Transduction	Genetic transfer via viral vectors	T cell, MSC	Oncolytic virus	Myeloma ovarian, carcinoma	76, 77
	Transfection	Genetic transfer via non-viral vectors	MSC	TRAILs	Lung cancer metastases	78
			MSC	SPIO@AuNPs	Head and neck cancer	79

Passive targeting: RES/non-RES targeting

The first rational living cell-based targeted drug delivery strategy is passive targeting, including RES and non-RES effects. For example, as RBCs age, they are recognized by the RES and subsequently undergo cell membrane lysis by lysosomal enzymes, hemoglobin degradation, and eventually lose their flexibility, integrity, and functionality[117-118]. Taking advantage of the metabolic pathways in physiologic conditions, RBCs-based drug/particle complexes can easily and efficiently target RES-related regions such as liver, spleen and bone marrow. RES clearance can be accelerated by altering the properties of RBCs during drug-loading processes in order to enhance the targeting efficiency. This strategy has been demonstrated in drug delivery for over 30 years[117, 119]. Its applications range from RES organ targeting to targeting specific cell types, including hepatic Kupffer cells, alveolar macrophages, peritoneal macrophages and peripheral blood monocytes, for delivering drugs to treat diverse diseases such as infections, inflammations, leukemia and cancers[120-125].

However, rapid clearance by the RES significantly limits the employment of RBC carriers in non-RES targeting. To address this problem, several approaches have been developed to permit the survival of heterologous or even xenogeneic RBCs in the circulation. A straightforward strategy to modify RBCs is by conjugating with nonimmunogenic materials onto the surfaces of RBCs in order to generate immunocamouflaged cells. Polyethylene glycol (PEG) is a bioinert and hydrophilic polymer that is widely used for cell stealth purposes. Using the same technique, methoxy(polyethylene glycol) (mPEG) was attached on the surface of RBCs and the immunogenity of modified RBCs was adjusted by varying the surface density and chain size of the coating polymer[126]. The mPEG-modified RBCs still retained normal morphology, osmotic fragility, oxygen transportation, and high cell viability, but became resistant to RES clearance with prolonged circulation time[127-128]. An alternative polymer, HPG, has similar antigen "mask" effects[98]. HPG is a hyperbranched polymer with the capability of further modification to achieve controlled release. Furthermore, stimuli-responsive RBCs, such as ultrasound-responsive RBCs[129] and magnetic-responsive RBCs[130], can be directed to disease sites, providing a platform for non-RES targeting. However, cell-based passive targeting strategies are still similar to the passive targeting of bare NPs, which typically cannot reach the desired targeting and drug delivery efficiency. Additionally, RBCs are mostly used in passive targeting pathways, which are less favorable for many diseases.

Live cell-mediated targeting strategies

Many circulating cells such as neutrophils, monocytes/macrophages, and stem cells present targeting capabilities to many diseases, including cancer, wounds, and ischemic tissues. For example, tumor-homing, one of the most studied targeting mechanisms of living cells, suggests that various cells can not only sense solid tumors, but can also track circulating cancer cells in the blood flow and reach the primary and metastatic tumor sites. Targeting solid tumors, tumor microenvironments, and cancer metastases will be reviewed, respectively here.

Leukocytes are the cells involved in immune response so they are essential for inflammation, disease development, regulation, and healing processes. The general paradigm of neutrophil and monocyte extravasation includes tethering, rolling, adhesion, crawling, and eventually transmigration[131-132]. For many diseases, an injured endothelium is activated directly by pathogens or indirectly by pro-inflammation[133-134]. Neutrophils are well recognized as the first cells to be recruited by chemoattractants (e.g. platelet-activating factor, interleukin 8, interleukin 17, formyl-methionyl-leucyl-phenylalanine, and complement component 5a), cytokines (e.g. TNF-α), growth factors (e.g. granulocyte-macrophage colony-stimulating factor), and bacterial products (e.g. formylated peptides and lipopolysaccharide)[135-136]. With the increasing expression of adhesion molecules, P-selectin and E-selectin in particular, neutrophils tether and slowly roll onto the surface of the endothelium[137-139]. Mediated by chemokines and adhesion molecules, neutrophils firmly attach on the endothelium and subsequently crawl towards endothelial adjunctions[140-143]. Finally, neutrophils cross the endothelium, basement membrane, pericytes, and reach disease sites[144]. Monocytes are also recruited by similar chemokines, cytokines and growth factors[132, 145]. Afterwards, monocytes differentiate into highly specialized macrophages to participate in the following immune responses[146]. Similarly, lymphocytes can infiltrate tumor sites and display a series of immune responses[147-148]. Therefore, the tumor-homing mechanisms of leukocytes provide active and "living" approaches for cell-mediated drug delivery. Doxorubicin (Dox) encapsulated liposome-carrying macrophages accumulated in lung tumors and released Dox after 24 hours of administration, but a smaller amount of Dox was observed without using macrophage carries (Figure 5A)[50]. Adhesion molecule-associated extravasation also provides targeting options. PEM functionalized T cells can migrate to ICAM-coated surface[62]. A recent study has demonstrated that RGD modified single-walled carbon nanotubes can target monocytes and enhance the delivery efficiency of monocyte carriers via integrin mediation (Figure 5B)[149].

A particularly important application of leukocyte-mediated drug/nanoparticle delivery is the brain drug delivery across the BBB. The BBB, which is a tight assembly of endothelium cells and astrocytes, separates the brain from the circulatory system[118]. The barrier is therefore highly selective and only allows the passage of essential nutrition such as water, soluble lipid molecules, amino acids, and glucose while blocking other substances. The BBB protects the brain from harmful substances; however, it also establishes a huge obstacle for delivering therapeutic agents to the brain. Interestingly, monocytes/macrophages are able to transmigrate across the BBB. Therefore, they can potentially serve as carriers for sheathed brain drug delivery. Bone-marrow-derived macrophages were found to be able to infiltrate into the brain and deliver nanozyme in a Parkinson's disease model[150]. MRI contrast agent-labeled murine monocytes/macrophages also targeted and accumulated in rat brain tumors[151]. A similar concept was applied in FluoSpheres® NeutrAvidin® labeled microspheres delivered to brain metastasis of breast cancer via monocytes/macrophages (Figure 5C)[54].

The tumor microenvironment, which involves a vast amount of cells, vessels, and stroma, is another potential target for circulating cell-mediated therapeutic delivery. Solid tumors generally contain hypoxic regions with low oxygen levels due to poor angiogenesis. Hypoxia is associated with necrosis and facilitates tumor mutation[152]. Due to the lack of blood vessels, hypoxic cells are resistant to traditional drug delivery approaches. Promisingly, monocytes, tumor-associated macrophages, and tumor infiltrating lymphocytes are frequently found within tumor microenvironments including hypoxic and necrotic regions, suggesting that leucocytes could access in the depth of tumors. Choi *et al.* utilized monocytes/macrophages as "Trojan Horses" to deliver gold nanoshells to the hypoxic regions of cancer and their results suggested that malignant cells were completely killed within these areas, where free nanoshells could not access[53].

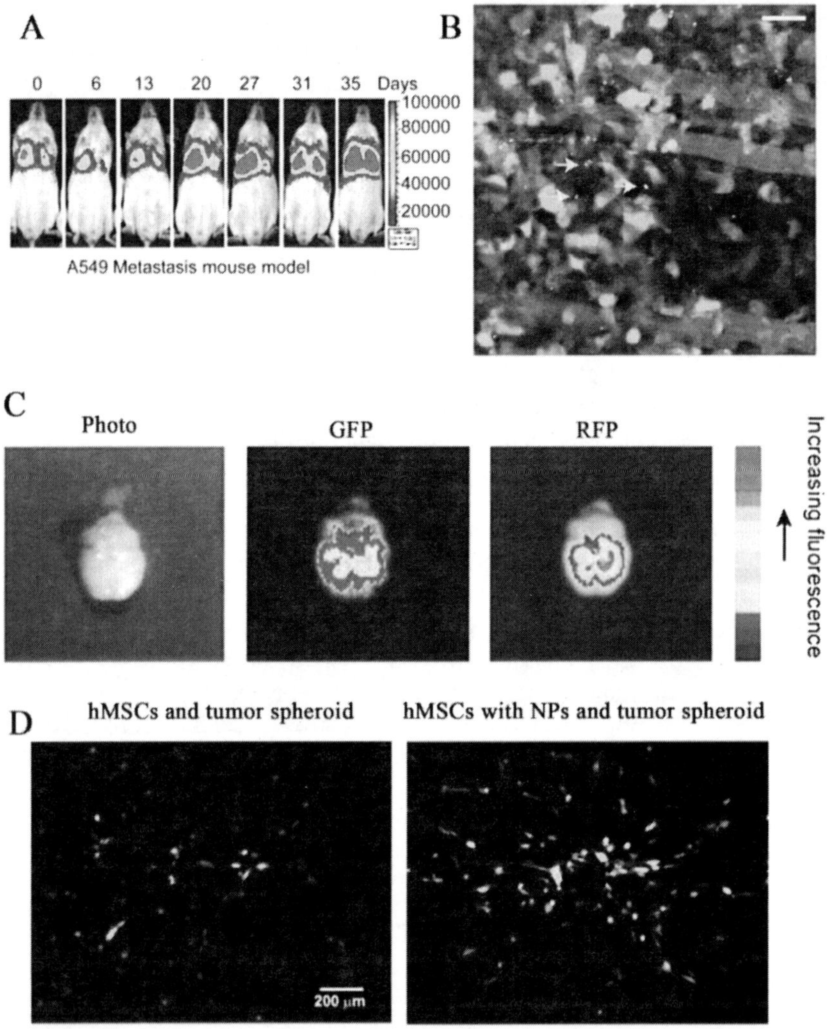

Figure 5. Cell-mediated targeting approaches to different disease sites via leukocytes and stem cells that retain their intrinsic capacities of crossing biological barriers and homing to tumor after nanoparticle assembly. (A): Bioluminescence images of macrophages ...

Stem cells can also naturally migrate towards injured tissues as part of the healing process. In addition, stem cells display tumor-homing and BBB-infiltrating properties [153]. Emerging evidences suggest that various types of stem cells have exclusive tropism for tumors and can thus mediate drug delivery. Previous attempts have emphasized on producing a large number of therapeutic agents in situ via genetically-modified stem cells. For example, interferon-β gene transduced/transfected hMSCs preferentially accumulated in gliomas, melanomas, and pulmonary metastases, then locally released interferon-β, and induced tumor regression[154-156]. In another study, neural stem cells migrated toward intracranial gliomas and delivered the oncolytic drug, 5-fluorouracil, to reduce about 80% of the tumor mass within 2 weeks[38]. Although stem cell-mediated gene therapy showed great outcomes, recent studies have put more attention on developing drug/biomaterial-laden cell carriers to reduce the potential risk of gene modification. NeutrAvidin nanoparticle-coated hMSCs were found be able to sense liver tumor spheroids *in vitro* (Figure 5D)[103]. MSCs surface-modified with Dox-loaded silica nanorattles also successfully tracked glioma cells and released Dox to kill cancer cells[67]. Recently, multi-functional theranostic NPs, such as SPIO-loaded gold NPs and photosensitizer silica NPs, were successfully delivered to different tumors via MSC-mediated approaches[157-158].

Cancer cells are particularly lethal when they reach the metastatic state. Unfortunately, delivering therapeutics to metastatic circulating tumor cells (CTCs) has been a significant challenge for decades due to their scarcity in the blood circulation. CTCs are capable of leaving a primary tumor, seeding metastases in distant sites. Effective therapies that directly remove CTCs from the blood circulation have not been found yet since CTCs circulate in the bloodstream with a low concentration (average 8 cells/ml) and are difficult to detect[159]. Interestingly, studies have indicated that tumor cells exhibit interactions with leukocytes and endothelial cells via adhesion molecules, which provide a platform for targeting CTCs through leukocytes. Inspired by selectin-dependent rolling along the vessel, human E-selectin, and cancer-specific TRAIL functionalized leukocytes were used to target CTCs, which displayed sialylated carbohydrate ligands for selectin under blood flow resulting in effective tumor killing outcome[160]. Studies have also found that tumor cells express adhesion molecules such as intercellular adhesion molecule 1 (ICAM-1, also known as CD54) that can bind to lymphocyte function-associated antigen (LFA-1, also known as CD11a) and macrophage-1 antigen (MAC-1, also known as CD11b) in the circulation[11, 161].

Therefore, circulating cells have shown promise as a powerful drug delivery system with an innate ability to detect diseases, cross biological barriers, and release drugs to disease sites. Compared to conventional drug delivery and nanomedicine, utilizing the natural ability of different cells is a completely different targeting mechanism, in which living cells act as guiders and drivers for delivering therapeutics and NPs. This strategy has also opened up new avenues for treatments of diseases that cannot be easily reached by traditional DDS.

Engineered targeting/binding

To maximize the targeting efficiency, engineered unnatural targeting/binding strategies have been developed for living cell-mediated DDS, allowing flexibility and diversity in binding pathways. Antibody decoration is one of the most popular techniques, by which cell carriers spontaneously search for specific antigens and even trigger the clearance of certain cells. Lipophilic anti-CD45 and anti-CD20 (Rituximab) surface painted RBCs effectively depleted CD45 positive leukocytes and CD19 +/CD20 +/CD45 + human lymphoma cells[162].

An alternative approach is to use an external magnetic field to enhance the accumulation of cell carriers in targets. Magnetic nanomaterials can enable remote control of the location of drug vehicles including circulating cells by an external magnetic field. SPION NPs have been widely explored due to their biocompatibility, ease of handling, and good MRI contrast. SPIO NPs loaded RBCs successfully targeted to tumor cells under magnetic stimulation, further reducing the RES aggregation[163]. The combination of molecular binding and cell-mediated drug delivery is a new and emerging field. The techniques in cellular engineering and biomaterial modifications offer significant opportunities for targeted drug and nanoparticle delivery.

CONTROLLED DRUG RELEASE

The controlled release of therapeutics is desired in medicine[164]. It is even more critical for circulating cell-mediated drug delivery, since it requires not only maximum treatment effects, but also limited drug release in carrier cells. Generally, the encapsulation or conjugation of free drugs in cell carriers is not ideal due to the cytotoxicity of drugs and drug-cell interactions. Drugs are thus protected in vehicles made with various materials forming new complexes to ultimately control the payload release. In this section, strategies for controlled drug release from circulating cells will be reviewed.

Degradation controlled drug release

Biodegradable polymers are the most widely used biomaterials for controlled drug release, since the release can be controlled by degradation mechanisms. Hydrolytic and enzymatic degradations are typical approaches to control drug release rates after being uptaken by cells. Abundant enzymes are located in the cytoplasm of phagocytic cells and able to facilitate the breakdown of engulfed substances. The degradation can be fine-tuned to obtain minimal release in circulating cells and release at desired disease sites. Liposomes are popular drug vehicles that have been used in many commercialized pharmaceutical formulations. Through elegant engineering designs, the desired release of drugs from liposomes can also be achieved after being delivered by living cells. For example, when liposome-Dox-loaded macrophages arrived at the tumor sites, drugs were released via macrophage death cause by leaky Dox from liposome degradation[50]. Biodegradable polymeric particles, for instance, polylactic acid (PLA) NPs, have manageable degradation profiles that are ideal for controlled drug release[165]. The slow degradation may ensure safety to circulating carrier cells and programmed release upon reaching disease sites. However, degradation controlled release may not achieve the ultimate safety and therapeutic effects due to the complicity of the *in vivo* environment and the fact that drug diffusion is also somewhat independent from material degradation. Thus, stimuli-triggered release from a smart drug delivery system can be advantageous.

Stimulated drug release by cells and local microenvironment

Environmental responsive/smart drug release has gained much attention in recent decades in the field of drug delivery, as payloads can be discharged by responding the local cellular/physiological changes, which may have resulted from disease development or inflammation[166-167]. It is also promising to combine the cell-mediated "live" targeting mechanism with a "smart" releasing mechanism. For instance, glutathione can reduce disulfide bonds and is available for responsive drug release within cytoplasm. An acute lymphoblastic leukemia drug, L-asparaginase, was covalently conjugated with cell-penetrating peptides (CPP) via disulfide linkages, followed by the encapsulation by RBCs without any membrane alteration[168]. The disulfide linkage disassociation occurred within the cytoplasm of RBCs, which contains glutathione, and resulted in the release of L-asparaginase release. RBCs spontaneously triggered drug release in the absence of external interventions, but underwent apoptosis to release the drug out of the cells[168].

Another pathway to facilitate drug release from cell carriers is discharging the drug-loaded nanoparticle as a whole. Generally, cells retain their equilibrium via constantly engulfing foreign substances and liberating engulfed particles. In contrast to endocytosis, exocytosis allows secreting drug vehicles out of cells, as a potential approach for nanoparticle delivery. In a recent study, bone-marrow-derived macrophages internalized catalase/polyethyleneimine-PEG nanozymes within 40–60 min and sustained released the cargos for 4-5 days[150]. And the release rate can be enhanced by phorbol myristate acetate treatment[52]. While the integrity of cell carriers can be maintained after the drug is released by taking advantages of exocytosis, the hemostasis may occur and the chances of off-site targeting may rise.

External stimulation triggered drug release

External stimulation is an attractive alternative to trigger the localized release of therapeutic agents after reaching the disease sites, if the local microenvironmental change is not available or not strong enough. Particularly, stimuli-responsive polymers have unique characteristics that respond to small changes in the microenvironment and are therefore being widely used in smart drug delivery[166, 169]. Drugs are securely protected by polymeric vehicles until external stimulation, which ensures rapid, transient and precise drug release is applied. Again, combining external stimuli-responsive release with living cell-mediated targeting is a potential area to be explored.

Until now, numerous smart drug vehicles have been developed that respond to external stimuli, such as temperature, light, pressure, light, or ultrasound. In some cases, drugs can be released by multiple stimulations to ensure an on-demand administration. For example, Swiston *et al*. functionalized lymphocytes with PEM coatings by combining both pH-sensitive and temperature-sensitive polymers. And these lymphocytes were able to regulate drug release by dual factors[62]. Overall, stimuli-responsive drug release can be crucial for a cell based DDS, even more significant than for the conventional DDS.

CONCLUSIONS AND FUTURE PROSPECTS

Circulating cell-mediated drug delivery has emerged as a new concept that focuses on the targeted delivery of therapeutics by living cells instead of passive factors or surface markers. A summary of comparison between conventional DDS and cell-mediated DDS is shown in Table 3. As we have discussed, various cell types in the circulation, including RBCs, leukocytes, and stem cells, are available for carrying drugs and particles to targeted diseases. Generally, three steps are considered in terms of design strategies: drugs/particles loading by cell carriers, cell carriers targeted to desired locations, and controlled drug release/treatment. Compared to traditional targeted drug delivery and nanomedicine strategies, additional factors must be considered for cell-mediated drug delivery. The drugs/particles must not be toxic to the cell carrier and should prevent cell alteration after drug loading. Thus, more attention has been paid to anchoring therapeutics on the surface of living cells. Certainly, high drug/particle loading or binding efficiency is favorable. We have elaborated on the feasibility and variety of cell targeting pathways to target different diseases. Rational selection of appropriate cells is necessary to achieve high targeting efficiency for new cell-mediated drug delivery system. Finally, it is imperative to employ a controlled drug releasing mechanism to ensure safety and maximize therapeutic effects. New treatment methods, such as photothermal and magnetic hyperthermia therapies, can also benefit from living cell-based delivery systems. The development of suitable biomaterials and biomedical engineering techniques is particularly important in creating new and effective cell-mediated drug delivery systems.

Table 3. A comparison of conventional DDS and cell-mediated DDS.

	Conventional DDS	Cell-mediated DDS
Drug Encapsulation	Physical or chemical methods	Biological, physical or chemical methods
Circulating Time	Short, easy to be eliminated via RES system; Stealth coating generally used	Longer, cells are capable of circulating in blood
Targeting	Decorated with targeting ligands and molecules; Mostly depend on EPR effects; Difficult in transmembrane penetration; Access to vascular areas only	Natural homing ability; Possess cell signaling; Penetrate through biological barriers; Reach to non-vascular areas
Drug Release	Controlled drug release via diffusion, degradation or external stimulation	Controlled drug release via exocytosis, degradation or external stimulation
Safety	May have off-site drug release	
Biocompatibility	May cause unwanted immune response;	Natural biocompatibility; May cause unwanted immune response after modification.

Although novel cell-mediated drug/nanomedicine delivery holds promises in improving diagnosis and therapeutic effects, it is debatable whether this strategy is safe enough for clinical practices. Many challenges still remain for this state-of-art approach. Generally, candidate cells are isolated and functionalized *in vitro*, followed by re-injection into the circulation or disease sites. Although current techniques allow effective isolation and expansion of certain cells *in vitro*, cell-carriers are susceptible to contamination and other risks. Alternatively, drugs and drug-engulfed vehicles can be directly injected into the circulation to bind with cell carriers *in vivo*. However, the drug loading efficiency may be compromised due to immune surveillance and drug vehicles' limited binding ability. Potential therapeutics leaking from the carrying cells is another concern. Additionally, a wide spectrum of cells playing multiple roles in disease development and the healing process may participate in the drug delivery procedure. Introducing cell carriers may interrupt the balance of natural physiological conditions. For example, leukocytes are attracted to inflammation sites to deliver medicine in the short term, but they can also cause chronic inflammation issues later. Since the mechanism and development of many diseases are not well understood, foreign cells with drug complexes may also exacerbate pathological conditions. In the meantime, more evidence is needed to verify the applicability of cell-mediated DDS in various diseases, including, but not limited to, cancers, wound healing, brain diseases, and cardiovascular diseases.

Regardless of these challenges, cell-mediated drug delivery holds great potential to revolutionize current diagnostic and medical techniques. Advances in biomaterials may open up the door to a new paradigm of DDS designs. Engineering tools for manipulating living cells and fabricating drug vehicles, such as NPs, will create efficient and intelligent targeting, releasing, and imaging strategies. Combining living circulating cells and nanotechnology will be the new direction for controlled drug delivery, while offering a powerful technique to improve overall diagnostic and therapeutic outcomes.

ACKNOWLEDGMENT

The authors are grateful for the funding support from a National Institutes of Health (NIH) Awards NIBIB EB012575, NCI CA182670, and NHLBI HL118498, and National Science Foundation (NSF) Awards DMR1313553, CMMI 1266116 and CBET-BME 1330663.

AUTHOR CONTRIBUTIONS

The manuscript was written through contributions of all authors. All authors have given approval to the final version of the manuscript.

REFERENCES

1. Allen TM, Cullis PR. Liposomal drug delivery systems: from concept to clinical applications. Adv. Drug Del. Rev. 2013;65:36–48.

2. Parveen S, Misra R, Sahoo SK. Nanoparticles: a boon to drug delivery, therapeutics, diagnostics and imaging. Nanomed. Nanotechnol. Biol. Med. 2012;8:147–166.

3. Kedar U, Phutane P, Shidhaye S, Kadam V. Advances in polymeric micelles for drug delivery and tumor targeting. Nanomed. Nanotechnol. Biol. Med. 2010;6:714–729.

4. Ashley GW, Henise J, Reid R, Santi DV. Hydrogel drug delivery system with predictable and tunable drug release and degradation rates. Proceedings of the National Academy of Sciences. 2013;110:2318–2323.

5. Hamidi M, Azadi A, Rafiei P. Hydrogel nanoparticles in drug delivery. Adv. Drug Del. Rev.2008;60:1638–1649.

6. Cui W, Zhou Y, Chang J. Electrospun nanofibrous materials for tissue engineering and drug delivery.Science and Technology of Advanced Materials. 2010;11:014108.

7. Yoo HS, Kim TG, Park TG. Surface-functionalized electrospun nanofibers for tissue engineering and drug delivery. Adv. Drug Del. Rev. 2009;61:1033–1042.

8. Prausnitz MR, Langer R. Transdermal drug delivery. Nat. Biotechnol. 2008;26:1261–1268.

9. Keefe AD, Pai S, Ellington A. Aptamers as therapeutics. Nature Reviews Drug Discovery. 2010;9:537–550.

10. Torchilin VP. Multifunctional nanocarriers. Adv. Drug Del. Rev. 2012;64:302–315.

11. Brannon-Peppas L, Blanchette JO. Nanoparticle and targeted systems for cancer therapy. Adv. Drug Del. Rev. 2012;64(Supplement):206–212.

12. Muro S. Challenges in design and characterization of ligand-targeted drug delivery systems. J. Control. Release. 2012;164:125–137.

13. Snyder EL, Dowdy SF. Cell penetrating peptides in drug delivery. Pharm. Res. 2004;21:389–393.

14. Torchilin VP, Trubetskoy VS. Which polymers can make nanoparticulate drug carriers long-circulating?Adv. Drug Del. Rev. 1995;16:141–155.

15. Greenwald RB, Choe YH, McGuire J, Conover CD. Effective drug delivery by PEGylated drug conjugates. Adv. Drug Del. Rev. 2003;55:217–250.

16. Alexis F, Pridgen E, Molnar LK, Farokhzad OC. Factors affecting the clearance and biodistribution of polymeric nanoparticles. Mol. Pharm. 2008;5:505–515.

17. Fang J, Nakamura H, Maeda H. The EPR effect: unique features of tumor blood vessels for drug delivery, factors involved, and limitations and augmentation of the effect. Adv. Drug Del. Rev.2011;63:136–151.

18. Tatum JL. Hypoxia: importance in tumor biology, noninvasive measurement by imaging, and value of its measurement in the management of cancer therapy. Int. J. Radiat. Biol. 2006;82:699–757.

19. Qin J, Yang X, Zhang RX, Luo YX, Li JL, Hou J, Zhang C, Li YJ, Shi J, Lu L, Wang JX, Zhu WL. Monocyte mediated brain targeting delivery of macromolecular drug for the therapy of depression.Nanomedicine. 2014

20. Berezina TL, Zaets SB, Morgan C, Spillert CR, Kamiyama M, Spolarics Z, Deitch EA, Machiedo GW. Influence of storage on red blood cell rheological properties. J. Surg. Res. 2002;102:6–12.

21. Muzykantov VR. Drug delivery by red blood cells: vascular carriers designed by mother nature. Expert opinion on drug delivery. 2010;7:403–427.

22. Hollowell JG, van Assendelft OW, Gunter EW, Lewis BG, Najjar M, Pfeiffer C. Vital Health Stat. Vol. 11. United States Bureau of the Census; 2005. Hematological and Iron-Related Analytes-Reference Data for Persons Aged 1 Year and Over: United States, 1988-94; pp. 1–156.

23. Amulic B, Cazalet C, Hayes GL, Metzler KD, Zychlinsky A. Neutrophil function: from mechanisms to disease. Annu. Rev. Immunol. 2012;30:459–489.

24. Smith JA. Neutrophils, host defense, and inflammation: a double-edged sword. J. Leukoc. Biol.1994;56:672–686.

25. Pillay J, den Braber I, Vrisekoop N, Kwast LM, de Boer RJ, Borghans JAM, Tesselaar K, Koenderman L. In vivo labeling with 2H2O reveals a human neutrophil lifespan of 5.4 days. Blood. 2010;116:625–627.

26. Adams DH, Rlloyd A. Chemokines: leucocyte recruitment and activation cytokines. The Lancet.1997;349:490–495.

27. McGuinness C, Humphries J, Waltham M, Burnand K, Collins M, Smith A. Recruitment of labelled monocytes by experimental venous thrombi. THROMBOSIS AND HAEMOSTASIS-STUTTGART.2001;85:1018–1024.

28. Bingle L, Brown NJ, Lewis CE. The role of tumour-associated macrophages in tumour progression: implications for new anticancer therapies. J. Pathol. 2002;196:254–265.

29. Sica A, Allavena P, Mantovani A. Cancer related inflammation: the macrophage connection. Cancer Lett. 2008;267:204–215.

30. Kunkel EJ, Butcher EC. Chemokines and the tissue-specific migration of lymphocytes. Immunity.2002;16:1–4.

31. Sakaguchi S, Yamaguchi T, Nomura T, Ono M. Regulatory T cells and immune tolerance. Cell.2008;133:775–787.

32. Mauri C, Bosma A. Immune regulatory function of B cells. Annu. Rev. Immunol. 2012;30:221–241.

33. Wagers AJ, Weissman IL. Plasticity of adult stem cells. Cell. 2004;116:639–648.

34. Bianco P, Robey PG. Stem cells in tissue engineering. Nature. 2001;414:118–121.

35. Kim SU, de Vellis J. Stem Cell-Based Cell Therapy in Neurological Diseases: A Review. J. Neurosci. Res. 2009;87:2183–2200.

36. Reagan MR, Kaplan DL. Concise Review: Mesenchymal Stem Cell Tumor Homing: Detection Methods in Disease Model Systems. Stem Cells. 2011;29:920–927.

37. Tu S-M, Lin S-H, Logothetis CJ. Stem-cell origin of metastasis and heterogeneity in solid tumours. The lancet oncology. 2002;3:508–513.

38. Aboody KS, Brown A, Rainov NG, Bower KA, Liu S, Yang W, Small JE, Herrlinger U, Ourednik V, Black PM. Neural stem cells display extensive tropism for pathology in adult brain: Evidence from intracranial gliomas. Proceedings of the National Academy of Sciences. 2000;97:12846–12851.

39. Pastan I, Willingham MC. Endocytosis. Springer; 1985. The pathway of endocytosis; pp. 1–44.

40. Canton I, Battaglia G. Endocytosis at the nanoscale. Chemical Society Reviews. 2012;41:2718–2739.

41. Ben-Bassat I, Bensch KG, Schrier SL. Drug-induced erythrocyte membrane internalization. J. Clin. Invest. 1972;51:1833.

42. Ginn FL, Hochstein P, Trump BF. Membrane Alterations in Hemolysis: Internalization of Plasmalemma Induced by Primaquine. Science. 1969;164:843–845.

43. Gopal V, Kumar AR, Usha N, Karthik A, Udupa N. Effective drug targeting by erythrocytes as carrier systems. Curr. Trends Biotechnol. Pharm. 2007;1:18–33.

44. Matovcik LM, Junga IG, Schrier SL. Drug-induced endocytosis of neonatal erythrocytes. Blood.1985;65:1056–1063.

45. Ben-Bassat I, Bensch KG, Schrier SL. Drug-induced erythrocyte membrane internalization. The Journal of clinical investigation. 1972;51:1833–44.

46. Schrier SL, Zachowski A, Devaux PF. Mechanisms of amphipath-induced stomatocytosis in human erythrocytes. Blood. 1992;79:782–6.

47. Aderem A, Underhill DM. Mechanisms of phagocytosis in macrophages. Annu. Rev. Immunol.1999;17:593–623.

48. Owens DE, III, Peppas NA. Opsonization, biodistribution, and pharmacokinetics of polymeric nanoparticles. Int. J. Pharm. 2006;307:93–102.

49. Beduneau A, Ma Z, Grotepas CB, Kabanov A, Rabinow BE, Gong N, Mosley RL, Dou H, Boska MD, Gendelman HE. Facilitated Monocyte-Macrophage Uptake and Tissue Distribution of Superparmagnetic Iron-Oxide Nanoparticles. PLoS One. 2009;4:e4343.

50. Choi J, Kim H-Y, Ju EJ, Jung J, Park J, Chung H-K, Lee JS, Lee JS, Park HJ, Song SY, Jeong S-Y, Choi EK. Use of macrophages to deliver therapeutic and imaging contrast agents to tumors. Biomaterials.2012;33:4195–4203.

51. Nowacek AS, Miller RL, McMillan J, Kanmogne G, Kanmogne M, Mosley RL, Ma Z, Graham S, Chaubal M, Werling J. NanoART synthesis, characterization, uptake, release and toxicology for human monocyte-macrophage drug delivery. Nanomedicine. 2009;4:903–917.

52. Batrakova EV, Li S, Reynolds AD, Mosley RL, Bronich TK, Kabanov AV, Gendelman HE. A macrophage-nanozyme delivery system for Parkinson's disease. Bioconjug. Chem. 2007;18:1498–1506.

53. Choi M-R, Stanton-Maxey KJ, Stanley JK, Levin CS, Bardhan R, Akin D, Badve S, Sturgis J, Robinson JP, Bashir R. A cellular Trojan Horse for

delivery of therapeutic nanoparticles into tumors. Nano Lett.2007;7:3759–3765.

54. Choi M-R, Bardhan R, Stanton-Maxey KJ, Badve S, Nakshatri H, Stantz KM, Cao N, Halas NJ, Clare SE. Delivery of nanoparticles to brain metastases of breast cancer using a cellular Trojan horse. Cancer Nanotechnol. 2012;3:47–54.

55. Doshi N, Mitragotri S. Macrophages Recognize Size and Shape of Their Targets. PLoS One.2010;5:e10051.

56. Champion J, Walker A, Mitragotri S. Role of Particle Size in Phagocytosis of Polymeric Microspheres.Pharm. Res. 2008;25:1815–1821.

57. Champion JA, Mitragotri S. Role of target geometry in phagocytosis. Proc. Natl. Acad. Sci. U. S. A.2006;103:4930–4934.

58. Beningo KA, Wang Y.-l. Fc-receptor-mediated phagocytosis is regulated by mechanical properties of the target. J. Cell Sci. 2002;115:849–856.

59. Fujiwara M, Baldeschwieler JD, Grubbs RH. Receptor-mediated endocytosis of poly(acrylic acid)-conjugated liposomes by macrophages. Biochim. Biophys. Acta. 1996;1278:59–67.

60. Juliano RL, Stamp D. The effect of particle size and charge on the clearance rates of liposomes and liposome encapsulated drugs. Biochem. Biophys. Res. Commun. 1975;63:651–658.

61. Champion J, Mitragotri S. Shape Induced Inhibition of Phagocytosis of Polymer Particles. Pharm. Res.2009;26:244–249.

62. Swiston AJ, Cheng C, Um SH, Irvine DJ, Cohen RE, Rubner MF. Surface functionalization of living cells with multilayer patches. Nano Lett. 2008;8:4446–4453.

63. Vasconcellos FC, Swiston AJ, Beppu MM, Cohen RE, Rubner MF. Bioactive polyelectrolyte multilayers: Hyaluronic acid mediated B lymphocyte adhesion. Biomacromolecules. 2010;11:2407–2414.

64. Swiston AJ, Gilbert JB, Irvine DJ, Cohen RE, Rubner MF. Freely Suspended Cellular "Backpacks" Lead to Cell Aggregate Self-Assembly. Biomacromolecules. 2010;11:1826–1832. [PMC free article]

65. Doshi N, Swiston AJ, Gilbert JB, Alcaraz ML, Cohen RE, Rubner MF, Mitragotri S. Cell-Based Drug Delivery Devices Using Phagocytosis-Resistant Backpacks. Advanced Materials. 2011;23:H105–H109.

66. Fernandes Stefanello T, Szarpak-Jankowska A, Appaix F, Louage B, Hamard L, De Geest BG, van der Sanden B, Nakamura CV, Auzély-Velty R. Thermoresponsive hyaluronic acid nanogels as hydrophobic drug carrier to macrophages. Acta Biomater. 2014;10:4750–4758.

67. Li L, Guan Y, Liu H, Hao N, Liu T, Meng X, Fu C, Li Y, Qu Q, Zhang Y, Ji S, Chen L, Chen D, Tang F. Silica Nanorattle–Doxorubicin-Anchored Mesenchymal Stem Cells for Tumor-Tropic Therapy. ACS nano. 2011;5:7462–7470.

68. Zaitsev S, Danielyan K, Murciano J-C, Ganguly K, Krasik T, Taylor RP, Pincus S, Jones S, Cines DB, Muzykantov VR. Human complement receptor type 1-directed loading of tissue plasminogen activator on circulating erythrocytes for prophylactic fibrinolysis. Blood. 2006;108:1895–1902. [PMC free article]

69. Kontos S, Hubbell JA. Improving protein pharmacokinetics by engineering erythrocyte affinity. Mol. Pharm. 2010;7:2141–7.

70. Kontos S, Kourtis IC, Dane KY, Hubbell JA. Engineering antigens for in situ erythrocyte binding induces T-cell deletion. Proceedings of the National Academy of Sciences. 2013;110:E60–E68.

71. Stephan MT, Stephan SB, Bak P, Chen J, Irvine DJ. Synapse-directed delivery of immunomodulators using T-cell-conjugated nanoparticles. Biomaterials. 2012;33:5776–5787.

72. Naldini L, Blömer U, Gallay P, Ory D, Mulligan R, Gage FH, Verma IM, Trono D. In vivo gene delivery and stable transduction of nondividing cells by a lentiviral vector. Science. 1996;272:263–267.

73. Tjuvajev JG, Stockhammer G, Desai R, Uehara H, Watanabe K, Gansbacher B, Blasberg RG. Imaging the expression of transfected genes in vivo. Cancer Res. 1995;55:6126–6132.

74. Chiocca EA. Oncolytic viruses. Nature Reviews Cancer. 2002;2:938–950.

75. Komarova S, Kawakami Y, Stoff-Khalili MA, Curiel DT, Pereboeva L. Mesenchymal progenitor cells as cellular vehicles for delivery of oncolytic adenoviruses. Mol. Cancer Ther. 2006;5:755–766.

76. Pereboeva L, Komarova S, Mikheeva G, Krasnykh V, Curiel DT. Approaches to utilize mesenchymal progenitor cells as cellular vehicles. Stem Cells. 2003;21:389–404.

77. Ong HT, Hasegawa K, Dietz AB, Russell SJ, Peng KW. Evaluation of T cells as carriers for systemic measles virotherapy in the presence of antiviral antibodies. Gene Ther. 2006;14:324–333.

78. Hu YL, Huang B, Zhang TY, Miao PH, Tang GP, Tabata Y, Gao JQ. Mesenchymal stem cells as a novel carrier for targeted delivery of gene in cancer therapy based on nonviral transfection. Mol. Pharm.2012;9:2698–2709.

79. Melancon MP, Lu W, Zhong M, Zhou M, Liang G, Elliott AM, Hazle JD, Myers JN, Li C, Jason Stafford R. Targeted multifunctional gold-based nanoshells for magnetic resonance-guided laser ablation of head and neck cancer. Biomaterials. 2011;32:7600–7608.

80. Huang D-M, Hsiao J-K, Chen Y-C, Chien L-Y, Yao M, Chen Y-K, Ko B-S, Hsu S-C, Tai L-A, Cheng H-Y, Wang S-W, Yang C-S, Chen Y-C. The promotion of human mesenchymal stem cell proliferation by superparamagnetic iron oxide nanoparticles. Biomaterials. 2009;30:3645–3651.

81. Sanz S, Lizano C, Luque J, Pinilla M. In vitro and in vivo study of glutamate dehydrogenase encapsulated into mouse erythrocytes by a hypotonic dialysis procedure. Life Sci. 1999;65:2781–9.

82. Kwant W, Seeman P. The erythrocyte ghost is a perfect osmometer. The Journal of general physiology.1970;55:208–219.

83. DeLoach J, Droleskey R, Andrews K. Encapsulation by hypotonic dialysis in human erythrocytes: a diffusion or endocytosis process. Biotechnol. Appl. Biochem. 1991;13:72–82.

84. Deloach JR. Carrier erythrocytes. Med. Res. Rev. 1986;6:487–504.

85. De Flora A, Zocchi E, Guida L, Polvani C, Benatti U. Conversion of encapsulated 5-fluoro-2′-deoxyuridine 5′-monophosphate to the antineoplastic drug 5-fluoro-2′-deoxyuridine in human erythrocytes.Proc. Natl. Acad. Sci. U. S. A. 1988;85:3145–9.

86. Alvarez FJ, Jordan JA, Herraez A, Diez JC, Tejedor MC. Hypotonically loaded rat erythrocytes deliver encapsulated substances into peritoneal macrophages. J. Biochem. 1998;123:233–9.

87. Pérez MT, Alvarez FJ, García-Pérez A-I, Lucas L, Tejedor MC, Sancho P. Heterogeneity of hypotonically loaded rat erythrocyte populations as detected by counter-current distribution in aqueous polymer two-phase systems. Journal of Chromatography B: Biomedical Sciences and Applications.1996;677:45–51.

88. Chiarantini L, Cerasi A, Fraternale A, Andreoni F, Scarí S, Giovine M, Clavarino E, Magnani M. Inhibition of Macrophage iNOS by Selective Targeting of Antisense PNA† Biochemistry. 2002;41:8471–8477.

89. Deloach J, Ihler G. A dialysis procedure for loading erythrocytes with enzymes and lipids. Biochim. Biophys. Acta. 1977;496:136–45.

90. Foroozesh M, Hamidi M, Zarrin A, Mohammadi-Samani S, Montaseri H. Preparation and in-vitro characterization of tramadol-loaded carrier erythrocytes for long-term intravenous delivery. J. Pharm. Pharmacol. 2011;63:322–332.

91. Weaver JC. Electroporation of biological membranes from multicellular to nano scales. Dielectrics and Electrical Insulation, IEEE Transactions on. 2003;10:754–768.

92. Tsong TY. Electroporation of cell membranes. Biophys. J. 1991;60:297–306. [PMC free article]

93. Flynn G, Hackett TJ, McHale L, McHale AP. Encapsulation of the thrombolytic enzyme, brinase, in photosensitized erythrocytes: a novel thrombolytic system based on photodynamic activation. Journal of Photochemistry and Photobiology B-Biology. 1994;26:193–196.

94. Mangal PC, Kaur A. Electroporation of red blood cell membrane and its use as a drug carrier system.Indian J. Biochem. Biophys. 1991;28:219–21.

95. Lizano C, Perez MT, Pinilla M. Mouse erythrocytes as carriers for coencapsulated alcohol and aldehyde dehydrogenase obtained by electroporation in vivo survival rate in circulation, organ distribution and ethanol degradation. Life Sci. 2001;68:2001–16.

96. Chen I, Howarth M, Lin W, Ting AY. Site-specific labeling of cell surface proteins with biophysical probes using biotin ligase. Nat. Methods. 2005;2:99–104.

97. Ferguson MA, Williams AF. Cell-surface anchoring of proteins via glycosyl-phosphatidylinositol structures. Annu. Rev. Biochem. 1988;57:285–320.

98. Rossi NAA, Constantinescu I, Kainthan RK, Brooks DE, Scott MD, Kizhakkedathu JN. Red blood cell membrane grafting of multi-functional hyperbranched polyglycerols. Biomaterials. 2010;31:4167–4178.

99. Stephan MT, Moon JJ, Um SH, Bershteyn A, Irvine DJ. Therapeutic cell engineering with surface-conjugated synthetic nanoparticles. Nat. Med. 2010;16:1035–1041.

100. Diamandis EP, Christopoulos TK. The biotin-(strept) avidin system: principles and applications in biotechnology. Clin. Chem. 1991;37:625–636.

101. Parrott MB, Adams KE, Mercier GT, Mok H, Campos SK, Barry MA. Metabolically biotinylated adenovirus for cell targeting, ligand screening, and vector purification. Mol. Ther. 2003;8:688–700.

102. Suzuki T, Dale GL. Biotinylated erythrocytes: in vivo survival and in vitro recovery. Blood.1987;70:791–795.

103. Cheng H, Kastrup CJ, Ramanathan R, Siegwart DJ, Ma M, Bogatyrev SR, Xu Q, Whitehead KA, Langer R, Anderson DG. Nanoparticulate cellular patches for cell-mediated tumoritropic delivery. ACS nano. 2010;4:625–631.

104. Murciano J-C, Medinilla S, Eslin D, Atochina E, Cines DB, Muzykantov VR. Prophylactic fibrinolysis through selective dissolution of nascent clots by tPA-carrying erythrocytes. Nat Biotech. 2003;21:891–896.

105. Murciano J-C, Higazi AA-R, Cines DB, Muzykantov VR. Soluble urokinase receptor conjugated to carrier red blood cells binds latent pro-urokinase and alters its functional profile. J. Control. Release.2009;139:190–196.

106. Baskin JM, Bertozzi CR. Bioorthogonal click chemistry: Covalent labeling in living systems. QSAR & Combinatorial Science. 2007;26:1211–1219.

107. Chang PV, Prescher JA, Sletten EM, Baskin JM, Miller IA, Agard NJ, Lo A, Bertozzi CR. Copper-free click chemistry in living animals. Proceedings of the National Academy of Sciences. 2010;107:1821–1826.

108. Kolb HC, Sharpless KB. The growing impact of click chemistry on drug discovery. Drug Discov. Today. 2003;8:1128–1137.

109. Hong V, Steinmetz NF, Manchester M, Finn MG. Labeling Live Cells by Copper-Catalyzed Alkyne-Azide Click Chemistry. Bioconjug. Chem. 2010;21:1912–1916.

110. Prescher JA, Bertozzi CR. Chemistry in living systems. Nat. Chem. Biol. 2005;1:13–21.

111. Hein JE, Fokin VV. Copper-catalyzed azide-alkyne cycloaddition (CuAAC) and beyond: new reactivity of copper (I) acetylides. Chemical Society Reviews. 2010;39:1302–1315. [PMC free article]

112. Baskin JM, Prescher JA, Laughlin ST, Agard NJ, Chang PV, Miller IA, Lo A, Codelli JA, Bertozzi CR. Copper-free click chemistry for dynamic in vivo imaging. Proceedings of the National Academy of Sciences. 2007;104:16793–16797.

113. Xu L, Zolotarskaya OY, Yeudall WA, Yang H. Click Hybridization of Immune Cells and Polyamidoamine Dendrimers. Advanced Healthcare Materials. 2014;3:1430–1438. [PMC free article]

114. Iwasaki Y, Matsuno H. Metabolic Delivery of Methacryloyl Groups on Living Cells and Cell Surface Modification via Thiol Ene "Click" Reaction. Macromol. Biosci. 2011;11:1478–1483.

115. Dou H, Destache CJ, Morehead JR, Mosley RL, Boska MD, Kingsley J, Gorantla S, Poluektova L, Nelson JA, Chaubal M. Development of a macrophage-based nanoparticle platform for antiretroviral drug delivery. Blood. 2006;108:2827–2835.

116. Lizano C, Pérez MT, Pinilla M. Mouse erythrocytes as carriers for coencapsulated alcohol and aldehyde dehydrogenase obtained by electroporation: In vivo survival rate in circulation, organ distribution and ethanol degradation. Life Sci. 2001;68:2001–2016.

117. Lynch WE, Sartiano GP, Ghaffar A. Erythrocytes as carriers of chemotherapeutic agents for targeting the reticuloendothelial system. Am. J. Hematol. 1980;9:249–259.

118. Knutson M, Wessling-Resnick M. Iron metabolism in the reticuloendothelial system. Crit. Rev. Biochem. Mol. Biol. 2003;38:61–88.

119. Zocchi E, Tonetti M, Polvani C, Guida L, Benatti U, Flora A. In vivo liver and lung targeting of adriamycin encapsulated in glutaraldehyde treated murine erythrocytes. Biotechnol. Appl. Biochem.1988;10:555–562.

120. Mishra P, Jain N. Biotinylated methotrexate loaded erythrocytes for enhanced liver uptake.'A study on the rat' Int. J. Pharm. 2002;231:145–153.

121. Crinelli R, Antonelli A, Bianchi M, Gentilini L, Scaramucci S, Magnani M. Selective Inhibition of NF-kB Activation and TNF-α Production in Macrophages by Red Blood Cell-Mediated Delivery of Dexamethasone. Blood Cells Mol. Dis. 2000;26:211–222.

122. Deloach J, Peters S, Pinkard O, Glew R, Hiler G. Effect of glutaraldehyde treatment on enzyme-loaded erythrocytes. Biochimica et Biophysica Acta (BBA)-General Subjects. 1977;496:507–515.

123. Zocchi E, Tonetti M, Polvani C, Guida L, Benatti U, De Flora A. Encapsulation of doxorubicin in liver-targeted erythrocytes increases the therapeutic index of the drug in a murine metastatic model.Proceedings of the National Academy of Sciences. 1989;86:2040–2044.

124. Kwon YM, Chung HS, Moon C, Yockman J, Park YJ, Gitlin SD, David AE, Yang VC. L-Asparaginase encapsulated intact erythrocytes for treatment of acute lymphoblastic leukemia (ALL) J. Control. Release. 2009;139:182–189.

125. Rossi L, Serafini S, Antonelli A, Pierigé F, Carnevali A, Battistelli V, Malatesta M, Balestra E, Caliò R, Perno C-F. Macrophage depletion induced by clodronate-loaded erythrocytes. J. Drug Target. 2005;13:99–111.

126. Bradley AJ, Murad KL, Regan KL, Scott MD. Biophysical consequences of linker chemistry and polymer size on stealth erythrocytes: size does matter. Biochim. Biophys. Acta. 2002;1561:147–158.

127. Murad KL, Mahany KL, Brugnara C, Kuypers FA, Eaton JW, Scott MD. Structural and functional consequences of antigenic modulation of red blood cells with methoxypoly (ethylene glycol) Blood.1999;93:2121–2127.

128. Scott MD, Murad KL, Koumpouras F, Talbot M, Eaton JW. Chemical camouflage of antigenic determinants: Stealth erythrocytes. Proceedings of the National Academy of Sciences. 1997;94:7566–7571.

129. Price RJ, Skyba DM, Kaul S, Skalak TC. Delivery of colloidal particles and red blood cells to tissue through microvessel ruptures created by targeted microbubble destruction with ultrasound. Circulation.1998;98:1264–7.

130. Sprandel U, Lanz DJ, von Horsten W. Magnetically responsive erythrocyte ghosts. Methods Enzymol.1987;149:301–12.

131. Kolaczkowska E, Kubes P. Neutrophil recruitment and function in health and inflammation. Nature Reviews Immunology. 2013;13:159–175.

132. Schenkel AR, Mamdouh Z, Muller WA. Locomotion of monocytes on endothelium is a critical step during extravasation. Nat. Immunol. 2004;5:393–400.

133. Albelda SM, Smith CW, Ward P. Adhesion molecules and inflammatory injury. The FASEB Journal.1994;8:504–512.

134. Pober JS, Cotran RS. The role of endothelial cells in inflammation. Transplantation. 1990;50:537–544.

135. Luster AD, Alon R, von Andrian UH. Immune cell migration in inflammation: present and future therapeutic targets. Nat. Immunol. 2005;6:1182–1190.

136. Sampson A. The role of eosinophils and neutrophils in inflammation. Clin. Exp. Allergy. 2000;30:22–27.

137. Carlos TM, Harlan JM. Leukocyte-endothelial adhesion molecules. Blood. 1994;84:2068–2101.

138. Bevilacqua MP. Endothelial-leukocyte adhesion molecules. Annu. Rev. Immunol. 1993;11:767–804.

139. Diacovo T, Roth S, Buccola J, Bainton D, Springer T. Neutrophil rolling, arrest, and transmigration across activated, surface-adherent platelets via sequential action of P-selectin and the beta 2-integrin CD11b/CD18. Blood. 1996;88:146–157.

140. Lo SK, Lee S, Ramos RA, Lobb R, Rosa M, Chi-Rosso G, Wright SD. Endothelial-leukocyte adhesion molecule 1 stimulates the adhesive activity of leukocyte integrin CR3 (CD11b/CD18, Mac-1, alpha m beta 2) on human neutrophils. The Journal of experimental medicine. 1991;173:1493–1500. [PMC free article]

141. von Andrian UH, Chambers JD, McEvoy LM, Bargatze RF, Arfors K-E, Butcher EC. Two-step model of leukocyte-endothelial cell interaction in inflammation: distinct roles for LECAM-1 and the leukocyte beta 2 integrins in vivo. Proceedings of the National Academy of Sciences. 1991;88:7538–7542.

142. Phillipson M, Heit B, Colarusso P, Liu L, Ballantyne CM, Kubes P. Intraluminal crawling of neutrophils to emigration sites: a molecularly distinct process from adhesion in the recruitment cascade. The Journal of experimental medicine. 2006;203:2569–2575.

143. Williams MR, Azcutia V, Newton G, Alcaide P, Luscinskas FW. Emerging mechanisms of neutrophil recruitment across endothelium. Trends Immunol. 2011;32:461–469.

144. Ley K, Laudanna C, Cybulsky MI, Nourshargh S. Getting to the site of inflammation: the leukocyte adhesion cascade updated. Nature Reviews Immunology. 2007;7:678–689.

145. Shi C, Pamer EG. Monocyte recruitment during infection and inflammation. Nature Reviews Immunology. 2011;11:762–774.

146. Geissmann F, Manz MG, Jung S, Sieweke MH, Merad M, Ley K. Development of Monocytes, Macrophages, and Dendritic Cells. Science. 2010;327:656–661.

147. Pages F, Galon J, Dieu-Nosjean M, Tartour E, Sautes-Fridman C, Fridman W. Immune infiltration in human tumors: a prognostic factor that should not be ignored. Oncogene. 2009;29:1093–1102.

148. Kawakami Y, Eliyahu S, Delgado CH, Robbins PF, Sakaguchi K, Appella E, Yannelli JR, Adema GJ, Miki T, Rosenberg SA. Identification of a human melanoma antigen recognized by tumor-infiltrating lymphocytes associated with in vivo tumor rejection. Proceedings of the National Academy of Sciences.1994;91:6458–6462.

149. Smith BR, Ghosn EEB, Rallapalli H, Prescher JA, Larson T, Herzenberg LA, Gambhir SS. Selective uptake of single-walled carbon nanotubes by circulating monocytes for enhanced tumour delivery. Nature nanotechnology. 2014

150. Batrakova EV, Li S, Reynolds AD, Mosley RL, Bronich TK, Kabanov AV, Gendelman HE. A Macrophage–Nanozyme Delivery System for Parkinson's Disease. Bioconjug. Chem. 2007;18:1498–1506.

151. Valable S, Barbier EL, Bernaudin M, Roussel S, Segebarth C, Petit E, Rémy C. In vivo MRI tracking of exogenous monocytes/macrophages targeting brain tumors in a rat model of glioma. Neuroimage.2008;40:973–983.

152. Harris AL. Hypoxia—a key regulatory factor in tumour growth. Nature Reviews Cancer. 2002;2:38–47.

153. Roger M, Clavreul A, Venier-Julienne M-C, Passirani C, Sindji L, Schiller P, Montero-Menei C, Menei P. Mesenchymal stem cells as cellular vehicles for delivery of nanoparticles to brain tumors. Biomaterials.2010;31:8393–8401.

154. Studeny M, Marini FC, Champlin RE, Zompetta C, Fidler IJ, Andreeff M. Bone marrow-derived mesenchymal stem cells as vehicles for interferon-β delivery into tumors. Cancer Res. 2002;62:3603–3608.

155. Studeny M, Marini FC, Dembinski JL, Zompetta C, Cabreira-Hansen M, Bekele BN, Champlin RE, Andreeff M. Mesenchymal stem cells: potential precursors for tumor stroma and targeted-delivery vehicles for anticancer agents. J. Natl. Cancer Inst. 2004;96:1593–1603.

156. Nakamizo A, Marini F, Amano T, Khan A, Studeny M, Gumin J, Chen J, Hentschel S, Vecil G, Dembinski J. Human bone marrow-derived mesenchymal stem cells in the treatment of gliomas. Cancer Res.2005;65:3307–3318.

157. Zhao J, Vykoukal J, Abdelsalam M, Recio-Boiles A, Huang Q, Qiao Y, Singhana B, Wallace M, Avritscher R, Melancon MP. Stem cell-mediated delivery of SPIO-loaded gold nanoparticles for the theranosis of liver injury and hepatocellular carcinoma. Nanotechnology. 2014;25:405101. [PMC free article]

158. Cao B, Yang M, Zhu Y, Qu X, Mao C. Stem Cells Loaded with Nanoparticles as a Drug Carrier for In Vivo Breast Cancer Therapy. Advanced Materials. 2014

159. Allard WJ, Matera J, Miller MC, Repollet M, Connelly MC, Rao C, Tibbe AGJ, Uhr JW, Terstappen LWMM. Tumor Cells Circulate in the Peripheral Blood of All Major Carcinomas but not in Healthy Subjects or Patients With Nonmalignant Diseases. Clin. Cancer Res. 2004;10:6897–6904.

160. Mitchell MJ, Wayne E, Rana K, Schaffer CB, King MR. TRAIL-coated leukocytes that kill cancer cells in the circulation. Proceedings of the National Academy of Sciences. 2014;111:930–935.

161. Liang S, Slattery MJ, Wagner D, Simon SI, Dong C. Hydrodynamic shear rate regulates melanoma-leukocyte aggregation, melanoma adhesion to the endothelium, and subsequent extravasation. Ann. Biomed. Eng. 2008;36:661–671.

162. Mukthavaram R, Shi G, Kesari S, Simberg D. Targeting and depletion of circulating leukocytes and cancer cells by lipophilic antibody-modified erythrocytes. J. Control. Release. 2014;183:146–153.

163. Wang C, Sun X, Cheng L, Yin S, Yang G, Li Y, Liu Z. Multifunctional Theranostic Red Blood Cells For Magnetic-Field-Enhanced in vivo Combination Therapy of Cancer. Advanced Materials. 2014

164. Soppimath KS, Aminabhavi TM, Kulkarni AR, Rudzinski WE. Biodegradable polymeric nanoparticles as drug delivery devices. J. Control. Release. 2001;70:1–20.

165. Xie Z, Zhang Y, Liu L, Weng H, Mason RP, Tang L, Nguyen KT, Hsieh JT, Yang J. Development of Intrinsically Photoluminescent and Photostable Polylactones. Advanced Materials. 2014;26:4491–4496.

166. Schmaljohann D. Thermo-and pH-responsive polymers in drug delivery. Adv. Drug Del. Rev.2006;58:1655–1670.

167. Bajpai A, Shukla SK, Bhanu S, Kankane S. Responsive polymers in controlled drug delivery. Progress in Polymer Science. 2008;33:1088–1118.

168. He H, Ye J, Wang Y, Liu Q, Chung HS, Kwon YM, Shin MC, Lee K, Yang VC. Cell-penetrating peptides meditated encapsulation of protein therapeutics into intact red blood cells and its application. J. Control. Release. 2014;176:123–132.

169. Bawa P, Pillay V, Choonara YE, Du Toit LC. Stimuli-responsive polymers and their applications in drug delivery. Biomed Mater. 2009;4:022001.

Chapter 5

Recent Advances in Ophthalmic Drug Delivery

Uday B Kompella,[1,2] Rajendra S Kadam,[1] and Vincent HL Lee[3,]

[1] *Department of Pharmaceutical Sciences, University of Colorado Denver, Aurora, CO, USA*

[2] *Department of Ophthalmology, University of Colorado Denver, Aurora, CO, USA*

[3] *School of Pharmacy, Faculty of Medicine, The Chinese University of Hong Kong, Shatin, N.T., Hong Kong*

ABSTRACT

Topical ocular drug bioavailability is notoriously poor, in the order of 5% or less. This is a consequence of effective multiple barriers to drug entry, comprising nasolacrimal drainage, epithelial drug transport barriers and clearance from the vasculature in the conjunctiva. While sustained drug delivery to the back of the eye is now feasible with intravitreal implants such as Vitrasert™(~6 months), Retisert™(~3 years) and Iluvien™(~3 years), currently there are no marketed delivery systems for long-term drug delivery to the anterior segment of the eye. The purpose of this article is to summarize the resurgence in interest to prolong and improve drug entry from topical administration. These approaches include mucoadhesives, viscous polymer vehicles, transporter-targeted prodrug design, receptor-targeted functionalized nanoparticles, iontophoresis, punctal plug and contact lens delivery systems. A few of these delivery systems might be useful in treating diseases affecting the back of the eye. Their effectiveness will be compared against intravitreal implants (upper bound of effectiveness) and trans-scleral systems (lower bound of effectiveness). Refining the animal model by incorporating the latest advances in microdialysis and imaging technology is key to expanding the knowledge central to the design, testing and evaluation of the next generation of innovative ocular drug delivery systems.

The current ophthalmology market is primarily driven by age- and lifestyle-related diseases, such as macular degeneration, cataracts, diabetic retinopathy and glaucoma. Dry eye syndrome, inflammation and ocular allergies are also contributing significantly. Traditionally, antiglaucoma products held the major market share. Driven by demographic trends such as an increase in aged population and lifestyle factors, the overall market for ophthalmic therapeutics is expected to undergo considerable and consistent growth during the next decade. In the USA, cases of visual impairment and blindness due to age-related macular degeneration (ARMD) are expected to increase from 620,000 cases in 2010 to 1.7 million in 2050 [1]. While therapeutic agents are currently available for treating only the wet form of ARMD, new treatments are anticipated in the future for treating the more prevalent dry form of ARMD [2]. The global ophthalmic pharmaceutical market with registered sales of US$14 billion in 2009, is forecasted to grow at a double digit rate [3].

The principal challenge in advancing therapeutics for treating diseases of the front as well as back of the eye is attainment of effective drug concentration at the drug target for prolonged periods of time, whilst minimizing any side effects. In addition, enhancing patient adherence with drug therapy, especially for asymptomatic diseases such as glaucoma, poses another challenge. The main focus of this article is to update the noninvasive sustained delivery systems administered in the front of the eye. Reference to posterior ocular drug delivery will be made for those systems that show promise there. In this context, key advances in sustained back of the eye drug delivery using invasive delivery systems, such as intravitreal implants and drug suspensions, will be summarized.

OCULAR PHARMACOKINETIC STUDIES & BARRIERS TO OCULAR DRUG DELIVERY

It is routine for pharmaceutical manufacturers to assess drug pharmacokinetics following oral administration of the drug product in human subjects. This pharmacokinetic data is a critical part of the new drug application (NDA). However, for ocular drug products, there is no such requirement for pharmacokinetic studies in human subjects. This is because the relevant target or surrogate tissues, such as aqueous humor and vitreous humor, cannot be sampled serially in human subjects to assess pharmacokinetics. For the same reasons, even during drug product development, pharmacokinetic data are not obtained in humans. Instead, pharmacokinetic studies have routinely relied on the use of animal models with eye sizes comparable to humans, such as rabbit, dog, monkey and pig, with rabbit being the most commonly used species for pharmacokinetic studies. More recently, pharmacokinetic data are being collected in rodent models, since several rodent disease models are available [4–14]. It is now evident that topical ocular drug delivery differs between rabbit, monkey and human models, even at the level of tear concentrations [15]. Thus, it can be anticipated that intraocular concentrations between species will differ. However, with appropriate mathematical models and scaling approaches for the size of the eye and the animal model, pharmacokinetics in one species can probably be extrapolated to other species [7]. Even in the absence of such sophisticated models, animal models such as the rabbit are valuable in assessing the relative availability of a drug from various formulations during drug product development. Between species, it is likely that the relative pharmacokinetic performance of various formulations will hold true on several occasions.

A key challenge in ocular pharmacokinetics is serial sampling of target tissues such as aqueous humor, iris-ciliary body, retina and vitreous humor to plot the concentration–time profiles for estimating various pharmacokinetic parameters, including area under the curve (AUC), time to maximum tissue concentration (T_{max}) and peak tissue concentration (C_{max}). Unlike the systemic pharmacokinetic studies where serial sampling in surrogate tissue (plasma) is feasible, conventional ocular pharmacokinetic studies do not allow serial drug sampling from eye tissues. Therefore, as mentioned by Lee and Robinson [16], ocular pharmacokinetic experiments are expensive, time consuming and require a large number of animals to define the pharmacokinetics of a drug after a single topical dose. However, such pharmacokinetic data is essential and routinely presented to the US FDA in a NDA.

Given the aforementioned animal-intensive nature of preclinical pharmacokinetic analysis in the eye, there is a critical need to develop alternative approaches. Such approaches, albeit with limitations, include microdialysis and non-invasive pharmacokinetic analysis. Microdialysis techniques allow continuous sampling of aqueous and vitreous humors in the same eye, making it feasible to assess pharmacokinetic parameters in both fluids using one single animal [17]. Microdialysis is a probe-based sampling technique that allows the analysis of the sample from fluid compartments or tissue-interstitial fluid compartments to measure unbound or free drug concentrations. The basic principle of microdialysis is based on a capillary dialysis probe, wherein continuous transfer of soluble molecules occurs between tissue space and the fluid enclosed in the probe across a semipermeable membrane. Since only unbound drug can permeate the membrane, drug concentrations in the probe would be reflective of free drug concentrations in the tissue.

In ocular pharmacokinetic studies, the microdialysis probe is generally placed in liquid compartments of the eye, such as aqueous humor (anterior chamber microdialysis) and vitreous humor (posterior chamber microdialysis). For aqueous humor microdialysis, a linear probe is placed in the anterior chamber with the help of a 25-gauge (G) needle by inserting the needle across the cornea, above the limbus. A linear microdialysis probe is inserted through the needle and held on the other side. The needle is then retracted so that the probe is suspended in aqueous humor in the middle of the anterior chamber. The outlets of the probe are fixed to the tissue using surgical glue. For vitreous humor microdialysis, a concentric probe is used. This probe is placed in the middle of the vitreous cavity using a 22-G needle and a guiding canula. Samples can be collected from aqueous humor, vitreous humor or both compartments simultaneously. Perfusion of the above probes with isotonic buffer and collection of perfusate at regular time intervals allows serial sampling and continuous measurement of solute concentration in both aqueous humor and vitreous humor of the eye in a single animal [17]. These measurements, although routinely made in anesthetized animals [18], can also be made in conscious animals [19]. However, for assessing sustained drug delivery over several months, microdialysis may not be necessary, since sporadic sampling with a few animals is probably adequate. Furthermore, the microdialysis technique does not allow continuous drug level assessment in solid target tissues, such as the iris-ciliary body or the retina.

Currently, there is growing interest in non-invasive approaches for assessing pharmacokinetics in the eye. Some of these approaches are summarized in Table 1. Such approaches might eventually allow assessment of drug pharmacokinetics in human eyes. Key limitations of the various noninvasive techniques for pharmacokinetic analysis include low sensitivity and hence inability to detect the low drug concentrations anticipated in target tissues; poor depth resolution and, therefore, inability to distinguish target tissue layers such as choroid, retinal pigment epithelium (RPE) and neural retina and applicability to specific molecules or molecules with an imaging signal as opposed to any given drug molecule. While microdialysis allows analysis of readily diffusible low-molecular-weight drugs with good sensitivity in the fluid compartments of the eye, it is not suitable for drug analysis in target tissues such as the iris-ciliary body and retina. Traditional pharmacokinetic studies, on the other hand, require a large number of animals, while allowing drug analysis in multiple tissues with sensitive analytical methods. Due to the comprehensive nature of data provided by traditional pharmacokinetic studies, these are still the norm in NDAs related to ophthalmic products.

Table 1. Comparison of various techniques for noninvasive or invasive assessment of ocular drug delivery and pharmacokinetics.

Technique	Type	Drugs	Procedure	Sensitivity	Resolution	Errors	Human study
MRI	Noninvasive	Applicable only for contrast agents, fluorinated (^{19}F) compounds or drug molecules labeled with contrast agent	Continuous monitoring over time in live animals	Relatively low sensitivity	Spatial resolution is ~0.3 mm, in-plane. Does not allow the clear distinction of small ocular structure such as retina, ciliary body, and trabecular meshwork	Tissue microenvironment affects relaxation time and hence the signal may give false results. Mobility artifacts	Feasible with approved contrast agent and ^{19}F compound
Positron emission tomography	Noninvasive	Applicable for fluorinated (^{18}F) compounds or drugs labeled with ^{11}C or ^{15}O	Continuous monitoring over time in live animals	Relatively high sensitivity compared with MRI and single photon emission computed tomography but lower than traditional pharmacokinetics	Spatial resolution is ~1–2 mm Does not allow the clear distinction of small ocular structure such as retina, ciliary body, and trabecular meshwork		Feasible with fluorinated drugs or drugs labeled with with ^{11}C or ^{15}O
Single photon emission computed tomography	Noninvasive	Applicable only for ^{99m}Tc and ^{111}In-chelates	Continuous monitoring over time in live animals	Relatively low sensitivity	Spatial resolution is ~1–2 mm Does not allow the clear distinction of small ocular structures such as retina, ciliary body, and trabecular meshwork		
Raman spectroscopy	Noninvasive	Applicable only for drugs with distinct Raman signal	Continuous monitoring over time in live animals	Relatively low sensitivity and depends on molecular properties of the drug. Able to detect sub-milligram to microgram levels in tissues	Spatial resolution is ~0.25 mm, in-plane Does not allow the clear distinction of small ocular structures such as retina, ciliary body, and trabecular meshwork	Raman signal from tissue background may give artifacts and false positive results	Feasible with lower intensity light source
Fluorophotometry	Noninvasive	Applicable only for fluorescent probes or drug molecules labeled with fluorescent probe (more specifically for fluorescein and similar molecules)	Continuous monitoring over time in live animals	More sensitivity than MRI and Raman spectroscopy, but low sensitivity when compared with liquid chromatography-mass spectrometry. Can detect nanogram levels	Spatial resolution is ~0.25 mm in plane Does not allow the clear distinction of small ocular structures such as retina, ciliary body, and trabecular meshwork	Background auto fluorescence of the tissue may give false positive results	Feasible
Microdialysis	Invasive	Most suitable for low molecular weight drugs	Continuous monitoring over time in live animals	Detection limited by drug diffusion through the probe, which declines with an increase in molecular size	Allows quantification of free drug in the aqueous and vitreous humors of the eye	Potential perturbations in drug levels due to probe insertion related injury	Not feasible
Traditional pharmacokinetics	Invasive	All types of drug molecules	Sacrifice of animals at each time point	High sensitivity and capable of detecting nanogram or even picogram levels in tissues	Allows quantification of drug in small tissues and even in subdivisions of tissue such as epithelium and stroma following dissection	Cross contamination and redistribution among tissues during and prior to dissection	Not feasible

Once a suitable animal model, analytical method for the drug of interest and pharmacokinetic method are identified, it is feasible to estimate drug distribution from a delivery system or dosage form. In addition, drug bioavailability or the extent of drug absorption from a dosage form can potentially be estimated. Several previous reports on topical ocular drug delivery do not estimate the bioavailability in an appropriate manner. While the parameters AUC, T_{max} and C_{max} can be used to compare the relative bioavailability (relative extent of absorption) between various formulations, absolute bioavailability (actual fraction of the dose absorbed) can only be estimated based on direct drug dosing in the target tissue. For instance, absolute bioavailability in the aqueous humor can be estimated by comparing the dose-normalized AUCs between topical and intracameral doses, provided the drug clearance does not change between the two routes of administration and the doses used. Due to the intrusive nature of experimentation used for the assessment of ocular pharmacokinetics, very limited reports are available that actually estimate the absolute bioavailability of drug molecules in the aqueous humor after topical application. Microdialysis and noninvasive methods of pharmacokinetics (Table 1) might allow efficient estimation of such parameters with minimal use of animals. Rittenhouse *et al.* determined the AUC for propranolol HCl solution in the aqueous humor of dogs and rabbits using a microdialysis technique after topical and intracameral modes of administration [20]. The estimated absolute bioavailability of propranolol in the aqueous humor after topical application as an eye drop was 0.98 ± 0.55% in dogs and 4.5 ± 2.8% in New Zealand white rabbits. Thus, the bioavailability of even a lipophilic drug such as propranolol is very low in the aqueous humor following topical dosing. Despite the dedicated efforts of ophthalmic formulation scientists over the last four decades, the goal to increase ocular drug bioavailability from less than 5% to at least 15–20% has been met with only limited success. These efforts aim at either overcoming permeability barriers such as the cornea or overcoming drug loss such as precorneal clearance. The extent of delivery is expected to be much lower in the vitreous humor when compared with the aqueous humor after topical application. The absolute bioavailability in the vitreous humor after topical application can be estimated by comparing the dose-normalized AUCs between topical and intravitreal doses. In one study, the vitreal pharmacokinetics of cyclosporine A were evaluated in albino rabbits after intravitreal injection at a 10 μg dose [21]. In this study, the vitreous AUC_{0-48h} was estimated to be 103.3 μg.h/g tissue. In another study, Acheampong *et al.* measured the AUC_{0-96h} for cyclosporine in the vitreous humor after topical application of 100 μg of cyclosporine as an emulsion in albino rabbits [22]; the measured AUC was 0.022 μg.h/g tissue. Thus, the absolute bioavailability of cyclosporine in rabbit vitreous humor after topical application as an eye drop is, at most, 0.002%, since cyclosporine levels may have contributed more to the vitreous AUC between 48 and 96 h following intravitreal dosing.

Thus, drug bioavailability in the anterior and posterior segments is very limited. Key reasons for such low bioavailability include short precorneal residence time of an eye drop as well as multiple permeability barriers that a drug has to cross prior to reaching target eye tissues. Topically applied solutions clear rapidly from the rabbit eye surface, with a half-life of approximately 1.3 min [23]. As discussed later, delivery systems such as punctal plugs, contact lenses and scleral lenses are being designed to overcome the short precorneal residence time.

Permeability barriers are natural defense mechanisms of the eye against the entry of xenobiotics. As shown inFigure 1, several permeability barriers exist in the eye, limiting intraocular drug delivery. The cornea is a tight epithelial barrier that limits drug absorption owing to the presence of tight junctions and a multilayered structure that restricts movement of water and solutes [24]. Corneal permeability, however, is critical for topically applied drugs targeting tissues of the anterior segment of the eye, including aqueous humor, iris-ciliary body and the lens. The conjunctiva is another barrier that limits drug permeability due to the presence of tight junctions and multicellular architecture [25]. However, isolated conjunctiva is more permeable compared with the cornea [26]. In the living animal, drug entering this tissue might be cleared rapidly, due to the presence of blood circulation in the conjunctiva. In addition, episcleral blood supply can potentially clear the drug that crosses conjunctiva. The fraction of the dose escaping conjunctival and episcleral barriers can permeate across sclera, a fibrous, porous tissue that also allows permeability of macromolecules [27]. Key target tissues in the back of the eye include the choroid, RPE and retina. Key diseases afflicting the back of the eye include wet and dry ARMD and diabetic retinopathy. While trans-scleral transport places the drug in the proximity of choroid, to reach neural retina the drug has to cross the RPE, a monolayer of cells with tight junctions (outer blood–retinal barrier). For the drug to enter the retina from the systemic circulation, it has to cross the retinal blood vessels, with a tight monolayer of endothelial cells (inner blood–retinal barrier). Due to the presence of such formidable barriers, eye drops and systemic doses do not efficiently deliver a drug to the back of the eye tissues. However, when dosed at sufficiently high levels, some drugs may achieve therapeutic levels in tissues of the back of the eye after topical [28] and systemic [29] administrations. Intravitreal injections, however, are the norm to deliver drugs efficiently to the retina. Alternative less invasive approaches to the globe being investigated include periocular injections for trans-scleral delivery [27], as well as suprachoroidal administrations [30]. However, permeability barriers imposed by surface ocular tissues can potentially be overcome to some extent using prodrugs, penetration enhancers,**iontophoresis**, ultrasound, and nanoparticles, as discussed subsequently in this article.

A new generation of ophthalmic vehicles and delivery systems are summarized in Table 2. As evident, prodrugs of prostaglandins are being employed to enhance membrane permeability of topically administered drugs. Viscous and mucoadhesive vehicles are being employed in topical formulations to enhance precorneal drug retention. In order to treat back-of-the-eye disorders such as ARMD, intravitreally injectable aptamer (Macugen™) and antibody fragment (Lucentis™) formulations have been introduced. To reduce the frequency of intra-vitreal injections and to treat chronic back-of-the-eye disorders such as diabetic retinopathy and uveitis, injectable (Ozurdex™; Ileuvien™) or surgically placed (Vitrasert™; Retisert™) sustained release implants have been developed.

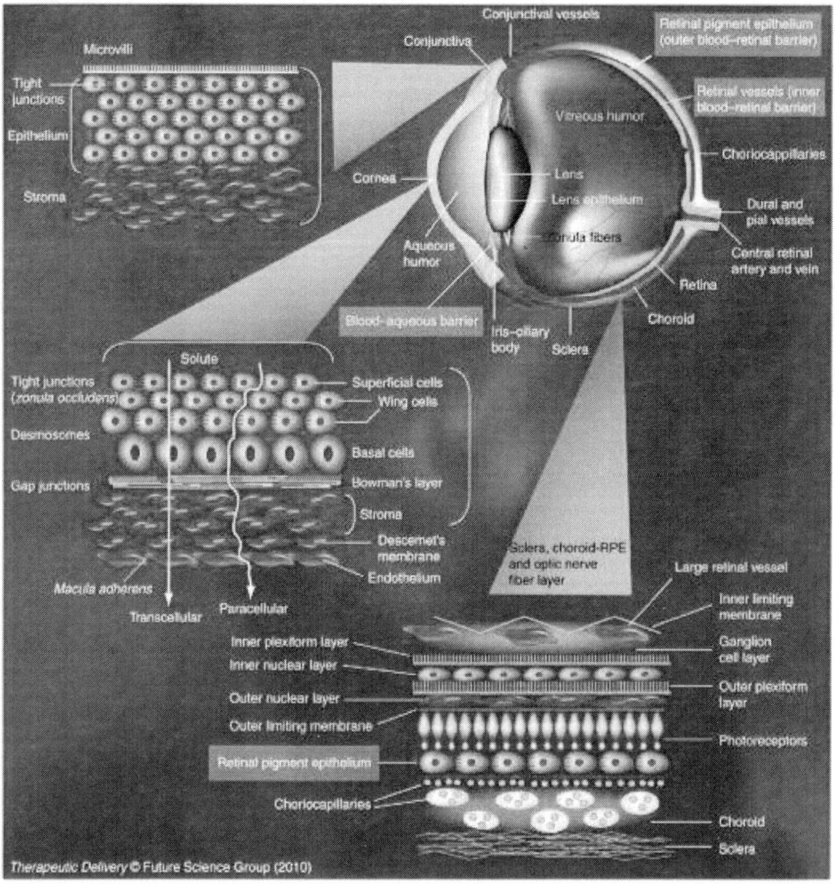

Figure 1. Permeability barriers for ocular drug delivery

KEY APPROACHES TO OVERCOME SHORT PRECORNEAL DRUG RESIDENCE

Mucoadhesive & viscosity-enhancing polymers

Topical formulations containing mucoadhesive or viscous materials can sustain and enhance drug delivery to target eye tissues. As shown in Table 2, several marketed topical formulations contain excipients that are capable of enhancing viscosity and/or allowing mucoadhesion. Examples of such excipients in ophthalmic formulations include gellan gum, polycarbophil, carbopol and poly(styrene-divinyl benzene) sulfonic acid. The viscosity of an ophthalmic formulation can range up to 20–30 cps [31]. Meseguer *et al.* reported that

Table 2. Some examples of ophthalmic drug-delivery systems approved by the US FDA.

Product category	Description	Drug	Year of approval	Indication
Topical delivery systems				
Ophthalmic solution (Acuvail™)	Preservative free 0.45% ketorolac tromethamine solution containing carboxymethylcellulose as viscosity enhancer	Ketorolac tromethamine	2009	NSAID for the treatment of pain and inflammation after cataract surgery
Ophthalmic solution (Lumigan™)	0.03% ophthalmic solution of bimatoprost with 0.005% benzalkonim chloride as preservative	Bimatoprost (prodrug)	2001	Reduction of elevated IOP
Ophthalmic solution (Travatan™)	0.004% ophthalmic solution of travoprost with 0.015% benzalkonim chloride as preservative	Travaprost (prodrug)	2001	Reduction of IOP
Ophthalmic solution (Xalatan™)	0.005% ophthalmic solution of latanoprost with 0.02% benzalkonim chloride as preservative	Latanoprost (prodrug)	1996	Reduction of IOP
Ophthalmic solution (Propine™)	0.1% ophthalmic solution of dipivefrin-HCl with benzalkonim chloride as preservative	Dipivefrin (prodrug)	1980	Control of IOP in chronic open-angle glaucoma
In situ gel-forming solution (Timoptic-XE™)	Sterile ophthalmic gel forming solution containing 0.25 or 0.5% of timolol equivalent timolol maleate. GELRITE gellan gum, tromethamine and mannitol, with 0.012% benzalkonim chloride as preservative	Timolol maleate	1993	Treatment of glaucoma
Ophthalmic solution (AzaSite™)	1% sterile aqueous topical ophthalmic solution of azithromycin formulated in DuraSite® (polycarbophil, disodium EDTA, sodium chloride) with 0.003% benzalkonim chloride as preservative	Azithromycin	2007	Bacterial conjunctivitis
Ophthalmic gel (Zirgan™)	Sterile topical ophthalmic gel containing 0.15% of ganciclovir formulated with carbopol and 0.0075% benzalkonim chloride as preservative	Ganciclovir	2009	Herpetic keratatis
Ophthalmic emulsion (Durezol™)	Sterile topical aqueous ophthalmic emulsion containing 0.05% of difluprednate formulated with boric acid, castor oil, glycerin, and polysorbate 80, with 0.1% sorbic acid as preservative	Difluprednate	2008	Steroid for treatment of inflammation and pain associated with ocular surgery
Ophthalmic emulsion (Restasis™)	Sterile preservative free ophthalmic emulsion containing 0.05% of	Cyclosporine	2002	Increase tear production in dry eye diseases

Product category	Description	Drug	Year of approval	Indication
	cyclosporine formulated with glycerin, castor oil, polysorbate-80 and carbomer1342.			
Ophthalmic suspension (Betoptic-S™)	0.25% ophthalmic suspension of betaxolol formulated with poly(styrene-divinyl benzene) sulfonic acid (Amberlite® IRP-69), Carbomer 934P, edetate disodium and 0.01% benzalkonim chloride as preservative	Betaxolol	1990	Treatment of glaucoma
Opthalmic inserts (Ocusert™)	Pilocarpine and alginic acid enclosed in thin ethylene-vinyl acetate membrane to control release of drug for 7 days	Pilocarpine	1974	Treatment of glaucoma
Intravitreal products				
Intravitreal implant (Ozurdex™)	Intravitreal implant containing 0.7 mg (700 μg) dexamethasone in the NOVADUR™ solid polymer drug delivery system	Dexamethasone	2009	Macular edema
Intra-vitreal implant (Retisert™)	Intravitreal implant containing 0.59 mg (590 μg) of fluocinolone acetonide formulated with microcrystalline cellulose, polyvinyl alcohol and magnesium stearate	Flucinolone acetonide	2005	Chronic noninfectious uveitis
Intra-vitreal implant (Vitrasert™)	Intravitreal implant containing 4.5 mg of ganciclovir formulated in the nonbiodegradable polymers, polyvinyl alcohol and ethylene-vinyl acetate	Ganciclovir	1996	Cytomegalovirus retinitis

precorneal drainage of eye drops can be reduced in rabbit models by choosing vehicles containing various excipients, including hydroxyethylcellulose (HEC), xan-than gum and gellan gum (Gelerite®) [32]. These viscous and mucoadhesive vehicles reduced the elimination rate constant for pilocarpine by a maximum of approximately tenfold [32]. Based on this prior observation, we simulated the influence of reduced precorneal loss by a similar magnitude on drug levels in tear fluid as well as in the aqueous humor (Figure 2). The first order aqueous humor elimination rate constant chosen for the simulation was fixed at 0.06 min^{-1}. This value is close to the 0.057 min^{-1} reported for the elimination of timolol from the aqueous humor in male Nippon albino rabbits [33]. From Figure 2, it is evident that a tenfold reduction in the precorneal loss rate constant sustains tear drug levels better, resulting in a 5.2-fold increase in the C_{max} of the drug in the aqueous humor.

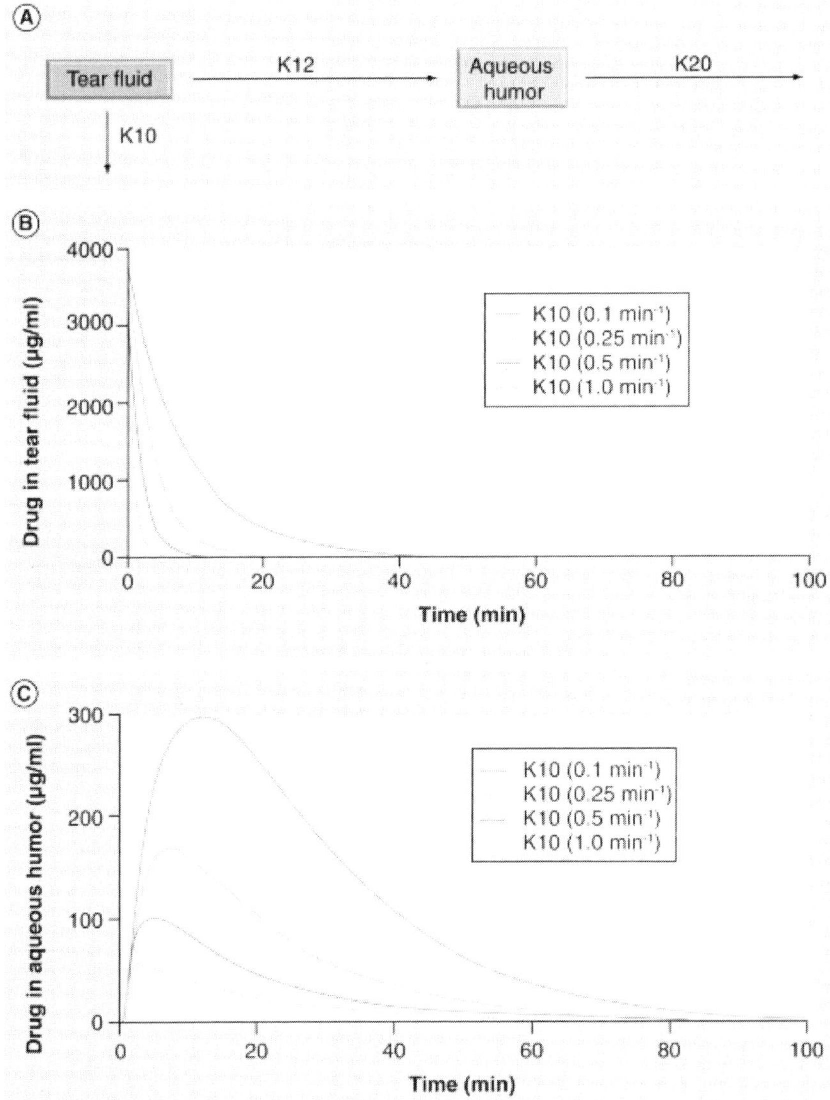

Figure 2. Choosing a mucoadhesive or viscous vehicle or carrier that reduces precorneal drug elimination can elevate tear fluid and aqueous humor drug levels

Formulations of azithromycin based on Durasite® [201] (polycarbophil USP – 90% polyacrylic acid cross-linked with 0.1–5% of divinyl glycol, ~1 × 10^6 Da, disodium ethylene diamine tetraacetic acid [EDTA], and sodium chloride) were recently shown to significantly enhance the ocular delivery of azithromycin in pigmented rabbits [34]. The AUC was 19.5-, 4.3-, and 2.8-fold higher in tears, cornea and aqueous humor, respectively, for the polycarbophil formulation when compared with the plain drug formulation. The mechanisms for enhanced drug delivery with polycarbophil can be twofold. First, this polymer is known to be a mucoadhesive agent in the eye. It was shown to reduce clearance from the eye surface in the rabbit model [32]. Second, evidence exists for the ability of polycarbophil to alter tight junctions, thereby elevating paracellular drug permeability [35]. While the elevations in AUCs observed in the azithromycin study are substantial, it is not clear whether such differences would translate to humans. A recent study reported that the topical delivery of besifloxacin (AUC_{0-24} [μg.h/g]) from an eye drop to the tear film follows the order: cynomolgus monkey (6440), pigmented rabbit (3240) and human (1263) [15].

Thus, viscous and/or mucoadhesive vehicles are useful in substantially enhancing topical drug delivery in animal models. However, it is unclear whether these benefits translate at the same level in humans with a blinking rate of 1000/h that is much higher than rabbits with a reported blinking rate of 3/h [36].

Contact lenses

Contact lens-mediated drug delivery to the eye dates back to 1960 when Wichterle demonstrated the use of soft contact lenses for topical drug delivery [37]. In the intervening 50 years there have been no FDA-approved products based on this platform. However, recent research articles and patent activities suggest renewed enthusiasm for this approach in sustaining ocular drug delivery by the topical route. There are multiple challenges in developing contact lenses as ocular drug delivery systems. The current challenges are incorporation of sufficient amounts of the drug into the lens matrix, sustaining the drug release for the desired time frame at a controlled rate, good optical clarity, patient comfort during prolonged wear and biocompatibility. Prolonged wear of contact lens is associated with risk of infection such as microbial keratitis and dry eye syndrome [38]. Wearing of contact lenses is contraindicated in various inflammatory conditions of anterior segment such as anterior uveitis, vernal conjunctivitis, microbial keratitis and dry eye syndrome, limiting the applicability of this delivery system. Several ocular diseases are more prevalent in geriatric patients, who might find the use of contact lenses and contact lens based delivery systems inconvenient. Various methods have been used for loading the drug into lens delivery systems (Figure 3). Some key approaches to prepare contact lens based drug-delivery systems are discussed below.

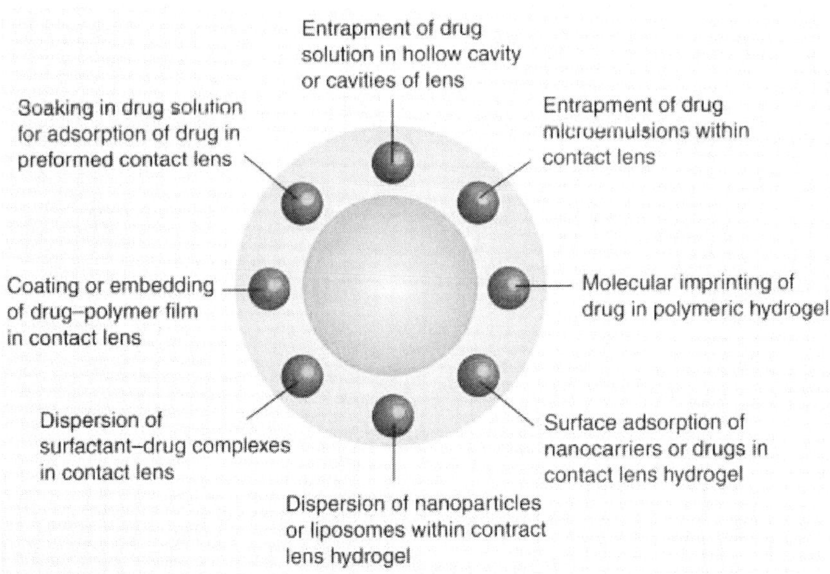

Figure 3. ontact lenses for ocular drug delivery

Soaking of lenses in drug solution

The traditional method is to soak the preformed contact lens in drug solution so that the drug is adsorbed into the polymeric lenses. Such presoaked lenses allow limited, slow release of the drug into the post-lens lacrimal fluid. Soft contact lenses made from poly-hydroxymethacrylate (pHEMA) hydrogels release the majority of their drug content in one day [39,40]. New contact lenses made from silicone–hydrogel hybrid showed similar uptake but lower release of drug when compared with pHEMA lenses [41,42]. The ciprofloxacin uptake in pHEMA hydrogel lenses (Polymacon) and silicone hydrogel lenses (Balafilcon A) was similar at approximately 1800 μg. However, pHEMA lenses released 200 μg of ciprofloxacin in 24 h, while silicone hydrogel lenses released only 80 μg of drug in 24 h. Although soaking of preformed contact lenses in drug solution is an easy way to incorporate the drug, this approach suffers from various limitations including low drug loading and fast diffusion of the drug from the lenses. However, recently, Kim *et al.* showed that modified silicon hydrogel material is able to release timolol and dexamethasone at near zero order for 120 days, with very low burst release (<5%) on the first day [43]. To increase the drug loading in contact lenses, Nakada and Sugiyama developed contact lenses characterized by a hollow cavity that binds two lenses [202]. The compounded lens with the hollow cavity was soaked in a drug solution and allowed to release the drug into the eye upon insertion. Although this method showed more drug loading, it suffers from an additional disadvantage of low O_2 and CO_2 permeability. An alternative to contact lenses is the use of collagen shields soaked in drug solutions as a delivery system. Haugen *et al.*demonstrated that collagen shields presoaked in gatifloxacin achieve effects similar to multiple eye drops of gatifloxacin in relieving endophthalmitis in a rabbit model [44].

Molecularly imprinted polymeric hydrogels

Another approach that has received recent attention is to use molecularly imprinted polymeric hydrogels for contact lens-mediated ophthalmic drug delivery. The polymeric content of the lens is molded to recognize the structural features and bonding preferences of target drug molecules. Such an approach allows high drug affinity, selectivity and greater drug loading capacity. Hiratani *et al.* showed that the uptake of timolol in molecularly imprinted contact lens is 4.9-fold higher (2.51 ± 0.01 mM) than the nonimprinted gels (0.51 ± 0.07 mM) [45]. The overall binding affinity of timolol to molecularly imprinted gel is 12.3-times higher compared with non-imprinted gels [45]. The molecularly imprinted soft contact lenses for norfloxacin exhibited low drug loading for timolol (2 µmol/g) when compared with norfloxacin (8 µmol/g), even though timolol has a spatial size similar to norfloxacin [46]. The difference in drug loading of imprinted hydrogel to norfloxacin and timolol is due to the specificity of the binding pocket in the imprinted gel used for preparing the contact lens [46]. Molecularly imprinted contact lenses exhibited a more prolonged drug release when compared with lenses soaked in drug solution. Molecularly imprinted contact lenses of ketotifen (diffusion coefficient: 5.57 ± 0.31 × 10 cm^2/s) showed a diffusion coefficient that was nine-times lower than non-imprinted lenses (50.2 ± 4.8 × 10 cm^2/s) [47]. For norfloxacin, non-imprinted contact lenses released almost all the drug content (>90%) within 12 h. On the other hand, norfloxacin-imprinted lenses released only 40% of the drug in 24 h with less than 10% burst release during the first 2 h [46].

Conjugation of nanoparticles or drug molecules to contact lens surface

Another approach to prepare contact lens drug delivery systems is to immobilize the drug or drug-loaded nanocarriers such as liposomes and nanoparticles on the surface of commercial soft contact lenses [48]. This can be achieved by surface functionalization of contact lenses to attach the drug or nanocarriers. Surface immobilization of levofloxacin liposomes on NeutrAvidin-coated Hioxifilcon-B contact lenses resulted in slower release than drug-soaked lenses. However, most of the drug (70%) was released within the first 5 h, with 30% remaining drug release occurring over the next 6 days [49]. Thus, the drawback of this approach is the relatively rapid detachment or disintegration of liposomes in the contact lenses. The lipid layer may also impede O_2 and CO_2 permeability.

Drug-polymer films integrated with contact lenses

Recent research from the Massachusetts Eye & Ear Infirmary (Needham, MA, USA) demonstrated that near zero-order, 4-week release of ciprofloxacin and fluorescein is feasible from a drug-polymer film (poly[lactide-co-glycolide] [PLGA] 65:35) coated with a hydrogel (pHEMA) contact lens [50]. *In vitro* release studies showed that the drug release rate can be reduced by using high-molecular-weight polymers or increasing the polymer-to-drug ratios. Sano *et al.*showed that the delivery of levofloxacin to the aqueous humor in albino rabbits was 15-times greater at the end of 4 h with the piggyback **contact lens delivery system** when compared with eye drops dosed every 30 min interval over 4 h [51]. However, a comparison of drug levels at earlier time points was not provided. One of the major drawbacks of this approach is the increase in the thickness of the contact lens, which might result in low patient acceptance and poor permeability of O_2 and CO_2. However, the use of some new materials such as silicone hydrogel with higher O_2 and CO_2 permeability, in conjunction with drug-polymer films of reduced thickness may make this option more suitable for sustained ocular drug delivery.

Liposome-loaded contact lenses

Another novel approach is the entrapment of liposomes, nanoparticles, microemulsions and surfactant–drug complexes in contact lenses during manufacturing. In this method, drug-laden nanoparticles, liposomes or surfactant are added to the polymerizing medium of hydrogel matrix followed by polymerization to form contact lenses dispersed with drug carriers [52–54]. The overall drug release can be regulated by controlling the release of the drug from the various carriers, followed by the lens matrix to the post-lens tear film. Contact lenses created with dispersion of lidocaine-loaded liposomes or nano-particles showed nonlinear release of drug for 7 days, with an initial 15–20% burst release over the first few hours followed by near zero-order release for the remaining 7 days [53]. Kapoor *et al.* showed near zero-order release of cyclosporine *in vitro* from surfactant-laden hydrogel lenses, with the drug-release rate being 1.5% per day over 50 days [55].

Scleral lens delivery systems

Scleral gas-permeable rigid lenses are contact lenses initially developed by Boston Foundation for Sight, the contact lens service at the Massachusetts Eye and Ear Infirmary. They were developed in order to mask the abnormal corneal astigmatism when traditional rigid gas-permeable corneal contact lenses fail or contraindicated for the management of severe ocular surface diseases such as keratoconjunctivitis. The scleral lenses were first approved by the FDA in 1994 for the management of ectasia and irregular astigmatism. A scleral lens rests on the sclera and creates a space over the cornea and limbus, acting as a reservoir for tear fluid between the inner surface of the scleral lens and the cornea to form an expanded tear film. This expanded tear film over the cornea acts as a liquid bandage in corneal surface irregularities.

Rosenthal and Jacobs investigated the application of scleral lenses made from a gas-permeable polymer (Boston IV polymer, Itafluorfocon B) for ocular drug delivery [203]. The scleral lens allowed improved retention of topically applied drug on the corneal surface in the expanded tear film, that is, the tear film or artificially introduced tears retained between the corneal surface and the scleral lens. The drug may be added to the scleral lens, which acts as a reservoir for sustained release of the drug, or administered externally as an eye drop, which is retained in the expanded precorneal tear film. The topical delivery of Avastin for the treatment of corneal neovascularization using **scleral lens delivery system** showed promising results in a clinical study with five patients [56]. One drop of 1% bevacizumab solution free from preservatives and penetration enhancers was applied twice daily over 3 months to the eyes of patients fitted with Boston ocular surface prosthesis (BOSP) lens in the fluid reservoir of BOSP (a reservoir for expanded tear film formed between the inner surface of the scleral lens and the cornea filled with artificial tears). Drug effects were evident during the 30 days of treatment with marked regression of pannus and improvement of vision from finger count to 20/40 in one patient. Another patient, whose prior vision was 20/400, improved to 20/200 4 months after treatment initiation and at the end of 6 months had vision of 20/70. In all patients, the effects continued for the subsequent 60 days after cessation of dosing and no recurrence was noted. There were no ocular adverse effects such as breakdown of the corneal epithelial integrity with the scleral lens delivery system in humans. On the other hand, Kim *et al.* demonstrated breakdown of corneal epithelial integrity in the human eye after treatment with topical bevacizumab eye drops [57].

The above observation with Avastin and another recent report indicating the delivery of an anti-TNF-α-antibody fragment (ESBA105) to the anterior and posterior eye tissues following high-frequency dosing (three drops every hour over 10 h) and 30 s blink-free opening of the rabbit eye indicates that given sufficient duration of exposure, even macromolecules might permeate ocular barriers to access target tissues [58].

Punctal plug delivery systems

Punctal plugs have been used for more than 20 years for the symptomatic relief of dry eye syndrome. Punctal plugs occlude the nasolacrimal duct, thereby reducing tear drainage through the upper and/or lower punctum in the eye. Depending on the material used for punctal plug, the duration of occlusion ranges from 7 to 180 days [59,60]. Punctal plugs made from silicone [204], teflon, hydroxyethyl methacrylate (HEM), polycaprolactone (PCL) or polydioxanone are intended for 180 day use, after which they need to be removed. On the other hand, punctal plugs made from animal collagen last for 7–10 days, during which time the plugs disintegrate [301]. Most punctal plugs are lost due to spontaneous extrusion from the puncta, and this risk is higher for plugs inserted in upper punctum compared with those inserted in lower punctum. Recently, punctal plugs made from thermosensitive, hydrophobic acrylic polymer were used to avoid extrusion problem (SmartPlug™, Medennium Inc.). This thermosensitive hydrophobic acrylic polymer is a solid rod at room temperature but becomes a soft cohesive gel when its temperature changes from room temperature to body temperature. Tuning of the material from a rigid rod into a gel-like plug conforms to the size and shape of the puncta, allowing retention of the plug.

Use of punctal plugs for the delivery of ophthalmic medications offers a novel approach for chronic treatment of patients with various eye diseases including glaucoma and dry eye syndrome. Punctal plugs have several potential advantages over eye drops, including dose reduction (and hence minimization of systemic side effects), enhanced efficacy possibly due to controlled release of the drug at an optimum rate and better patient compliance with drug therapy. Punctal plugs used for drug delivery can be made from various polymers in a variety of shapes and sizes, with the final dimensions being limited by the punctum. The punctal plugs are mainly composed of a cylindrical body containing the drug compound, an optional outer shell made up of material impermeable to the drug and the tear fluid, an optional cap material containing pores and an optional unit to retain the punctal plug over prolonged periods of time. The bottom end is tapered and/or narrower in most punctal plugs to allow easy insertion into the punctum. The other end, otherwise known as the head portion, is exposed to the tear film and configured to rest on the exterior of punctum. The cap may have one or more pores extended throughout the body for release of the drug. In some designs, additional canalicular extension are configured to improve plug retention in the punctum and to serve as an additional reservoir for the therapeutic compound.

Generally, a punctal plug drug delivery system is coated with a material that is impermeable to the drug and tear fluid on all sides except the head portion, through which the drug is released into the tear film. The release of the drug from a punctal plug is controlled by drug diffusion from the polymeric core to the tear fluid. The drug can be loaded in the central polymeric core as solution, suspension, microemulsions, nanoparticles, microparticles or liposomes with or without an additional polymer matrix. In some cases, punctal plugs are coated with a drug-containing polymeric matrix [204]. Alternatively, preformed plugs can be soaked in drug solution. However, these drug-loading approaches, when performed in the outer coat alone, result in limited drug loading. Most examples of punctal plugs showed near zero-order drug release rates for drug molecules [205,206]. For instance, cyclosporine-loaded punctal plugs made of HEMA core and coated with a silicone shell on all sides except the cylindrical head showed near zero-order release, with a release rate of approximately 3.0 μg/day over 45 days [206].

Punctal plug-mediated ocular delivery of latanoprost is in Phase II clinical study (QLT, Inc., BC, Canada). A Phase I study conducted in Mexico with five patients indicated safety and good tolerability and approximately 30% intraocular pressure (IOP) reduction over 3 months [302]. Current Phase II trials of latanoprost **punctal plug delivery system** (L-PPDS) are testing 44 and 81 μg doses in the United States. Preliminary results of ongoing Phase II trials with 44 μg L-PPDS showed a mean reduction in IOP by 3.5 mm Hg at the end of 4 weeks, with 36% of patients showing a decrease in IOP more than 5 mm Hg [302]. Out of the 60 patients enrolled, 48 patients (78%) completed the 4 week follow-up and retained L-PPDS in both eyes for 4 weeks. The L-PPDS was reasonably well tolerated over the testing period, with the overall adverse event incidence ranging from 1.7 to 11.7%. Commonly observed adverse effects included eye itchiness, eye irritation, increased lacrimation and ocular discomfort. Figure 4 summarizes the various possibilities for punctal plug designs in a single scheme.

Key approaches to overcome permeability barriers

Several approaches, including prodrug derivatization, drug–cyclodextrin complexation, penetration enhancers and driving forces such as electrical currents or ultrasound, have been used to increase the permeability of drug molecules across the cornea. All these approaches except ultrasound have been shown to be beneficial in human studies, as described below.

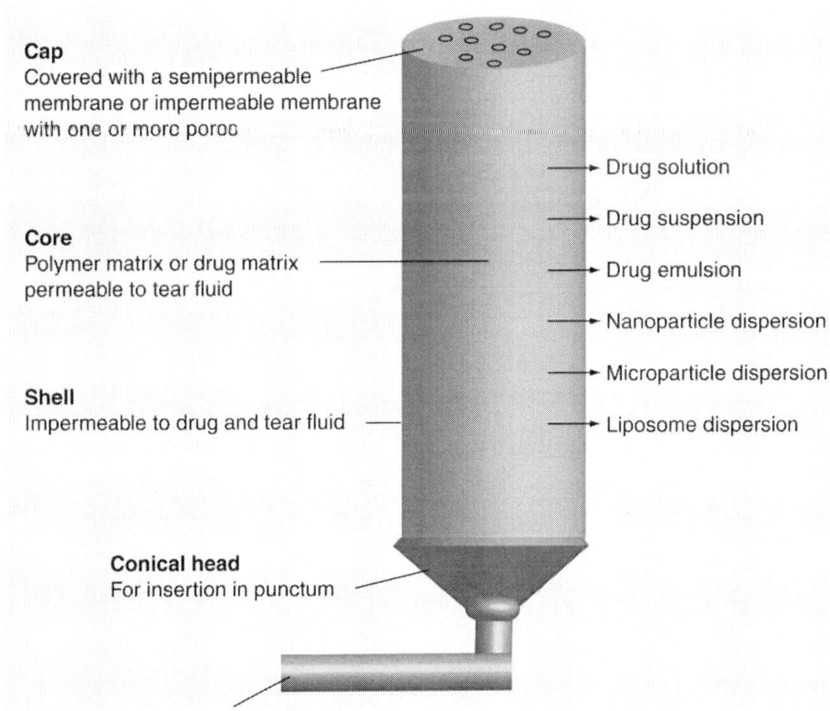

Figure 4. Punctal plug delivery systems with various components and drug-loading methods

Prodrugs

Prodrug derivatization can be used to enhance drug lipophilicity in order to overcome the permeability barrier. Examples of marketed prodrugs include dipivefrine, latanoprost, travoprost and bimatoprost. Dipivefrine, an ester prodrug of epinephrine, is 600-times more lipophilic and 17-fold more permeable across the cornea compared with epinephrine [61]. Ocular tissue distribution of dipivefrine and epinephrine in female New Zealand rabbits showed 15- and 11-times higher $AUC_{(0\text{-}t)}$ in cornea and aqueous humor, respectively, for the prodrug when compared with the parent molecule [61]. Dipivefrine comparative studies in humans showed that dipivefrine produces similar reduction in the IOP at 20-fold lower dose compared with epinephrine [62]. The prostaglandin analogs latanoprost, travoprost, bimatoprost are all prodrugs, with the former two being isopropyl ester prodrugs and the latter being an ethanolamine amide prodrug. These prodrugs increase drug lipophilicity and, hence, are expected to enhance drug absorption [63–66].

Alternatively, for a drug with dissolution rate-limited absorption, prodrugs can be used to enhance drug solubility and, therefore, the driving force to encourage passive drug absorption. Cyclosporine A (CsA) is a potent immunosuppressive drug that is of therapeutic value in treating various ocular disease including dry eye syndrome, uveitis and corneal graft rejection [67]. Poor aqueous solubility of CsA is a limiting factor in the development of topical ocular formulations [67,68]. An emulsion formulation of CsA (Restasis®) is currently being marketed for dry eye syndrome. However, blurred vision has been reported with this emulsion formulation [69,70]. The vehicles for such emulsions should be chosen with care since some oily vehicles can enhance corneal permeability of fluorescein by sevenfold in human eyes [71]. UNIL088 is a hydrophilic ester prodrug of CsA with 25,000-times higher solubility than CsA in isotonic phosphate buffered saline (PBS) (pH 7.0) [67]. Eye drops of this CsA prodrug are effective in treating corneal graft rejection in rats [72].

Currently, prodrugs to enhance transporter mediated drug absorption are being investigated for enhanced ocular drug delivery [73]. Evidence exists for the presence of organic anion, organic cation and peptide transporters in various ocular barriers including cornea, conjunctiva, and blood–retinal barriers [74–76]. In addition, evidence exists for efflux transporters such as P-glycoprotein and multidrug resistance-associated proteins (MRPs) in the ocular barriers [47,77–80]. Delivery of an anionic aldose reductase inhibitor to cultured human retinal pigment epithelial cells [13], as well as neural retina *in vivo* in a rat model [13], was elevated by probenecid treatment. This may be an outcome of drug efflux inhibition from eye tissues by probenecid.

Functionalized nanoparticles

One approach to enhance the epithelial uptake and transport of poorly permeable drugs is to employ nanoparticle formulations that allow superior cellular delivery. For instance, using a bovine *ex vivo*, intact eye model, Kompella *et al.* demonstrated that corneal epithelial uptake of 20-nm nanoparticles can be elevated from approximately 2–16% by surface coating or functionalizing nanoparticles [81]. The nano-particles did not perturb the tight junctional architecture or paracellular permeability of the cornea. Furthermore, the uptake of transferrin-**functionalized nanoparticles** could be reduced by excess free transferrin ligand. In this study, an elevation in particle uptake was observed as early as 5 min following eye drop instillation. Thus, functionalized nanoparticles allow rapid, receptor-mediated uptake of particles into epithelial cells of the eye.

Drug cyclodextrin complexes

Drug–cyclodextrin complexation is a useful way to increase the aqueous solubility of poorly water-soluble drugs without changing their molecular properties [82]. A key advantage of cyclodextrins over conventional membrane-perturbing penetration enhancers such as benzalkonium chloride (BAK) and EDTA, is that they increase ocular drug bioavailability predominantly by increasing drug solubility [83]. Various clinical and preclinical studies demonstrated the efficiency of this approach in increasing the ocular bioavailability of poorly water soluble drug molecules such as hydrocortisone, dexamethasone, cyclosporine, and acetazoloa-mide [84–87]. A clinical study of 0.32% dexamethasone-cyclodextrin-polymer complexes in humans showed 2.6-times higher AUC in the aqueous humor when compared with a dexamethasone suspension [88]. Interestingly, several eye drop products containing cyclodextrins are currently marketed in Europe. Examples include Voltaren® (diclofenac-Na with hydroxyl propyl-β-cyclodextrin [HPβCD]), Indocid® (indomethacin with HPβCD) and Clorocil® (chloramphenicol with methyl β-cyclodextrin [MβCD]). It is anticipated that ophthalmic products containing cyclodextrins may enter the US market in the near future.

In addition to the use of cyclodextrins, drug solubility can also be enhanced by including suitable salt forms of the drug, surfactants or co-solvents.

Penetration enhancers

Despite extensive research in the area of penetration enhancers to overcome epithelial barriers, no clinical ophthalmic formulation relies directly on the penetration-enhancing properties of its excipients. However, the fact that several ophthalmic formulations contain benzalkonium chloride and/or EDTA as preservatives is noteworthy. Both these excipients are well known to alter biological membranes, thereby enhancing drug permeability [89–91]. For a comprehensive review on the mechanisms of action of penetration enhancers, the readers are referred to an article by Lee *et al.* [92].

Iontophoresis

For charged drug molecules or vehicles, iontophoresis is a viable approach for enhanced drug delivery. Iontophoresis utilizes low currents to enhance the penetration of charged molecules across tissue barriers. The drug is applied using an electrode carrying the same charge as the drug. An electrode with the opposite charge placed elsewhere in the body completes the circuit. The ionized drug molecule penetrates the tissue by electric repulsion. Additionally, neutral molecules can potentially be delivered using iontophoresis on the basis of electro-osmosis or solute-associated fluid transport.

Eljarrat-Binstock and Domb prepared a comprehensive review on the application of iontophoresis for ocular drug delivery [93]. This noninvasive method can deliver drug molecules to both anterior and posterior segments of the eye. Several preclinical studies demonstrated the potential of iontophoresis in delivering antibiotics, antifungals, steroids and NSAIDs to both anterior and posterior segments using transcorneal and trans-scleral iontophoresis, respectively. Several investigators conducted clinical studies using iontophoresis [94–98]. The EyeGate® II system, designed for transcorneal drug delivery, consists of an ocular applicator, syringe, adaptor for transferring the drug product from reservoirs to the applicator and generator to provide consistent current to the electrode. FDA-granted orphan drug designation for the delivery of dexamethasone phosphate using the EyeGate-II system for corneal graft rejection therapy. Iontophoresis may prove an attractive option to patients who are not responsive to eye drop therapy. In addition, Phase II clinical trials are ongoing for the evaluation of safety and efficacy of EGP-437 (combination of the above drug and device) to treat uveitis and dry eye syndrome. Ocuphor (Iomed Inc., USA) and Visulex (Aciont Inc., USA) are the other ocular iontophoresis systems under investigation for trans-scleral iontophoresis. Harwath-Winter *et al.*demonstrated the superiority of ocular iontophoresis for the delivery of sodium iodide in treating dry eye syndrome in 28 patients [99]. The investigators observed no corneal edema or epithelial damage in this study. Parkinson *et al.* reported severe burning sensation and discomfort in two out of six patients at a current dose of 4 mA over 2 min during an ocular iontophoresis study [100]. Thus, one limitation of ocular iontophoresis is potential discomfort to patients. Another limitation of ocular iontophoresis is inadequate sustained delivery. Given the chronic nature of eye disorders, such as diabetic retinopathy, macular degeneration, glaucoma and uveitis, iontophoresis is required at a high frequency. The safety of prolonged iontophoretic drug delivery has yet to be established.

INVASIVE APPROACHES FOR DRUG DELIVERY TO THE BACK OF THE EYE

Trans-scleral drug delivery

In order to enhance drug delivery to the back of the eye, drug can be administered adjacent to the sclera by various periocular routes including subconjunctival, sub-Tenon, peribulbar and retrobulbar routes [27]. Ayalasomayajula and Kompella showed that posterior subconjunctival injection enhances celecoxib delivery to the retina by 54-fold compared with a systemic injection [11]. Since this route of administration is invasive, albeit less invasive to the globe than intravitreal injections, dosing at a low frequency is desirable. That is, sustained release dosage forms are required. This can be achieved by administering drug suspensions, *in situ*-forming drug suspensions, drug–polymer particles or implants, preferably those that allow unidirectional drug permeability towards sclera. Using a single periocular injection of slow-release polymeric microparticles of celecoxib, Amrite *et al.* demonstrated that drug levels as well as effects can be sustained for at least 2 months in a diabetic rat model [9,101]. Sclera is more permeable to drug molecules when compared with cornea and conjunctiva [102]. Kim *et al.* showed that the delivery of high-molecular-weight dextran–dye conjugate (70 kDa) to choroid and retina after subconjunctival injection in mice [103]. However, the choroid-Bruch's membrane can bind lipophilic drugs, hindering their transport [104]. For lipophilic molecules, drug partitioning or binding in tissue is significantly higher in choroid-RPE when compared with the retina. High, nonproductive binding to choroid-RPE minimizes drug entry into the retina during **trans-scleral drug delivery** [105]. The RPE, a tight monolayer of epithelial cells underlying choroid, is an additional barrier for drug delivery to the neural retina. Thus, although periocular administration improves drug delivery by overcoming the conjunctival permeability barrier, there are additional barriers that need to be overcome in facilitating retinal drug delivery. Table 3 summarizes some of the sustained trans-scleral drug delivery approaches assessed to date.

Intravitreal injections

Intravitreal injection of drug solution or suspension is the most efficacious method for targeted delivery of drug molecules to tissues of the posterior segment of eye. Drugs in solution disappear rapidly form the vitreous humor, requiring frequent injections to maintain therapeutic levels [106]. By contrast, drug suspensions prolong drug residence in the vitreous by allowing slow dissolution of the drug in the vitreous [106]. Durairaj *et al.* demonstrated that the elimination half-life for diclofenac acid suspension was 24 and 18 days in vitreous and choroid-retina, respectively, when compared with 2.9 and 0.9 h observed with diclofenac sodium solution after intravitreal injection in New Zealand white rabbits [107]. Intravitreal injection of triamcinolone acetonide suspension is used in clinical settings for the treatments of wet ARMD and diabetic macular edema [108,109]. The duration of drug effect depends on the dose administered and the pathological condition. In nonvitrectomized human eyes, the duration of the effect after single intravitreal injection of triamcinolone acetonide suspension was approximately 6–9 and 2–4 months for 20- and 4-mg doses, respectively [110]. Single intravitreal injection of triamcinolone acetonide suspension (4 mg) showed drug levels in vitreous up to 3 months in nonvitrectomized human eyes [111]. The elimination half-life from vitreous is twofold higher in vitrectomized eyes when compared with the nonvitrectomized eyes [112].

Table 3. Sustained trans-scleral drug delivery for treating retinal disorders.

Drug	Delivery system	Route of injection	Duration of observation	Ref.
Ovalbumin	Thermosetting gel	Subconjunctival injection	Drug levels were detected in choroid-retina for 14 days	[121]
Celecoxib	PLGA microparticles	Subconjunctival injection for posterior segment delivery	Drug levels were detected in retina, vitreous, lens and cornea for 14 days Significant reduction of diabetic retinopathy related changes at 14 days	[10]
Tetramethyl rhodamine-dextran	Solution	Subconjunctival injection	Drug levels were detected in sclera, choroid and retina for 70 h	[122]
Budesonide	PLA microparticles	Subconjunctival injection for posterior segment delivery	Drug levels were detected in retina, vitreous, and cornea for 14 days	[123]
Budesonide	PLA nanoparticles	Subconjunctival injection for posterior segment delivery	Drug levels were detected in retina, vitreous and cornea for 7 days	[123]
Celecoxib	PLGA microparticles	Subconjunctival injection for posterior segment delivery	Drug levels were detected in sclera-choroid, retina, vitreous and cornea after 2 months of injection Significant reduction of diabetic retinopathy-related changes after 2 months	[9]
Carboplatin	PAMAM dendrimeric nanoparticles	Subconjunctival injection for posterior segment delivery	Significant reduction in retinal tumor burden after 21 days	[124]
5-flurouracil	PLGA microspheres	Subconjunctival injection for anterior and posterior segment delivery	Drug levels were detected in aqueous humor, and sclera for 7 days	[125]
Carboplatin	Fibrin sealant	Subconjunctival injection for posterior segment delivery	Effect was observed for 21 days	[126]
PKC412	PLGA microspheres	Periocular injection for posterior segment delivery	Drug levels were detected in vitreous, retina and choroid after 20 days Marked reduction in choroidal neovascularization area after 20 days	[127]
PEDF protein	Protein solution	Subconjunctival injection for posterior segment delivery	Drug levels were detected in sclera and retina for 24 h	[85]

Intravitreal implants

Implants are biodegradable or nondegradable solid dosage forms typically made of polymers that slowly release the drug of interest over prolonged periods [113,114]. Depending on the biodegradation nature of the polymer used, implants are classified as nonbiodegradable and biodegradable implants. Drug release can be controlled more precisely using nondegradable implants. Due to changes in the properties of the system with polymer degradation, control of drug release from degradable systems is more difficult. While degradable systems offer the advantage of physiological clearance of all system components over time, nondegradable systems remain permanently in the body unless surgically removed. Currently, Vitrasert, Retisert and Ozurdex are three implants approved for intravitreal administration. Whilst the former two are nondegradable implants, the latter is degradable. Intravitreal implants are either sutured in the pars plana area or injected into the vitreous humor.

Implants have many potential advantages over traditional delivery systems including drug delivery closer to the target tissue in the posterior segment, dose reduction due to localized delivery (and hence minimization of systemic side effects)and reduced complications including infections due to low frequency dosing [115]. This results in reduced dosing frequency and enhanced efficacy possibly due to controlled release of the drug at an optimum rate. Although implants are currently approved for treating diseases of the posterior segment of the eye, they can potentially be used for treating diseases of the anterior segment as well. Subconjunctival and intrascleral implants may deliver greater quantities of the drug to the anterior segment, whereas suprachoroidal and intravitreal implants might deliver greater quantities of the drug to the posterior segment. Additionally, evidence exists for the usefulness of subcutaneously administered biodegradable implants in sustaining the retinal delivery and efficacy of an aldose-reductase inhibitor in treating diabetic retinopathy [14].

Vitrasert was the first nonbiodegradable intra-vitreal implants approved by the FDA in 1996 for the treatment of cytomegalovirus retinitis. Vitrasert is a reservoir implant device consisting of a drug pellet of ganciclovir coated with polyvinyl alcohol (PVA), a permeable polymer that regulates drug diffusion and ethylene vinyl acetate (EVA). EVA is an impermeable polymer that controls the area through which a drug is released [116]. This device needs surgical implantation in the pars plana region and also requires surgical removal once the drug is depleted. This device exhibits no burst release and sustains the drug release for 5–8 months after administration. However, the device is associated with risk of endopthalmitis and retinal detachment.

Retisert is another nonbiodegradable implant of flucinolone acetonide approved in 2005 for the treatment of chronic noninfectious uveitis. This implant contains 0.59 mg of drug coated with PVA and silicone laminates and provides sustained drug delivery over a period of 3 years. It delivers approximately 0.5 μg of drug/day at a constant rate. Retisert has shown significant efficacy but results in side effects, including cataracts and elevated IOP, which are associated with several corticosteroids such as fluocinolone acetonide [116]. Medidur/Iluvien is another nonbiodegradable implant of flucinolone acetonide that is undergoing Phase III clinical trials for the treatment of diabetic macular edema. Fluocinolone acetonide inhibits secretion of vascular endothelial growth factor, a pathological implicated in vascular leakage and neovascularization associated with diabetic retinopathy [117]. Medidur or Ileuvien is much smaller in size (3.0 mm long and 0.37 mm in diameter) and can be injected intravitreally using a 25 G needle [118]. Thus, surgical procedures are being obviated with the development of miniaturized, injectable nondegradable implants. Given the small size of these implants, and the already compromised state of eyes with back-of-the-eye disorders, these implants are not expected to cause any major vision disturbances due to their mere presence.

As an alternative to nondegradable systems, biodegradable implants, such as Ozurdex and Surodex, are being investigated for dexamethasone delivery. Dexamethasone, a corticosteroid similar to fluocinolone acetonide, is expected to alleviate vascular complications. Ozurdex is an intravitreal implant containing 0.7 mg of dexametahsone and was approved by the FDA in 2009 for the treatment of macular edema. Surodex is an anterior segment implant of dexamethasone for treating postoperative inflammation after cataract surgery. Biodegradable implants are typically made of degradable polymers such as PLGA, which are degraded in the body into water and CO_2. Therefore, these implants disappear with time and require no surgical removal. The release rate of the drug from biodegradable implants can be controlled by altering the ratios of lactide and glycolide in the polymers. Unlike nondegradable systems, zero-order release is difficult to achieve with degradable systems. Similar to nondegradable implants of corticosteroids, degradable systems also suffer from side effects including cataracts and elevated IOP. Intravitreal implants are revolutionizing the therapy of back-of-the-eye disorders by allowing effective drug delivery over prolonged periods. To reduce the side effects of corticosteroid implants, more efforts are required in proper dose selection and placement [119] of the devices in the vitreous cavity.

Challenges & future perspectives

The incidence of age-related chronic ocular diseases such as glaucoma, cataract, macular and other retinal degenerative disorders is expected to rise dramatically in the next two decades. Thus, there is an urgent need to develop novel ocular delivery systems that meet the multimedication needs of the elderly with pre-existing visual and motor impairment. The ideal ocular delivery system is one that is easy to manufacture, allows noninvasive self-administration, achieves and maintains effective drug concentrations at the target site for desired time intervals, minimizes systemic exposure and affords good patient comfort, acceptance and compliance. While eye drops may be ideal for treating anterior segment eye diseases, it may not be the most reliable drug delivery vehicle entrusted with targeted drug delivery to the posterior segment. By contrast, other systems discussed elsewhere in this article, such as implants, are less than ideal, they will remain the 'best' there is for the time being. Clearly, more innovative measures are needed to achieve breakthroughs in topical ocular drug delivery. These measures will be discussed in turn.

Noninvasive topical drug delivery to the back of the eye

Currently there is no product approved for treating disorders associated with the back of the eye utilizing noninvasive topical drug administration. Some potential approaches to enhance drug delivery to the back of the eye include delivery systems that allow prolonged precorneal retention, such as contact lens-type devices and mucoadhesive delivery systems. In addition, prodrugs targeting transporters present in conjunctiva and underlying barriers might be a viable option for enhanced drug delivery to the back of the eye using eye drops. Similarly, functionalized nanoparticles targeting the ocular surface and underlying barriers are potential options to allow noninvasive drug delivery to the back of the eye. Iontophoresis also allows enhanced drug delivery to the back of the eye from ocular surface.

Sustained drug delivery to the anterior segment of the eye

Although sustained drug delivery to the back of the eye is now feasible for up to 3 years with injectable or surgically placed intravitreal implants, sustained drug delivery to the tissues of the anterior segment of the eye for several days or longer periods is not feasible with any of the topical ophthalmic preparations that are currently available. The key challenge to overcome is short precorneal residence time. While contact lens-delivery systems are likely to sustain drug delivery for a week or several weeks, punctal plug delivery systems will likely allow sustained drug delivery for a few months. Implantable systems such as Surodex, although suitable for anterior segment delivery following placement in the anterior chamber, widespread patient acceptance of such visible systems has yet to be established.

Membrane & dynamic barriers to drug permeability

Cornea, conjunctiva, choroid layer and RPE are some of the barriers that a topically applied drug has to cross prior to reaching target tissues in the anterior or posterior segments of the eye. In addition, the drug has to overcome vascular clearance mechanisms in the conjunctiva, episclera and choroid during membrane transport prior to reaching the tissues of the back of the eye. It appears that if sufficient duration of drug contact is allowed on the eye surface, several of these barriers can be crossed to a significant extent even by macromolecules. Alternative approaches to overcoming membrane barriers are to design prodrugs targeted to transporters or design nanoparticles capable of crossing barriers efficiently.

Side effects associated with intravitreal implants

It is well known that intravitreal procedures can potentially cause retinal detachment and endophthalmitis, particularly when repeated. Thus, there has been a need for sustained drug delivery for several years since diseases such as diabetic retinopathy and ARMD require several years of drug therapy. While sustained delivery for approximately 3 years is currently feasible for fluocinolone acetonide, a small lipophilic molecule, such systems have yet to be developed for small or large hydrophilic drugs. In the future, it is anticipated that significant efforts will be made in this direction with innovative polymers and delivery systems. In addition, even at low dose delivery, corticosteroids in the vitreous humor result in cataracts and elevated IOP. To minimize these side effects, improved localization of the implant near the target site, further reduction in dose and pulsatile steroid delivery are some of the potential future options.

It is encouraging to see that several new approaches for sustained and enhanced topical drug delivery, notably punctal plugs, scleral lenses and iontophoresis, are in clinical trial (Table 4). With such innovation, significant advances are anticipated in drug delivery from the precorneal area similar to those achieved with novel delivery systems administered in the intravitreal space. In addition to the technical challenges discussed in this article, several regulatory and toxicology issues must also be considered in developing a safe and effective drug product for the eye. The reader is referred to an excellent review by Novack addressing these topics [120].

Table 4. Some novel ophthalmic drug delivery methods that are under clinical development.

Delivery system	Drug	Indication	Developmental stage
Iontophoresis	Dexamethasone phosphate	Dry eye syndrome, uveitis and corneal graft rejection	Clinical trials
Punctal plug	Latanoprost	Ocular hypertension and open angle glaucoma	Phase II clinical trial
Scleral lenses	Avastin	Corneal neovascularization	Clinical study

FOOTNOTES

For reprint orders, please contact erutuf@stnirper-moc.ecneics

Financial & competing interests disclosure

This work was supported, in part, by NIH grants EY018940 and EY017533. Uday Kompella is a consultant for Allergan, Genentech and Vistakon. The authors have no other relevant affiliations or financial involvement with any organization or entity with a financial interest in or financial conflict with the subject matter or materials discussed in the manuscript apart from those disclosed.

No writing assistance was utilized in the production of this manuscript.

BIBLIOGRAPHY

1. Rein DB, Wittenborn JS, Zhang X, Honeycutt AA, Lesesne SB, Saaddine J. Forecasting age-related macular degeneration through the year 2050: the potential impact of new treatments. Arch Ophthalmol.2009;127(4):533–540.

2. Nowak JZ. Age-related macular degeneration (AMD): pathogenesis and therapy. Pharmacol Rep.2006;58(3):353–363.

3. VisionGain. Visiongain Reports. 2010. World ophthalmic pharmaceutical market 2010–2025.

4. Cheruvu NP, Amrite AC, Kompella UB. Effect of diabetes on trans-scleral delivery of celecoxib. Pharm Res. 2009;26(2):404–414.

5. Amrite AC, Edelhauser HF, Singh SR, Kompella UB. Effect of circulation on the disposition and ocular tissue distribution of 20 nm nanoparticles after periocular administration. Mol Vis. 2008;14:150–160.

6. Cheruvu NP, Amrite AC, Kompella UB. Effect of eye pigmentation on trans-scleral drug delivery. Invest Ophthalmol Vis Sci. 2008;49(1):333–341.

7. Amrite AC, Edelhauser HF, Kompella UB. Modeling of corneal and retinal pharmacokinetics after periocular drug administration. Invest Ophthalmol Vis Sci. 2008;49(1):320–332.

8. Kador PF, Randazzo J, Babb T, et al. Topical aldose reductase inhibitor formulations for effective lens drug delivery in a rat model for sugar cataracts. J Ocul Pharmacol Ther. 2007;23(2):116–123.

9. Amrite AC, Ayalasomayajula SP, Cheruvu NP, Kompella UB. Single periocular injection of celecoxib-PLGA microparticles inhibits diabetes-induced elevations in retinal PGE2, VEGF, and vascular leakage.Invest Ophthalmol Vis Sci. 2006;47(3):1149–1160.

10. Ayalasomayajula SP, Kompella UB. Subconjunctivally administered celecoxib-PLGA microparticles sustain retinal drug levels and alleviate diabetes-induced oxidative stress in a rat model. Eur J Pharmacol.2005;511(2–3):191–198.

11. Ayalasomayajula SP, Kompella UB. Retinal delivery of celecoxib is several-fold higher following subconjunctival administration compared with systemic administration. Pharm Res. 2004;21(10):1797–1804.

12. Koushik K, Kompella UB. Preparation of large porous deslorelin-plga microparticles with reduced residual solvent and cellular uptake using a supercritical carbon dioxide process. Pharm Res.2004;21(3):524–535.

13. Sunkara G, Ayalasomayajula SP, Rao CS, Vennerstrom JL, Deruiter J, Kompella UB. Systemic and ocular pharmacokinetics of N-4-benzoylaminophenylsulfonylglycine (BAPSG), a novel aldose reductase inhibitor. J Pharm Pharmacol. 2004;56(3):351–358.

14. Aukunuru JV, Sunkara G, Ayalasomayajula SP, Deruiter J, Clark RC, Kompella UB. A biodegradable injectable implant sustains systemic and ocular delivery of an aldose reductase inhibitor and ameliorates biochemical changes in a galactose-fed rat model for diabetic complications. Pharm Res. 2002;19(3):278–285.

15. Proksch JW, Granvil CP, Siou-Mermet R, Comstock TL, Paterno MR, Ward KW. Ocular pharmacokinetics of besifloxacin following topical administration to rabbits, monkeys, and humans. J Ocul Pharmacol Ther. 2009;25(4):335–344.

16. Lee VH, Robinson JR. Topical ocular drug delivery: recent developments and future challenges. J Ocul Pharmacol. 1986;2(1):67–108.

17. Macha S, Mitra AK. Ocular pharmacokinetics in rabbits using a novel dual probe microdialysis technique. Exp Eye Res. 2001;72(3):289–299.

18. Macha S, Duvvuri S, Mitra AK. Ocular disposition of novel lipophilic diester prodrugs of ganciclovir following intravitreal administration using microdialysis. Curr Eye Res. 2004;28(2):77–84.

19. Dias CS, Mitra AK. Posterior segment ocular pharmacokinetics using microdialysis in a conscious rabbit model. Invest Ophthalmol Vis Sci. 2003;44(1):300–305.

20. Rittenhouse KD, Pelffer RL, Jr, Pollack GM. Evaluation of microdialysis sampling of aqueous humor forin vivo models of ocular absorption and disposition. J Pharm Biomed Anal. 1998;16(6):951–959.

21. Pearson PA, Jaffe GJ, Martin DF, et al. Evaluation of a delivery system providing long-term release of cyclosporine. Arch Ophthalmol. 1996;114(3):311–317.

22. Acheampong AA, Shackleton M, Tang-Liu DD, Ding S, Stern ME, Decker R. Distribution of cyclosporin a in ocular tissues after topical administration to albino rabbits and beagle dogs. Curr Eye Res.1999;18(2):91–103.

23. Wei G, Xu H, Ding PT, Li SM, Zheng JM. Thermosetting gels with modulated gelation temperature for ophthalmic use: the rheological and γ scintigraphic studies. J Control Release. 2002;83(1):65–74.

24. Klyce SD, Crosson CE. Transport processes across the rabbit corneal epithelium: a review. Curr Eye Res. 1985;4(4):323–331.

25. Kompella UB, Kim KJ, Lee VH. Active chloride transport in the pigmented rabbit conjunctiva. Curr Eye Res. 1993;12(12):1041–1048.

26. Wang W, Sasaki H, Chien DS, Lee VH. Lipophilicity influence on conjunctival drug penetration in the pigmented rabbit: a comparison with corneal penetration. Curr Eye Res. 1991;10(6):571–579.

27. Raghava S, Hammond M, Kompella UB. Periocular routes for retinal drug delivery. Expert Opin Drug Deliv. 2004;1(1):99–114.

28. Rao VR, Prescott E, Shelke NB, et al. Delivery of SAR 1118 to retina via ophthalmic drops and its effectiveness in reduction of retinal leukostasis and vascular leakiness in rat streptozotocin (STZ) model of diabetic retinopathy (DR) Invest Ophthalmol Vis Sci. 2010

29. Hariprasad SM, Mieler WF, Holz ER. Vitreous and aqueous penetration of orally administered gatifloxacin in humans. Arch Ophthalmol. 2003;121(3):345–350.

30. Olsen TW, Feng X, Wabner K, et al. Cannulation of the suprachoroidal space: a novel drug delivery methodology to the posterior segment. Am J Ophthalmol. 2006;142(5):777–787.

31. Felt O, Furrer P, Mayer JM, Plazonnet B, Buri P, Gurny R. Topical use of chitosan in ophthalmology: tolerance assessment and evaluation of precorneal retention. Int J Pharm. 1999;180(2):185–193.

32. Meseguer G, Gurny R, Buri P, Rozier A, Plazonnet B. γ scintigraphic study of precorneal drainage and assessment of miotic response in rabbits of various ophthalmic formulations containing pilocarpine. Int J Pharm. 1993;95:229–234.

33. Yamamura K, Sasaki H, Nakashima M, et al. Characterization of ocular pharmacokinetics of β-blockers using a diffusion model after instillation. Pharm Res. 1999;16(10):1596–1601.

34. Akpek EK, Vittitow J, Verhoeven RS, et al. Ocular surface distribution and pharmacokinetics of a novel ophthalmic 1% azithromycin formulation. J Ocul Pharmacol Ther. 2009;25(5):433–439.

35. Hornof MD, Bernkop-Schnurch A. In vitro evaluation of the permeation enhancing effect of polycarbophil-cysteine conjugates on the cornea of rabbits. J Pharm Sci. 2002;91(12):2588–2592.

36. Maurice D. The effect of the low blink rate in rabbits on topical drug penetration. J Ocul Pharmacol Ther.1995;11(3):297–304.

37. Witcherle O, Lim D. Hydrophilic gels for biological use. Nature. 1960;185:117–118.

38. Lim L, Loughnan MS, Sullivan LJ. Microbial keratitis associated with extended wear of silicone hydrogel contact lenses. Br J Ophthalmol. 2002;86(3):355–357.

39. Hehl EM, Beck R, Luthard K, Guthoff R, Drewelow B. Improved penetration of aminoglycosides and fluorozuinolones into the aqueous humour of patients by means of acuvue contact lenses. Eur J Clin Pharmacol. 1999;55(4):317–323.

40. Sedlacek J. Possibility of the application of ophthalmic drugs with the use of gel contact lenses. Cesk Oftalmol. 1965;21(6):509–512.

41. Karlgard CC, Jones LW, Moresoli C. Ciprofloxacin interaction with silicon-based and conventional hydrogel contact lenses. Eye Contact Lens. 2003;29(2):83–89.

42. Karlgard CC, Wong NS, Jones LW, Moresoli C. In vitro uptake and release studies of ocular pharmaceutical agents by silicon-containing and p-HEMA hydrogel contact lens materials. Int J Pharm.2003;257(1–2):141–151.

43. Kim J, Conway A, Chauhan A. Extended delivery of ophthalmic drugs by silicone hydrogel contact lenses. Biomaterials. 2008;29(14):2259–2269.

44. Haugen B, Werner L, Romaniv N, et al. Prevention of endophthalmitis by collagen shields presoaked in fourth-generation fluoroquinolones versus by topical prophylaxis. J Cataract Refract Surg. 2008;34(5):853–858.

45. Hiratani H, Alvarez-Lorenzo C. Timolol uptake and release by imprinted soft contact lenses made ofN,N-diethylacrylamide and methacrylic acid. J Control Release. 2002;83(2):223–230.

46. Alvarez-Lorenzo C, Yanez F, Barreiro-Iglesias R, Concheiro A. Imprinted soft contact lenses as norfloxacin delivery systems. J Control Release. 2006;113(3):236–244.

47. Ali M, Horikawa S, Venkatesh S, Saha J, Hong JW, Byrne ME. Zero-order therapeutic release from imprinted hydrogel contact lenses within in vitro physiological ocular tear flow. J Control Release.2007;124(3):154–162.

48. Danion A, Brochu H, Martin Y, Vermette P. Fabrication and characterization of contact lenses bearing surface-immobilized layers of intact liposomes. J Biomed Mater Res A. 2007;82(1):41–51.

49. Danion A, Arsenault I, Vermette P. Antibacterial activity of contact lenses bearing surface-immobilized layers of intact liposomes loaded with levofloxacin. J Pharm Sci. 2007;96(9):2350–2363.

50. Ciolino JB, Hoare TR, Iwata NG, et al. A drug-eluting contact lens. Invest Ophthalmol Vis Sci.2009;50(7):3346–3352.

51. Sano K, Tokoro T, Imai Y. A new drug delivery system utilizing piggyback contact lenses. Acta Ophthalmol Scand. 1996;74(3):243–248.

52. Gulsen D, Chauhan A. Dispersion of microemulsion drops in HEMA hydrogel: a potential ophthalmic drug delivery vehicle. Int J Pharm. 2005;292(1–2):95–117.

53. Gulsen D, Li CC, Chauhan A. Dispersion of DMPC liposomes in contact lenses for ophthalmic drug delivery. Curr Eye Res. 2005;30(12):1071–1080.

54. Gulsen D, Chauhan A. Ophthalmic drug delivery through contact lenses. Invest Ophthalmol Vis Sci.2004;45(7):2342–2347.

55. Kapoor Y, Thomas JC, Tan G, John VT, Chauhan A. Surfactant-laden soft contact lenses for extended delivery of ophthalmic drugs. Biomaterials. 2009;30(5):867–878.

56. Lim M, Jacobs DS, Rosenthal P, Carrasquillo KG. The boston ocular surface prosthesis as a novel drug delivery system for bevacizumab. Seminars in Ophthalmology. 2009;24(3):149–155.

57. Kim SW, Ha BJ, Kim EK, Tchah H, Kim TI. The effect of topical bevacizumab on corneal neovascularization. Ophthalmology. 2008;115(6):e33–38.

58. Ottiger M, Thiel MA, Feige U, Lichtlen P, Urech DM. Efficient intraocular penetration of topical anti-TNF-α single-chain antibody (ESBA105) to anterior and posterior segment without penetration enhancer.Invest Ophthalmol Vis Sci. 2009;50(2):779–786.

59. Calonge M. The treatment of dry eye. Surv Ophthalmol. 2001;45(Suppl 2):S227–S239.

60. Balaram M, Schaumberg DA, Dana MR. Efficacy and tolerability outcomes after punctal occlusion with silicone plugs in dry eye syndrome. Am J Ophthalmol. 2001;131(1):30–36.

61. Wei CP, Anderson JA, Leopold I. Ocular absorption and metabolism of topically applied epinephrine and a dipivalyl ester of epinephrine. Invest Ophthalmol Vis Sci. 1978;17(4):315–321.

62. Kass MA, Mandell AI, Goldberg I, Paine JM, Becker B. Dipivefrin and epinephrine treatment of elevated intraocular pressure: a comparative study. Arch Ophthalmol. 1979;97(10):1865–1866.

63. Rautio J, Kumpulainen H, Heimbach T, et al. Prodrugs: design and clinical applications. Nat Rev Drug Discov. 2008;7(3):255–270.

64. Hellberg MR, Ke TL, Haggard K, Klimko PG, Dean TR, Graff G. The hydrolysis of the prostaglandin analog prodrug bimatoprost to 17-phenyl-trinor PGF2α by human and rabbit ocular tissue. J Ocul Pharmacol Ther. 2003;19(2):97–103.

65. Netland PA, Landry T, Sullivan EK, et al. Travoprost compared with latanoprost and timolol in patients with open-angle glaucoma or ocular hypertension. Am J Ophthalmol. 2001;132(4):472–484.

66. Susanna R, Jr, Chew P, Kitazawa Y. Current status of prostaglandin therapy: latanoprost and unoprostone. Surv Ophthalmol. 2002;47(Suppl 1):S97–S104.

67. Lallemand F, Perottet P, Felt-Baeyens O, et al. A water-soluble prodrug of cyclosporine a for ocular application: a stability study. Eur J Pharm Sci. 2005;26(1):124–129.

68. Lallemand F, Felt-Baeyens O, Rudaz S, et al. Conversion of cyclosporine a prodrugs in human tears vs rabbits tears. Eur J Pharm Biopharm. 2005;59(1):51–56.

69. Liu H, Wang Y, Li S. Advanced delivery of ciclosporin a: present state and perspective. Expert Opin Drug Deliv. 2007;4(4):349–358.

70. Olivero DK, Davidson MG, English RV, Nasisse MP, Jamieson VE, Gerig TM. Clinical evaluation of 1% cyclosporine for topical treatment of keratoconjunctivitis sicca in dogs. J Am Vet Med Assoc.1991;199(8):1039–1042.

71. Benitez Del Castillo JM, Del Aguila C, Duran S, Hernandez J, Garcia Sanchez J. Influence of topically applied cyclosporine a in olive oil on corneal epithelium permeability. Cornea. 1994;13(2):136–140.

72. Bourges JL, Lallemand F, Agla E, et al. Evaluation of a topical cyclosporine a prodrug on corneal graft rejection in rats. Mol Vis. 2006;12:1461–1466.

73. Dey S, Anand BS, Patel J, Mitra AK. Transporters/receptors in the anterior chamber: pathways to explore ocular drug delivery strategies. Expert Opin Biol Ther. 2003;3(1):23–44.

74. Janoria KG, Gunda S, Boddu SH, Mitra AK. Novel approaches to retinal drug delivery. Expert Opin Drug Deliv. 2007;4(4):371–388.

75. Hosoya K, Tachikawa M. Inner blood-retinal barrier transporters: role of retinal drug delivery. Pharm Res. 2009;26(9):2055–2065.

76. Ueda H, Horibe Y, Kim KJ, Lee VH. Functional characterization of organic cation drug transport in the pigmented rabbit conjunctiva. Invest Ophthalmol Vis Sci. 2000;41(3):870–876.

77. Karla PK, Pal D, Mitra AK. Molecular evidence and functional expression of multidrug resistance associated protein (MRP) in rabbit corneal epithelial cells. Exp Eye Res. 2007;84(1):53–60.

78. Mannermaa E, Vellonen KS, Ryhanen T, et al. Efflux protein expression in human retinal pigment epithelium cell lines. Pharm Res. 2009;26(7):1785–1791.

79. Prasanna G, Fortner J, Xiang C, et al. Ocular pharmacokinetics and hypotensive activity of PF-04475270, an EP4 prostaglandin agonist in preclinical models. Exp Eye Res. 2009;89(5):608–617.

80. Gehlbach P, Demetriades AM, Yamamoto S, et al. Periocular injection of an adenoviral vector encoding pigment epithelium-derived factor inhibits choroidal neovascularization. Gene Ther. 2003;10(8):637–646.

81. Kompella UB, Sundaram S, Raghava S, Escobar ER. Luteinizing hormone-releasing hormone agonist and transferrin functionalizations enhance nanoparticle delivery in a novel bovine ex vivo eye model. Mol Vis. 2006;12:1185–1198.

82. Loftssona T, Jarvinen T. Cyclodextrins in ophthalmic drug delivery. Adv Drug Deliv Rev.1999;36(1):59–79.

83. Stefansson E, Loftssona T. Cyclodextrins in eye drop formulations. J Incl Phenomena Macrocyclic Chem. 2002;44:23–27.

84. Bary AR, Tucker IG, Davies NM. Considerations in the use of hydroxypropyl-β-cyclodextrin in the formulation of aqueous ophthalmic solutions of hydrocortisone. Eur J Pharm Biopharm. 2000;50(2):237–244.

85. Loftsson T, Frithriksdottir H, Stefansson E, Thorisdottir S, Guthmundsson O, Sigthorsson T. Topically effective ocular hypotensive acetazolamide and ethoxyzolamide formulations in rabbits. J Pharm Pharmacol.1994;46(6):503–504.

86. Cheeks L, Kaswan RL, Green K. Influence of vehicle and anterior chamber protein concentration on cyclosporine penetration through the isolated rabbit cornea. Curr Eye Res. 1992;11(7):641–649.

87. Loftsson T, Fridriksdottir H, Thorisdottir S, Stefansson E. The effect of hydroxypropyl methylcellulose on release of dexamethazone from aqueous 2-hydroxypropyl-β-cyclodextrin formulations. Int J Pharm.1994;104:181–184.

88. Kristinsson JK, Fridriksdottir H, Thorisdottir S, Sigurdardottir AM, Stefansson E, Loftsson T. Dexamethasone–cyclodextrin– polymer co-complexes in aqueous eye drops. Aqueous humor pharmacokinetics in humans. Invest Ophthalmol Vis Sci. 1996;37(6):1199–1203.

89. Ashton P, Podder SK, Lee VH. Formulation influence on conjunctival penetration of four β blockers in the pigmented rabbit: a comparison with corneal penetration. Pharm Res. 1991;8(9):1166–1174.

90. Lee YH, Lee VH. Formulation influence on ocular and systemic absorption of topically applied atenolol in the pigmented rabbit. J Ocul Pharmacol. 1993;9(1):47–58.

91. Scholz M, Lin JE, Lee VH, Keipert S. Pilocarpine permeability across ocular tissues and cell cultures: influence of formulation parameters. J Ocul Pharmacol Ther. 2002;18(5):455–468.

92. Lee VH, Yamamoto A, Kompella UB. Mucosal penetration enhancers for facilitation of peptide and protein drug absorption. Crit Rev Ther Drug Carrier Syst. 1991;8(2):91–192.

93. Eljarrat-Binstock E, Domb AJ. Iontophoresis: a noninvasive ocular drug delivery. J Control Release.2006;110(3):479–489.

94. Nurieva SM. Effect on the eye vessels of vasodilator agents after their administration by different electrophoretic methods (clinical study) Vestn Oftalmol. 1978;(2):71–73.

95. Polunin GS, Kas'ianova NS. Fluorophotometry in the clinical aspects of eye diseases. Vestn Oftalmol.1979;(5):52–57.

96. Bourne WM, Brubaker RF. Decreased endothelial permeability in the iridocorneal endothelial syndrome.Ophthalmology. 1982;89(6):591–595.

97. Maurice DM. Iontophoresis of fluorescein into the posterior segment of the rabbit eye. Ophthalmology.1986;93(1):128–132.

98. Meyer DR, Linberg JV, Vasquez RJ. Iontophoresis for eyelid anesthesia. Ophthalmic Surg.1990;21(12):845–848.

99. Horwath-Winter J, Schmut O, Haller-Schober EM, Gruber A, Rieger G. Iodide iontophoresis as a treatment for dry eye syndrome. Br J Ophthalmol. 2005;89(1):40–44.

100. Parkinson TM, Ferguson E, Febbraro S, Bakhtyari A, King M, Mundasad M. Tolerance of ocular iontophoresis in healthy volunteers. J Ocul Pharmacol Ther. 2003;19(2):145–151.

101. Amrite A, Pugazhenthi V, Cheruvu N, Kompella U. Delivery of celecoxib for treating diseases of the eye: influence of pigment and diabetes. Expert Opin Drug Deliv. 2010;7(5):631–645.

102. Prausnitz MR, Noonan JS. Permeability of cornea, sclera, and conjunctiva: a literature analysis for drug delivery to the eye. J Pharm Sci. 1998;87(12):1479–1488.

103. Kim TW, Lindsey JD, Aihara M, Anthony TL, Weinreb RN. Intraocular distribution of 70-kDa dextran after subconjunctival injection in mice. Invest Ophthalmol Vis Sci. 2002;43(6):1809–1816.

104. Cheruvu NP, Kompella UB. Bovine and porcine trans-scleral solute transport: influence of lipophilicity and the choroid-bruch's layer. Invest Ophthalmol Vis Sci. 2006;47(10):4513–4522.

105. Kadam RS, Kompella UB. Influence of lipophilicity on drug partitioning into sclera, choroid-retinal pigment epithelium, retina, trabecular meshwork, and optic nerve. J Pharmacol Exp Ther. 2010;332(3):1107–1120. Retracted

106. Durairaj C, Shah JC, Senapati S, Kompella UB. Prediction of vitreal half-life based on drug physicochemical properties: quantitative structure–pharmacokinetic relationships (QSPKR) Pharm Res.2009;26(5):1236–1260.

107. Durairaj C, Kim SJ, Edelhauser HF, Shah JC, Kompella UB. Influence of dosage form on the intravitreal pharmacokinetics of diclofenac. Invest Ophthalmol Vis Sci. 2009;50(10):4887–4897.

108. Danis RP, Ciulla TA, Pratt LM, Anliker W. Intravitreal triamcinolone acetonide in exudative age-related macular degeneration. Retina. 2000;20(3):244–250.

109. Jonas JB, Kreissig I, Sofker A, Degenring RF. Intravitreal injection of triamcinolone for diffuse diabetic macular edema. Arch Ophthalmol. 2003;121(1):57–61.

110. Jonas JB. Intravitreal triamcinolone acetonide: a change in a paradigm. Ophthalmic Res.2006;38(4):218–245.

111. Mason JO, 3rd, Somaiya MD, Singh RJ. Intravitreal concentration and clearance of triamcinolone acetonide in nonvitrectomized human eyes. Retina. 2004;24(6):900–904.

112. Chin HS, Park TS, Moon YS, Oh JH. Difference in clearance of intravitreal triamcinolone acetonide between vitrectomized and nonvitrectomized eyes. Retina. 2005;25(5):556–560.

113. Choonara YE, Pillay V, Danckwerts MP, Carmichael TR, Du Toit LC. A review of implantable intravitreal drug delivery technologies for the treatment of posterior segment eye diseases. J Pharm Sci.2009;99(5):2219–2239.

114. Short BG. Safety evaluation of ocular drug delivery formulations: techniques and practical considerations. Toxicol Pathol. 2008;36(1):49–62.

115. Bourges JL, Bloquel C, Thomas A, et al. Intraocular implants for extended drug delivery: therapeutic applications. Adv Drug Deliv Rev. 2006;58(11):1182–1202.

116. Ramchandran RS, Fekrat S, Stinnett SS, Jaffe GJ. Fluocinolone acetonide sustained drug delivery device for chronic central retinal vein occlusion: 12-month results. Am J Ophthalmol. 2008;146(2):285–291.

117. Ayalasomayajula SP, Ashton P, Kompella UB. Fluocinolone inhibits VEGF expression via glucocorticoid receptor in human retinal pigment epithelial (ARPE-19) cells and TNF-α-induced angiogenesis in chick chorioallantoic membrane (CAM) J Ocul Pharmacol Ther. 2009;25(2):97–103.

118. Del Amo EM, Urtti A. Current and future ophthalmic drug delivery systems. A shift to the posterior segment. Drug Discov Today. 2008;13(3–4):135–143.

119. Kompella UB, Li X, Zhan G, Lee SS. Influence of intravitreal implant location on ocular disposition of fluocinolone acetonide. ARVO Annual Meeting. 2009;5971:A5594.

120. Novack GD. Ophthalmic drug delivery: development and regulatory considerations. Clin Pharmacol Ther. 2009;85(5):539–543.

121. Rieke ER, Amaral J, Patricia Becerra S, Lutz RJ. Sustained subconjunctival protein delivery using a thermosetting gel delivery system. J Ocul Pharmacol Ther. 2010 (In Press)

122. Kim EY, Gao ZG, Park JS, Li H, Han K. RHEGF/HP-β-CD complex in poloxamer gel for ophthalmic delivery. Int J Pharm. 2002;233(1–2):159–167.

123. Kompella UB, Bandi N, Ayalasomayajula SP. Subconjunctival nano- and microparticles sustain retinal delivery of budesonide, a corticosteroid capable of inhibiting VEGF expression. Invest Ophthalmol Vis Sci.2003;44(3):1192–1201.

124. Kang SJ, Durairaj C, Kompella UB, O'Brien JM, Grossniklaus HE. Subconjunctival nanoparticle carboplatin in the treatment of murine retinoblastoma. Arch Ophthalmol. 2009;127(8):1043–1047.

125. Chiang CH, Tung SM, Lu DW, Yeh MK. In vitro and in vivo evaluation of an ocular delivery system of 5-fluorouracil microspheres. J Ocul Pharmacol Ther. 2001;17(6):545–553.

126. Van Quill KR, Dioguardi PK, Tong CT, et al. Subconjunctival carboplatin in fibrin sealant in the treatment of transgenic murine retinoblastoma. Ophthalmology. 2005;112(6):1151–1158.

127. Saishin Y, Silva RL, Saishin Y, et al. Periocular injection of microspheres containing PKC412 inhibits choroidal neovascularization in a porcine model. Invest Ophthalmol Vis Sci. 2003;44(11):4989–4993.

128. Sunkara G, Kompella UB. Membrane transport processes in the eye. In: Mitra AK, editor. Ophthalmic Drug Delivery Systems. Marcel Dekker Inc; NY, USA: 2003. pp. 13–58.

Chapter 6

Nanocarrier Systems for Transdermal Drug Deliver

José Juan Escobar-Chávez[1], Isabel Marlen Rodríguez-Cruz[2], Clara Luisa Domínguez-Delgado[3], Roberto Díaz- Torres[4], Alma Luisa Revilla-Vázquez[5] and Norma Casas Aléncaster[6]

[1] *Unidad de Investigación Multidisciplinaria, Laboratorio 12: Sistemas Transdérmicos y Materiales Nanoestructurados, Facultad de Estudios Superiores Cuautitlán-Universidad Nacional Autónoma de México, Estado de México*

[2] *Departamento de Enseñanza e Investigación, Hospital Regional de Alta Especialidad de Zumpango, Zumpango, Estado de México*

[3] *Unidad de Investigación Multidisciplinaria, Laboratorio 9, Facultad de Estudios Superiores Cuautitlán-Universidad Nacional Autónoma de México, Estado de México*

[4] *Unidad de Investigación Multidisciplinari, Laboratorio 9: Sistemas Transdérmicos y Materiales Nanoestructurado, Facultad de Estudios Superiores Cuautitlán-Universidad Nacional Autónoma de México, Estado de México*

[5] *Laboratorio de Desarrollo de Métodos Analíticos, Facultad de Estudios Superiores Cuautitlán-Universidad Nacional Autónoma de México, Cuautitlán Izcalli, Estado de México*

[6] *Laboratorio de Propiedades Reológicas de Alimentos, Facultad de Estudios Superiores Cuautitlán-Universidad Nacional Autónoma de México, Cuautitlán Izcalli, Estado de México, Estado de México*

1. INTRODUCTION

The nanomedicine which is the application of technologies on the scale of 1 to 500 nm to diagnose and treat diseases, it has become a very relevant topic nowadays. During the last century, there has been a lot of new research and patents regarding nanomedicine in health sciences [1]. The objective of nanomedicine is to diagnose and preserve the health without side effects with noninvasive treatments. To reach these goals, nanomedicine offers a lot of new tools and capabilities. The manipulation that nanomedicine provides to the drugs and other materials in the nanometer scale can change the basic properties and bioactivity of materials. The solubility, increment in surface area, control release and site-targeted delivery are some characteristics that nanotechnology can manipulate on drug delivery systems.

Nanotechnology applied to health sciences contains new devices used in surgery, new chips for better diagnostics, new materials for substituting body structures and some structures capable to carry drugs through the body for treatment of a lot of diseases. These structures can be made of a lot of different materials and they are very different in structure and chemical nature. All these nanostructures are called nanocarriers and they can be administrated into the organisms by topical and transdermal routes [2]. Nanocarriers are a powerful weapon against a lot of illnesses since they are so small to be detected by immune system and they can deliver the drug in the target organ. For that reason, drug doses using nanocarriers and side effects decrease a lot.

The idea for using these tiny systems is not as new as we think but the use of nanocarriers in pharmaceutical products is not frequent, since the technology is expensive for certain types of nanoparticles and because nanocarriers need to be evaluated for demonstrating they do not have toxic effects. Nowadays the controversy of biological effects due to nanostructures is an open discussion, in one hand, the nanotechnologist continue making new and more sophisticated nanocarriers and in the other hand, toxicologist continue evaluating possible damaging effects.

Whatever it happens, nanotechnology is the new era and nanomedicine cannot be taking off. New nanocarriers will be created and the entire scientist working in nanomedicine bet for it to be the cure of diseases that in this moment are difficult to deal with [3]

The application of preparations to the skin for medical purposes is as old as the history of medicine itself, with references to the use of ointments and salves found in the records of Babylonian and Egyptian medicine. The historical development of permeation research is well described by Hadgraft & Lane [4]. Over time, the skin has become an important route for drug delivery in which topical, regional or systemic effects are desired. Nevertheless, skin constitutes an excellent barrier and presents difficulties for the transdermal delivery of therapeutic agents, since few drugs possess the characteristics required to permeate across the stratum corneum in sufficient quantities to reach a therapeutic concentration in the blood. In order to enhance drug transdermal absorption different methodologies have been investigated developed and patented [5,6]. Improvement in physical permeation-enhancement technologies has led to renewed interest in transdermal drug delivery. Some of these novel advanced transdermal permeation enhancement technologies include: iontophoresis, electroporation, ultrasound, microneedles to open up the skin and more recently the use of transdermal nanocarriers [3,7-10].

A number of excellent reviews that have been published contain detailed discussions concerning many aspects of transdermal nanocarriers [11-17]. The present chapter shows an updated overview of the use of submicron particles and other nanostructures in the pharmaceutical field, specifically in the area of topical and transdermal drugs. This focus is justified due to the magnitude of the experimental data available with the use of these nanocarriers. The development of submicron particles and other nanostructures in the pharmaceutical and cosmetic fields has been emerged in the last decades for designing best formulations for application through the skin [18-21].

2. THE SKIN

The skin is the largest organ of the body [22-24], accounting for more than 10% of body mass, and the one that enables the body to interact more intimately with its environment. Essentially, the skin consists of four layers: The SC, that is the outer layer of the skin (non-viable epidermis), and forms the rate-controlling barrier for diffusion for almost all compounds. It is composed of dead flattened, keratin-rich cells, the corneocytes. These dense cells are surrounded by a complex mixture of intercellular lipids, namely, ceramides, free fatty acids, cholesterol, and cholesterol sulphate. Their most important feature is that they are structured as ordered bilayer arrays [25-28]. The other layers are: the remaining layers of the epidermis (viable epidermis), the dermis, and the subcutaneous tissue (Figure 1). There are also several associated appendages: hair follicles sweat ducts, glands and nails [29,30].

Many agents are applied to the skin either deliberately or accidentally, with either beneficial or deleterious outcomes. The main interest in dermal absorption assessment is related to: a) Local effects in dermatology (e.g., corticosteroids for dermatitis); b) transport through the skin seeking a systemic effect (e.g., nicotine patches, hormonal drug patches, etc.) [31]; c) surface effects (e.g., sunscreens, cosmetics, and anti-infectives) [32,33]; d) targeting of deeper tissues (e.g., nonsteroidal anti-inflammatory agents) [34-37]; and e) unwanted absorption (e.g., solvents in the workplace, pesticides or allergens) [38,39].

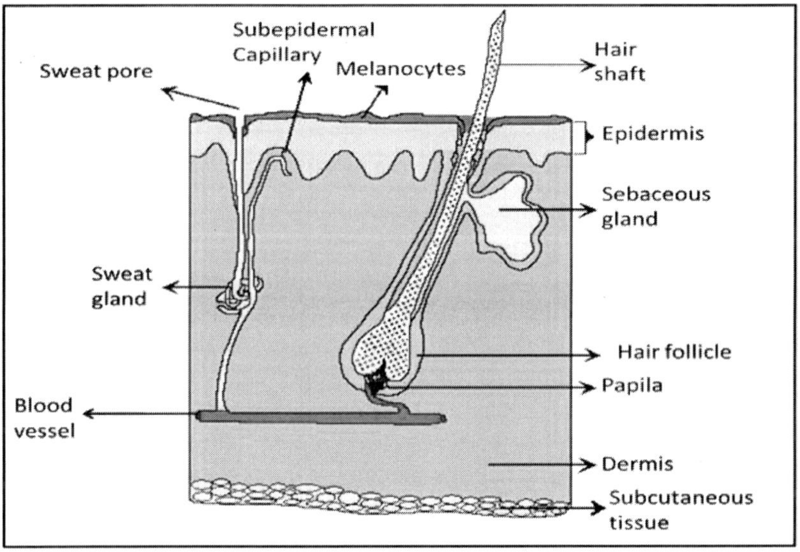

Figure 1. Schematic representation of skin layer.

2.1. EPIDERMIS

2.1.1. STRATUM CORNEUM

The stratum corneum is the heterogeneous outermost layer of the epidermis and is approximately 10-20 μm thick. The stratum corneum consists of about 15 to 25 layers of flattened, stacked, hexagonal, and cornified cells embedded in an intercellular matrix of lipids. These lipid domains form a continuous structure so they are considered to play a crucial role in the maintenance of the skin barrier that helps avoid transepidermal water loss. Each cell is approximately 40 μm in diameter and 0.5 μm thick [40].

The stratum corneum barrier properties may be partly related to its very high density (1.4 g/cm^3 in the dry state) and its low hydration of 15–20 %, compared with the usual 70 % for the body. Each stratum corneum cell is composed mainly of insoluble bundled keratins (70 %) and lipid (20 %) encased in a cell envelope, accounting for about 5% of the stratum corneum weight. The permeability barrier is located within the lipid bilayers in the intercellular spaces of the stratum corneum [6-8] and consists of ceramides (40–50%), fatty acids (15–25%), cholesterol (20–25%) and cholesterol sulphate (5–10 %) [41-45].

The barrier function is further facilitated by the continuous desquamation of this horny layer with a total turnover of the stratum corneum occurring once every 2–3 weeks. The stratum corneum functions as a barrier are to prevent the loss of internal body components, particularly water, to the external environment. The cells of the stratum corneum originate in the viable epidermis and undergo many morphological changes before desquamation. Thus, the epidermis consists of several cell strata at varying levels of differentiation.

The origins of the cells of the epidermis lie in the basal lamina between the dermis and viable epidermis. In this layer there are melanocytes, Langerhans cells, Merkel cells, and two major keratinic cell types: the first functioning as stem cells having the capacity to divide and produce new cells; the second serving to anchor the epidermis to the basement membrane [46]. The basement membrane is 50–70 nm thick and consists of two layers, the lamina densa and lamina lucida, which comprise mainly proteins, such as type IV collagen, aminin, nidogen and fibronectin. Type IV collagen is responsible for the mechanical stability of the basement membrane, whereas laminin and fibronectin are involved with the attachment between the basement membrane and the basal keratinocytes. The cells of the basal lamina are attached to the basement membrane by hemidesmosomes, which are found on the ventral surface of basal keratinocytes [47]. Hemidesmosomes appear to comprise three distinct protein groups: two of which are bullous pemphigoid antigens (BPAG1 and BPAG2), and the other epithelial cellspecific integrins [48-50]. BPAG1 is associated with the organization of the cytoskeletal structure and forms a link between the hemidesmosome structure and the keratin intermediate filaments. The integrins are transmembrane receptors that mediate attachment between the cell and the extracellular matrix. Human epidermal basal cells contain integrins $\alpha_2\beta_1$, $\alpha_3\beta_1$ and $\alpha_6\beta_4$. Integrin $\alpha_6\beta_4$ and BPAG2 appear to be the major hemidesmosomal protein contributors to the anchoring of the keratinocyte, spanning from the keratin intermediate filament, through the lamina lucida, to the lamina densa of the basement membrane [51]. In the lamina densa, these membrane-spanning proteins interact with the protein laminin-5 which, in turn, is linked to collagen VII, the major constituent of the anchoring fibrils within the dermal matrix. It has also been suggested that both BPAG2 and integrin $\alpha_6\beta_4$ mediate in the signal transductions required for hemidesmosome formation and cell differentiation and proliferation. Integrin $\alpha_3\beta_1$ is associated with actin and may be linked with laminin-5. Epidermal wounding results in an up-regulation of these proteins that appears to be involved with cell motility and spreading. The importance of maintaining a secure link between the basal lamina cells and the basement membrane is obvious, and the absence of this connection results in chronic blistering diseases such as pemphigus and epidermolysis bullosa.

2.2.DERMIS

The dermis is about 0.1–0.5 cm thick and consists of collagenous (70 %) and elastin fibres. In the dermis, glycosaminoglycans or acid mucopolysaccharides, are covalently linked to peptide chains to form proteoglycans, the ground substance that promotes the elasticity of the skin.The main cells present are the fibroblasts, which produce the connective tissue components of collagen, laminin, fibronectin and vitronectin; mast cells, which are involved in the immune and inflammatory responses; and melanocytes involved in the production of the pigment melanin [51]. Nerves, blood vessels and lymphatic vessels are also present in the dermis.

Contained within the dermis is an extensive vascular network providing for the skin nutrition, repair, and immune responses for the rest of the body, heat exchange, immune response, and thermal regulation. Skin blood vessels derive from those in the subcutaneous tissues (hypodermis), with an arterial network supplying the papillary layer, the hair follicles, the sweat and apocrine glands, the subcutaneous area, as well as the dermis itself. These arteries feed into arterioles, capillaries, venules, and, thence, into veins. Of particular importance in this vascular network is the presence of arteriovenous anastomoses at all levels in the skin. These arteriovenous anastomoses, which allow a direct shunting of up to 60% of the skin blood flow between the arteries and veins, thereby avoiding the fine capillary network, are critical to the skin's functions of heat regulation and blood vessel control. Blood flow changes are most evident in the skin in relation to various physiological responses and include psychological effects, such as shock (''draining of color from the skin'') and embarrassment (''blushing''), temperature effects, and physiological responses to exercise, hemorrhage, and alcohol consumption.

The lymphatic system is an important component of the skin in regulating its interstitial pressure, mobilization of defense mechanisms, and in waste removal. It exists as a dense, flat meshwork in the papillary layers of the dermis and extends into the deeper regions of the dermis. Also present in the dermis are a number of different types of nerve fibers supplying the skin, including those for pressure, pain, and temperature [52]. Epidermal appendages such as hair follicles and sweat glands are embedded in the dermis [53].

2.3. HYPODERMIS

The deepest layer of the skin is the subcutaneous tissue or hypodermis. The hypodermis acts as a heat insulator, a shock absorber, and an energy storage region. This layer is a network of fat cells arranged in lobules and linked to the dermis by interconnecting collagen and elastin fibers. As well as fat cells (possibly 50% of the body's fat); the other main cells in the hypodermis are fibroblasts and macrophages. One of the major roles of the hypodermis is to carry the vascular and neural systems for the skin. It also anchors the skin to underlying muscle. Fibroblasts and adipocytes can be stimulated by the accumulation of interstitial and lymphatic fluid within the skin and subcutaneous tissue [54]. The total thickness of skin is about 2–3 mm, but the thickness of the stratum corneum is only about 10–15 μm.

2.4. SKIN APPENDAGES

There are four skin appendages: the hair follicles with their associated sebaceous glands, eccrine and apocrine sweat glands, and the nails, but these occupy only about 0.1 % of the total human skin surface.

The pilosebaceous follicles have about 10 to 20 % of the resident flora and cannot be decontaminated by scrubbing. The hair follicles are distributed across the entire skin surface with the exception of the soles of the feet, the palms of the hand and the lips. A smooth muscle, the erector pilorum, attaches the follicle to the dermal tissue and enables hair to stand up in response to fear. Each follicle is associated with a sebaceous gland that varies in size from 200 to 2000 μm in diameter. The sebum secreted by this gland consisting of triglycerides, free fatty acids, and waxes, protects and lubricates the skin as well as maintaining a pH of about 5. Sebaceous glands are absent on the palms, soles and nail beds. Sweat glands or eccrine glands respond to temperature via parasympathetic nerves, except on palms, soles and axillae, where they respond to emotional stimuli via sympathetic nerves [51]. The eccrine glands are epidermal structures that are simple, coiled tubes arising from a coiled ball, of approximately 100 μm in diameter, located in the lower dermis. It secretes a dilute salt solution with a pH of about 5, this secretion being stimulated by temperature-controlling determinants, such as exercise and high environmental temperature, as well as emotional stress through the autonomic (sympathetic) nervous system. These glands have a total surface area of about 1/10,000 of the total body surface. The apocrine glands are limited to specific body regions and are also coiled tubes. These glands are about ten times the size of the eccrine ducts, extend as low as the subcutaneous tissues and are paired with hair follicles.

Nail function is considered as protection. Nail plate consists of layers of flattened keratinized cells fused into a dense but elastic mass. The cells of the nail plate originate in the nail matrix and grow distally at a rate of about 0.1 mm/day. In the keratinization process the cells undergo shape and other changes, similar to those experienced by the epidermal cells forming the stratum corneum. This is not surprising because the nail matrix basement membrane shows many biochemical similarities to the epidermal basement membrane [55,56]. Thus, the major components are highly folded keratin proteins with small amounts of lipid (0.1–1.0%). The principal plasticizer of the nail plate is water, which is normally present at a concentration of 7–12 %.

3. SKIN FUNCTIONS

Many of the functions of the skin can be classified as essential to survival of the body bulk of mammals and humans in a relatively hostile environment. In a general context, these functions can be classified as a protective, maintaining homeostasis or sensing. The importance of the protective and homeostatic role allows the survival of humans in an environment of variable temperature; water content (humidity and bathing); and the presence of environmental dangers, such as chemicals, bacteria, allergens, fungi and radiation. In a second context, the skin is a major organ for maintaining the homeostasis of the body, especially in terms of its composition, heat regulation, blood pressure control, and excretory roles. It has been argued that the basal metabolic rate of animals differing in size should be scaled to the surface area of the body to maintain a constant temperature through the skin's thermoregulatory control [57]. Third, the skin is a major sensory organ in terms of sensing environmental influences, such as heat, pressure, pain, allergen, and microorganism entry. Finally, the skin is an organ that is in a continual state of regeneration and repair. To fulfill each of these functions, the skin must be tough, robust, and flexible, with effective communication between each of its intrinsic components mentioned above.

The stratum corneum also functions as a barrier to prevent the loss of internal body components, particularly water, to the external environment. The epidermis plays a role in temperature, pressure, and pain regulation.

Appendage functions are following: hair follicle and sebaceous gland fulfill with protect (hair) and lubricate (sebum), eccrine and apocrine glands have the functions of cooling and vestigial secondary sex gland, respectively; and nails has the function of to protect.The hypodermis acts as a heat insulator, a shock absorber and an energy storage region. One of the major roles of the hypodermis is to carry the vascular and neural systems for the skin.

4. ROUTES OF DRUG PENETRATION THROUGH THE SKIN

The determination of penetration pathways of topically applied substances into the skin is the subject of several investigations. The permeation of drugs through the skin includes the diffusion through the intact epidermis y through the skin appendages. These skin appendages are hair follicles and sweat glands which form shunt pathways through the intact epidermis, occupying only 0.1%

of the total human skin [58]. It is known drug permeation through the skin is usually limited by the stratum corneum. Two pathways through the intact barrier may be identified, the intercellular and transcellular route (Figure 2):

a. The intercellular lipid route is between the corneocytes.

Interlamellar regions in the stratum corneum, including linker regions, contain less ordered lipids and more flexible hydrophobic chains. This is the reason of the non-planar spaces between crystalline lipid lamellae and their adjacent cells outer membrane. Fluid lipids in skin barrier are crucially important for transepidermal diffusion of the lipidic and amphiphilic molecules, occupying those spaces for the insertion and migration through intercellular lipid layers of such molecules [59,60]. The hydrophilic molecules diffuse predominantly "laterally" along surfaces of the less abundant, water filled inter-lamellar spaces or through such volumes; polar molecules can also use the free space between a lamella and a corneocyte outer membrane to the same end [61].

1. The transcellular route contemplates the crossing through the corneocytes and the intervening lipids [24].

Intracellular macromolecular matrix within the stratum corneum abounds in keratin, which does not contribute directly to the skin diffusive barrier but supports mechanical stability and thus intactness of the stratum corneum. Transcellular diffusion is practically unimportant for transdermal drug transport [62]. The narrow aqueous transepidermal pathways have been observed using confocal laser scanning microscopy (CLSM). Here regions of poor cellular and intercellular lipid packing coincide with wrinkles on skin surface and are simultaneously the sites of lowest skin resistance to the transport of hydrophilic entities. This lowest resistance pathway leads between clusters of corneocytes at the locations where such cellular groups show no lateral overlap.

The better sealed and more transport resistant is the intra-cluster/inter-corneocyte pathway [63]. Hydrophilic conduits have openings between ≥5 μm (skin appendages) and ≤10 nm (narrow inter-corneocyte pores). So sweat ducts (≥50 μm), pilosebaceous units (5–70 μm), and sebaceous glands (5–15 μm) represent the largest width/lowest resistance end of the range. Junctions of corneocytes-clusters and cluster boundaries fall within the range [64]. It was determined that the maximally open hydrophilic conduits across skin are approximately 20–30 nm wide, including pore penetrant/opener thickness [63]. Another studies revealed the width of the negatively charged hydrophilic transepidermal pores expanded by electroosmosis to be around of 22–48 nm [65]. Lipophilic cutaneous barrier is governed by molecular weight and distribution coefficient rather than molecular size [66]. The relative height of cutaneous lipophilic barrier consequently decreases with lipophilicity of permeant, but molecules heavier than 400–500 Da are so large permeants to find sufficiently wide defects in the intercellular lipidic matrix to start diffusing through the lipidic parts of cutaneous barrier [64, 66,67].

The contribution to transdermal drug transport can increases with the pathways widening or multiplication, for example such that is caused by exposing the stratum corneum to a strong electrical (electroporation/iontophoresis), mechanical (sonoporation/sonophoresis), thermal stimulus, or suitable skin penetrants [59].

Recently, follicular penetration has become a major focus of interest due to the drug targeting to the hair follicle is of great interest in the treatment of skin diseases. However due to follicular orifices only occupying 0.1% of the total skin surface area, it was assumed as a non important route. But a variety of studies shown the hair follicles as could be a way to trough the skin [68- 73]. Such follicular pathway also has been proposed for topical administration of nanoparticles and microparticles and it has been investigated in porcine skin, because in recent studies the results have confirmed the *in vitro*penetration into the porcine hair follicles might be considered similar to those on humans *in vivo.* After topical application of dye sodium fluorescein onto porcine skin mounted in Franz diffusion cells with the acceptor compartment beneath the dermis, the fluorescence was detected on the surface, within the horny layer, and in most of the follicles confirming the similarity in the penetration between porcine and human skin [72]. So nanoparticles have been studied in porcine skin revealing in the surface images that polystyrene nanoparticles accumulated preferentially in the follicular openings, this distribution was increased in a time-dependent manner, and the follicular localization was favored by the smaller particle size [74]. In other investigations, it has been shown by differential stripping the influence of size microparticles in the skin penetration. It can act as efficient drug carriers or can be utilized as follicle blockers to stop the penetration of topically applied substances [73].

It has already been postulated that certain molecules can hydrogen bond to groups present on the surfaces of follicular pores [75]. However, more studies have to be made in order to identify all the molecular properties that influence drug penetration into hair follicles.

Nowadays, there are currently a number of methods available for quantifying drugs localized within the skin or various layers of the skin. To date, a direct, non-invasive quantification of the amount of topically applied substance penetrated into the follicles had not been possible. Therefore, stripping techniques, tape stripping and cyanoacrylate skin surface biopsy have been used to remove the part of the stratum corneum containing dye topically applied [76]. Thus, the "differential stripping" has been shown as a new method that can be used to study the penetration of topically applied substances into the follicular infundibula non-invasively and selectively [20,26,30,76,77].

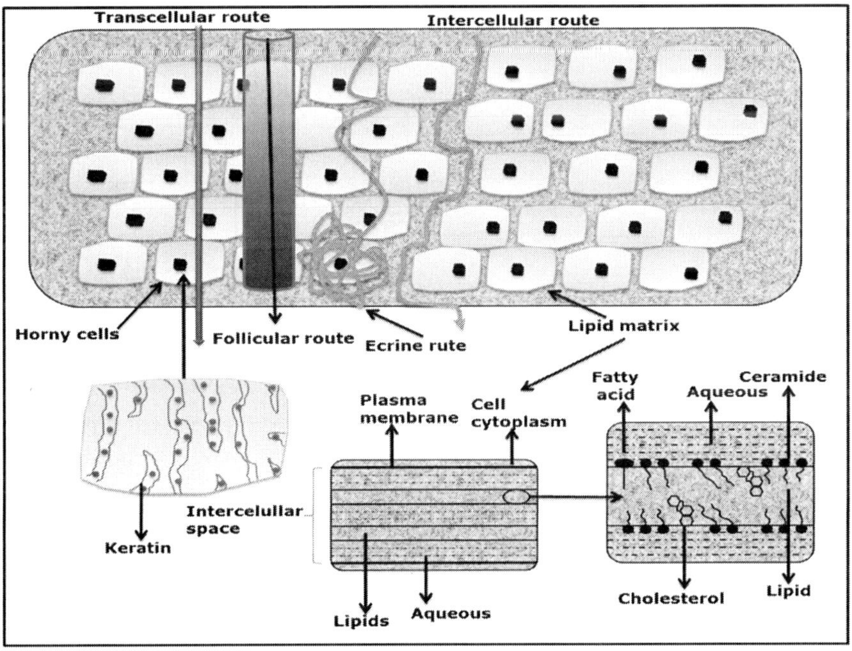

Figure 2. Schematic representation of penetration routes of drugs throughout the skin.

5. ADVANTAGES AND DISADVANTAGES OF TRANSDERMAL DRUG DELIVERY

Transdermal drug delivery systems offer several important advantages over more traditional approaches, in addition to the benefits of avoiding the hepatic first-pass effect, and higher patient compliance, the additional advantages and the disadvantages [78-80] that transdermal drug delivery offers can be summarized as follows in Table 1.

Table 1. Main advantages and disadvantages of transdermal drug delivery

Advantages of transdermal drug delivery	Disadvantages of transdermal drug delivery
Longer duration of action	Possibility of local irritation at the site of application
Reduction in dosing frequency	Erythema, itching, and local edema can be caused by the drug, the adhesive, or other excipients in the patch formulation
More uniform plasma levels	The skin's low permeability limits the number of drugs that can be delivered in this manner
Useful for drugs that require relatively consistent plasma levels	
It is an alternative route of administration to accommodate patients who cannot tolerate oral dosage forms (specially for nauseated or unconscious patients)	
Improved bioavailability	
Reduction of side effects	
Flexibility of terminating the drug administration by simply removing the patch from the skin	

6. NANOCARRIER SYSTEMS

Nanocarriers have demonstrated increased drug absorption, penetration, half-life, bioavailability, stability, etc. Nanocarriers are so small to be detected by immune system and they can deliver the drug in the target organ using lower drug doses in order to reduce side effects. Nanocarriers can be administrated into the organisms by all the routes; one of them is the dermal route. The nanocarriers most used and investigated for topical/transdermal drug delivery in the pharmaceutical field are shown in Figure 3 and Table 2.

6.1. NANOPARTICLES

Nanoparticles are smaller than 1,000 nm. Nowadays, it is possible to insert many types of materials such as drugs, proteins, peptides, DNA, etc. into the nanoparticles. They are constructed from materials designed to resist pH, temperature, enzymatic attack, or other problems [81]. Nanoparticles can be classified as nanospheres or nanocapsules (See Figure 4). Nanospheres are solid-core structures and nanocapsules are hollow-core structures. Nanoparticles can be composed of polymers, lipids, polysaccharides and proteins [82,83]. Nanoparticles preparation techniques are based on their physicochemical properties. They are made by emulsification-diffusion by solvent displacement, emulsification-polymerization, in situ-polymerization, gelation, nanoprecipitation, solvent evaporation/extraction, inverse salting out, dispersion polymerization and other derived from these one.

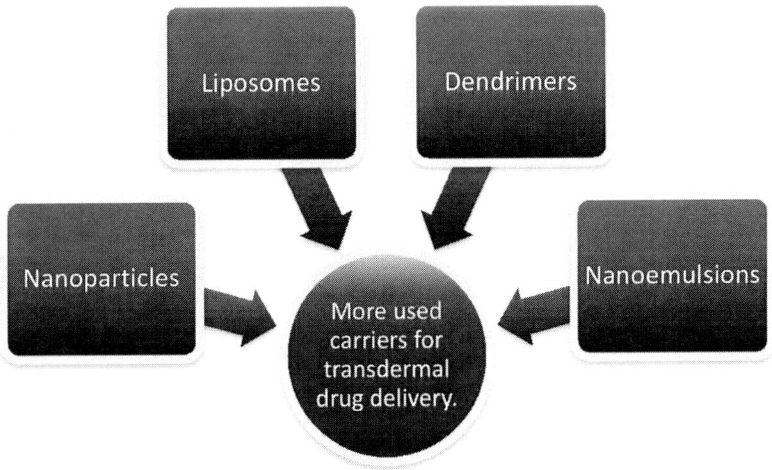

Figure 3. More used transdermal nanocarriers

6.2. NANOEMULSIONS

Nanoemulsions are isotropic dispersed systems of two non miscible liquids, normally consisting of an oily system dispersed in an aqueous system (o/w nanoemulsion), or an aqueous system dispersed in an oily system but forming droplets or other oily phases of nanometric sizes (100 nm). They can be stable (methastable) for long times due to the extremely small sizes and the use of adequate surfactants. Nanoemulsions can use hydrophobic and hydrophilic drugs because it is possible to make both w/o or o/w nanoemulsions [84]. They are non-toxic and non-irritant systems and they can be used for skin or mucous membranes, parenteral and non parenteral administration in general and they have been used in the cosmetic field. Nanoemulsions can be prepared by three methods mainly: high-pressure homogenization, microfluidization and phase inversion temperature. Transdermal delivery using nanoemulsions has been reduced due to the stability problems inherent to this dosage form.

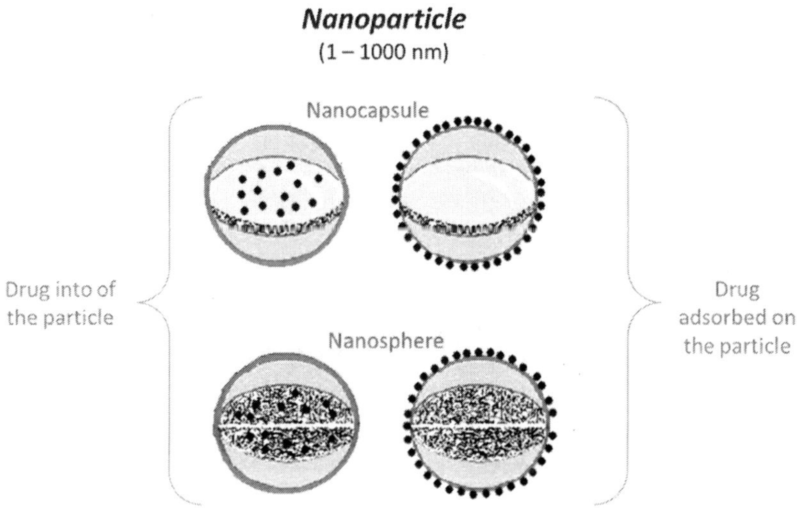

Figure 4. Nanospheres and nanocapsules are small vesicles used to transport drugs. Nanospheres are typically solid polymers with drugs embedded in the polymer matrix. Nanocapsules are a shell with an inner space loaded with the drug of interest. Both systems are useful for controlling the release of a drug and protecting it from the surrounding environment.

6.3. LIPOSOMES

Liposomes are hollow lipid bilayer structures (Figure 5) that can transport hydrophilic drugs inside the core and hydrophobic drugs between the bilayer [85]. They are structures made of cholesterol and phospholipids. They can have different properties depending on the excipients included and the process of their elaboration. The nature of liposomes makes them one of the best alternatives for drug delivery because they are non-toxic and remain inside the bloodstream for a long time. Liposomes can be surface-charged as neutral, negative or positive, depending on the functional groups and pH medium. Liposomes can encapsulate both lipophilic and hydrophilic drugs in a stable manner, depending on the polymer added to the surface [86]. There are small unilamellar vesicles (25 nm to 100nm), medium-sized unilamellar vesicles (100 nm and 500nm), large unilamellar vesicles, giant unilamellar vesicles, oligolamellar vesicles, large multilamellar vesicles and multivesicular vesicles (500 nm to microns). The thickness of the membrane measures approximately 5 to 6 nm. These shapes and sizes depend of the preparation technique, the lipids used and process variables. Depending on these parameters, the behavior both in vivo and in vitro can change and opsonization processes, leakage profiles, disposition in the body and shelf life are different due to the type of liposome [86].

Liposomes preparation techniques follow three basic steps with particular features depending on safety, potential scale up and simplicity: 1) Lipid must be hydrated, 2) Liposomes have to be sized and 3) Nonencapsulated drug has to be removed. The degree of transdermal drug penetration is affected by the lamellarity, lipid composition, charge on the liposomal surface, mode of application and the total lipid concentrations [87].

Figure 5. Liposomes are spherical vesicles that comprise one or more lipid bilayer structures enclosing an aqueous core. They protect encapsulated drugs from degradation. Liposomes can also be functionalized to improve cell targeting and solubility

6.4. DENDRIMERS

Dendrimers are monodisperse populations that are structurally and chemically uniform (Figure 6). They allow conjugation with numerous functional groups due to the nature of their branches. The amount of branches increases exponentially and dendrimers growth is typically about 1 nm per generation [88]. The dendrimers classification is based on the number of generations. After the creation of a core, the stepwise synthesis is called first generation; after that, every stepwise addition of monomers creates the next generation. This approach allows an iterative synthesis, providing the ability to control both molecular weight and architecture.

The kind of polymer chosen to construct the dendrimer by polimerization is very important with regard to the final architecture and features. In addition, the use of branched monomers has the peculiarity of providing tailored loci for site-specific molecular recognition and encapsulation. Notably, 3D and fractal architecture, as well as the peripheral functional groups, provide dendrimers with important characteristic physical and chemical properties. In comparison with linear polymers, dendritic structures have "dendritic voids" that give these molecules important and useful features. These spaces inside dendrimers can mimic the molecular recognition performed by natural proteins. Furthermore, dendrimers have a high surface-charge density due to ionizable groups that help them to attach drugs by electrostatic forces, regardless of the stoichimetry. This dendrimer-drug association provides drugs with better solubility, increasing their transport through biological membranes and sometimes increasing drug stability. The number of molecules that can be incorporated into dendrimers is related to the number of surface functional groups; therefore, later-generation dendrimers are more easily incorporated into dendritic structure. However, not all the functional groups are available for interaction due to steric volume, molecule rotation or stereochemistry effects. Dendrimers can have positive and negative charges, which allows them to complex different types of drugs [89].

6.5. ADVANTAGES AND LIMITATIONS OF USING NANOCARRIERS FOR TRANSDERMAL DRUG DELIVERY

As it is has been mentioned before, the search for new strategies able to enhance the topical and transdermal penetration of drugs has become essential [61]. Different carrier systems have been proposed in an attempt to favour the transport of drugs through the skin, enabling drug retention and in some cases allowing a controlled release. Skin penetration is essential to a number of current concerns, for example, contamination by microorganisms and chemicals, drug delivery to skin (dermatological treatments) and through skin (transdermal treatments), and skin care and protection (cosmetics) [103-106].

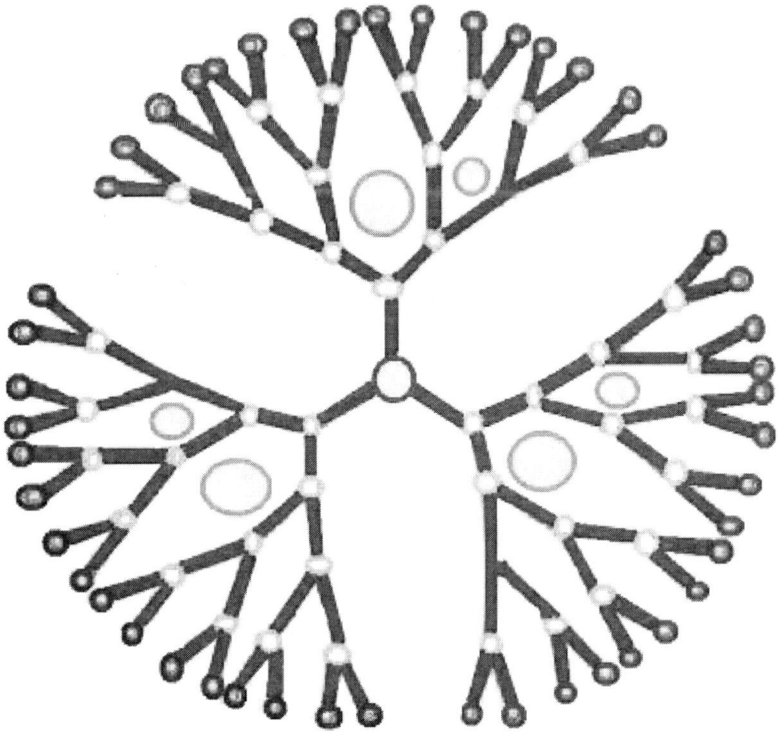

Figure 6. Dendrimers are highly branched polymers with a controlled three-dimensional structure around a central core. They can accommodate more than 100 terminal groups.

Follicular penetration has become a major focus of interest due to the drug targeting to the hair follicle is of great interest in the topical treatment of skin diseases.However due to follicular orifices only occupying ~0.1% of the total skin surface area, it was assumed as a non important route. But recently, a variety of studies have shown that the hair follicles represent a important way to trough the skin and some techniques have been used to test the penetration of drugs loaded nanoparticles/microsparticles [70,72,73,107-109]. Confocal laser scanning microscopy which permits optical sectioning of thick tissues and cells and their subsequent computerized three-dimensional reconstruction, has been used to study the entry of drugs through the skin [110]. It was visualized in the fresh human scalp skin on-line the diffusion processes of a model fluorophore into the hair follicle at different depths [72]. Such follicular pathway also has been proposed for topical administration of nanoparticles and microparticles using porcine skin. Recent studies have confirmed that the *in vitro* penetration into the porcine hair follicles might be considered similar to those on humans *in vivo*[73]. Studies in porcine skin revealed in the surface images that polystyrene nanoparticles accumulated preferentially in the follicular openings, this distribution was increased in a time-dependent manner, and the follicular localization was favored by the smaller particle size [76]. In other investigations, it has been shown by differential stripping the influence of size microparticles in the skin penetration. It can act as efficient drug carriers or can be utilized as follicle blockers to stop the penetration of topically applied substances [75]. An alternative technique is multiphoton microscopy (MPM) especially two-photon excitation microscopy has been widely used in imaging biological specimens treated with nanoparticles [111-114]. The near-infrared light used in the two-photon microscope can penetrate deeper in highly scattering tissues such as *in vivo* human skin than confocal microscopes operated with ultraviolet excitation [115]. Futhermore, this technique provides both cellular and extracellular structural information, with subcellular resolution helpful for clinical dermatological diagnosis, both *ex vivo* and*in vivo*. In addition, it can be used to characterize stratum corneum structures, visualize and quantify transcutaneous drug delivery, detect skin cancers, explore collagen structural transitions, and watch laser–skin interactions [116,117]. A common method used for quantifying drugs localized within the skin or various layers of the skin are the tape stripping and cyanoacrylate skin surface biopsy techniques, which have been used to remove the part of the stratum corneum containing dye topically applied. Thus, the "differential stripping" has been shown as a new method that can be used to study the penetration of topically applied substances into the follicular infundibula non-invasively and selectively [118].It has been reported in a previous *in vitro* permeation studies using tape stripping, that triclosan-loaded nanoparticles penetrated into the skin and their retention favoured a local effect. Moreover, polymeric nanoparticles are expected to be able to form a depot in the hair follicles, providing a targeted controlled drug delivery [33,119].

In general, the principal advantages of microparticles and nanoparticles over conventional formulations such as creams, solutions, ointments, lotions, gels, and foams, is that the second ones have different absorption characteristics and aesthetic properties, and they also have some major limitations, such as poor penetration and uncontrolled drug release. Furthermore, tolerability and safety end points, such as irritation, dryness, erythema, itching, stinging and burning will be key factors in determining its usefulness. It happens because using a traditional system, drug delivery is sometimes rapid and topical or plasmatic concentrations can result in toxic effects. However, for the case of nanoparticles a smaller amount of the drug is necessary and due to the targeted nature of delivery [120]. Topical or transdermal drug deliveries have many advantages over the other routes: fewer side effects, increased patient compliance, controlled release, and the lack of a hepatic first pass [121].

TABLE 2. Examples of nanocarriers used for drug delivery.

Nanocarrier	Size range	Preparation methods	Characteristics	References
Polymeric nanoparticles	10-1000 nm	-In situ polymerization -Emulsification-evaporation, -Emulsification-diffusion, -Emulsification-diffusion by solvent displacement.	Solid or hollow particles wich have entraped, binded or encapsulated drugs.	[33]
Solid lipid nanoparticles	50-1000 nm	-High-pressure homogenization.	Similar to polymeric nanoparticles but made of solid lipids.	[90]
Inorganic Nanoparticles	<50nm	-Sol-gel technique	Nanometric particles, made up of inorganic compounds such as silica, titania and alumina.	[91]
Liposomes	25 nm-100 μm	-Sonication -Extrusion, -Mozafari method	Vesicles composed of one or more concentric lipid bilayers, separated by water or aqueous buffer compartments.	[92]
Dendrimers	3–10 nm	-Polymerization	Macromolecular high branched structures.	[93]
Quantum dots	2-10nm	-Colloidal assembly, viral assembly, -Electrochemical assembly.	Made up of organic surfactants, precursors and solvents.	[94]
Lipid globules	1-100 nm	-Emulsification espontaneous systems.	Multicomponent fluid made of water, a hydrophobic liquid, and one or several surfactants resulting in a stable system.	[95]
Lipid microcylinders	<1 μm	-Self emulsification	Self organizing system in which surfactants crystallize into tightly packed bilayers that spontaneously form cylinders.	[96]

It has been reported that nanoencapsulation of drugs (nano-medicines) increases their efficacy, specificity, tolerability and therapeutic index [122-124]. These nano-formulations are reported to be superior to traditional medicine with respect to controlled release, targeted delivery and therapeutic impact. The targeting capabilities of nanomedicines are influenced by particle size, surface charge, surface modification, and hydrophobicity. Of these, nanoparticle size distribution is an important factor in determining the interaction with the cell membrane and their penetration across physiological barriers, being dependent on the tissue, target site and circulation [125]. Example of this are the nanostructured lipid carriers (NLC) by their structure (lipid nanoparticles with solid matrix) increase in loading capacity, physical and chemical long-term stability, triggered release and potentially supersaturated topical formulations with respect to solid lipid nanoparticles (SLN). Other advantages of NLC include improvement in stabilisation of incorporated compounds, controlled release, occlusivity, film formation on skin including in vivo effects on the skin. Lipid nanoparticles have been observed as a good option for transdermal delivery because they can be prepared in different sizes and it is possible to modify surface polarity in order to improve skin penetration [126,127]. From the upper skin, nanoparticles can reach deeper skin regions because they exhibit mechanical flexion [128].

Additionaly, transdermal nanocarriers are able to reach target organs because they can be attached to antibodies, antigens, vitamins and other molecules to be more specific. Nanoparticles can travel largely undetected by the immune system depending of the nanocarriers size of the antigen added as well as its composition. So, by hiding functional groups or protecting these groups with other molecules, drugs can be released specifically in the target organ. Consecuently, nanoparticles can even travel from the skin to lymph nodes, representing a promising tool for immunomodulation [129]. One of the first strategies for transdermal delivery were the liposomes. The nature of liposomes makes them one of the best alternatives for drug delivery because they are non-toxic and remain inside the bloodstream for a long time [130-133]. Nevertheless, some factors affect the degree of transdermal drug penetration such as the lamellarity, the lipid composition, the charge on the liposomal surface, the mode of application and the total lipid concentrations [89,134]. For that reason, flexible vesicles called transfersomes or transformable liposomes have been compared with those rigid vesicles to enhance penetration [135-142]. The lipids present in the liposome bilayer can interact with lipids present in the stratum corneum changing the structure of the upper skin. This change is beneficial for the penetration of lipophilic drugs into the stratum corneum [143]. Some liposomes may have a deformable structure and pass through the stratum or may accumulate in the channel-like regions in the stratum corneum, depending upon their composition [144,145]. In order to obtain transformable liposomes more flexible, they are prepared using surfactants or alcohol (ethosomes) in the lipid bilayer, to be able to deform them when a pressure is applied in the transdermal route.

Some limitations for nanocarriers are the important tests and regulations that should be carried out to ensure an adequate characterization, analytical evaluation, toxicological and pharmacological assessment, which is necessary to determine the efficacy of using these nanostructures in therapies and diagnosis because of their tiny size, their high surface energy, their composition, their architecture, their attached molecules, etc. Those things are frequently reviewed for the dendrimers. One of the main advantages is that they have multivalency and it is possible to get control of the functional groups on the surface [146,147]. Due to their form and size (1–10 nm), these molecules can carry drugs, imaging agents, and can interact with lipids present in membranes, because it was reported a better permeation in cell cultures and intestinal membranes. They also increased the permeation of lipophilic drugs instead of hydrophilic drugs. The main problems with this kind of transdermal carrier are poor biodegradation and inherent cytotoxicity [148]. To obtain dendrimers less toxic, dendrimers have been linked to peptides. Dendrimers-peptides are formed from amino acids linked via peptide-amide bonds to the branches of dendrimers in the core or on the surface to get down the toxicity. Then, they are bio-transformed to produce amino-acid derivatives. Besides, the synthesis of these structures is less expensive and purification does not present any difficulty [149,150].

It is suggested in future research to elucidate the interactions between nanocarriers and other molecules as well as interactions between nanocarriers and biological entities. The toxicology of nanostructures is also a current concern. Materials behave very differently when they are diminished to nanosizes. Traditional laws do not work at this "meso-scale" in the same way as they function at the macro-scale. On the macro scale, bulk properties in a material predominate over surface properties. At the micro-scale, surface properties tend to dominate. At the meso-scale, both types of properties play significant roles [151,152]. Furthermore, the effects of metabolized/altered nanostructures on the biological system are difficult to predict. Regulatory agencies are taking action to assess new Nanotechnology-based products.

In addition, the fabrication of nanocarriers scaling up from the lab at the industrial production is difficult and the materials used to prepare nanocarriers are very expensive in the majority of the cases.

7.APPLICATIONS OF NANOCARRIER SYSTEMS IN TOPICAL/TRANSDERMAL DELIVERY

Nanocarriers as drug delivery systems were first intended for use in parenteral or oral routes of administration and as such still continue to be the focus of many studies [153]. However skin application of these nanocarriers, and especially for liposomes, polymeric and lipidic nanoparticles, also makes sense when considering surface effects (film formation and occlusive effects), local effects in the skin (drug delivery in the epidermis and dermis) and systemic effects (deeper drug permeation and transdermal delivery). In potential uses apart from those concerned with surface effects the nanocarrier has to overcome the SC barrier in order to deliver the drug more or less deeply into skin layers. Recent advances in the study of penetration mechanisms deal with the control of the intercellular penetration route by the crystalline state of lipids, and the penetration through skin appendages (the follicular pathway) that appears to contribute much more than was previously thought. Applications dependent on skin penetration that have received special attention include transdermal delivery of nano- and microparticles by hair follicles, especially for nanoparticles which penetrate hair follicles very efficiently targeting the skin immune system in order to develop new vaccination strategies, and problems relating to skin diseases [154,155].

Options for topical and transdermal delivery are the solid lipid nanoparticles (SLNs) and nanostructured lipid carriers (NLCs) [156]. Some drugs such as triptolide, triamcinolone acetonide acetate, cyclosporin A have been used to be entrapped in SLN [157-159]. SLN can be admixed to an already commercially available and established topical formulation, e.g. a cosmetic day cream. Admixing the SLN leads to an increase in occlusivity while still maintaining the 'light character' of the day cream and avoiding the glossiness of more occlusive night creams. This phenomena is explained in Figure 7. However, having a highly occlusive night cream already, addition of SLN will have little or no effect [156].

Lipid nanoparticles are other options to load arthemeter and econazole nitrate [160,161]. Celecoxib [162]. It was compared the permeability of coenzima Q 10 incorporated in NLC and in an emulsion with the same lipid contain. The occlusion effect of the cream was also investigated. The result showed a higher permeability of the molecule and a higher occlusive effect for the NLC than for the emulsion as it could be observed in Figures 8 and 9 [163].

Different studies shown lipid nanoparticles were able to enhance the chemical stability of compounds sensitive to light, oxidation and hydrolysis. Enhancement of chemical stability after incorporation into lipid nanocarriers was proven for many cosmetic actives, e.g. coenzyme Q10 [164-166], ascorbyl palmitate [164,167], tocopherol (vitamin E) [165] and retinol (vitamin A) [168-170].

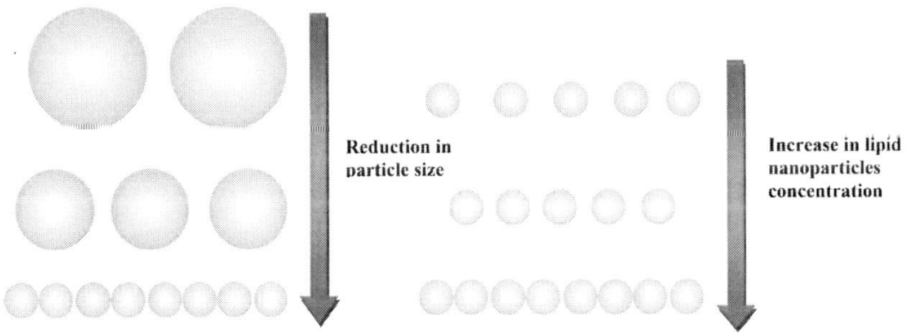

Figure 7. The occlusion factor of lipid nanoparticles depends on various factors: at identical lipid content, reducing the particle size leads to an increase in particle number, the film becomes denser (left) and therefore the occlusion factor increases. At a given particle size, increasing the lipid concentration increases particle number and density of the film (right) which also leads to a higher occlusion factor.

Three vitamin derivatives including vitamin C (ascorbyl tetraisoplamitate), vitamin E (tocopherol acetate) and vitamin A (retinyl palmitate) were also loaded in PLGA nanospheres, for skin whitening and anti-wrinkles/aging applications due vitamin C suppresses the blemishes because it limits the activity of tyrosinase, which promotes melanin production. Furthermore, it increases collagen formation to reduce wrinkles, and prevents cell oxidation by eliminating active oxygen. As to vitamin E and A, they also act as antioxidant and collagen promoter, respectively. They were able to reach the target areas in a stable form, sustain the pharmacological effect for a long time and be effective to reduce the wrinkles and produce a whitening effect [171]. In a sense, the idea of nanoparticle design for drug delivery systems in cosmetics applications is important.

As it was described before, the follicles are deep invaginations inside skin where the SC is thinner, the vascularisation is denser, and there are several targets of interest along a follicle structure from both cosmetic and pharmaceutical viewpoints [172]. Here, minoxidil which is an antihypertensive has been introduced in a block of copolymer poly (ε-caprolactone)-block-poly(ethyleneglycol) to treat the alopecia areata disorder, by widening blood vessels and opening potassium channels, it allows more oxygen, blood, and nutrients to the follicle. This disorder is an inflammatory condition, often reversible hair loss affecting mainly children and young adults. Clinically, round hairless patches appear on the scalp while hair follicles remain intact. This skin

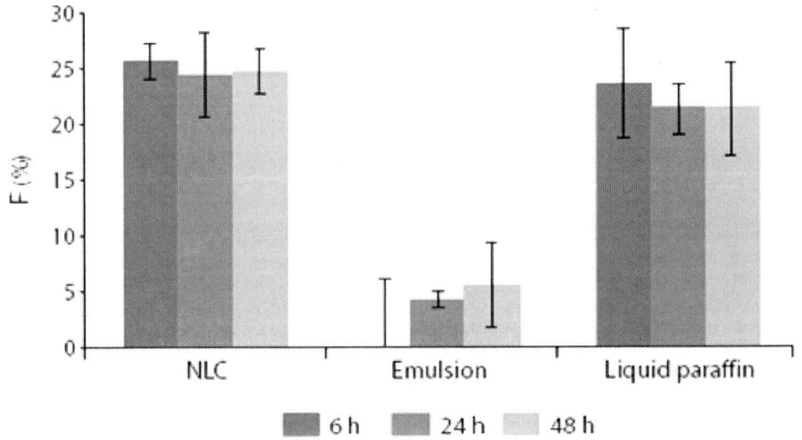

Figure 8. Occlusion factor (F) of NLC, emulsion and liquid paraffin (with permission from authors) [163].

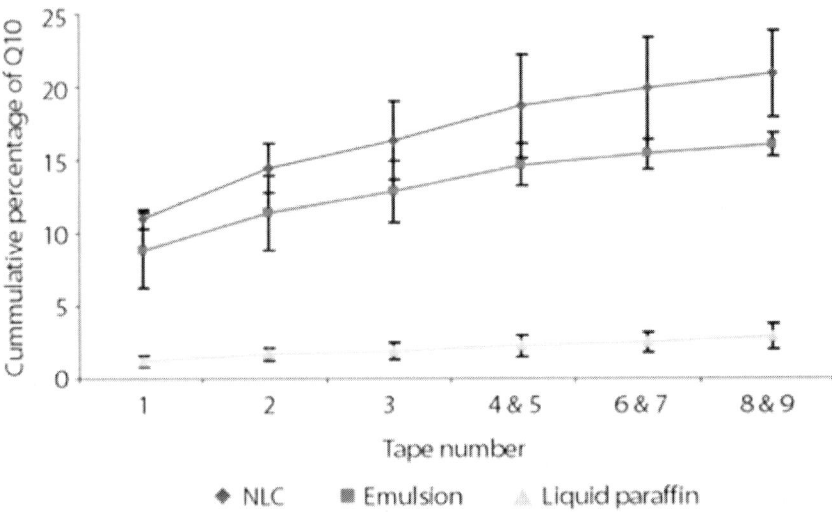

Figure 9. Tape stripping test: summation of coenzyme Q10 found in the tapes related to the applied amount (with permission from authors) [163].

disorder is related with the distal part of the human hair follicle immune system, especially with the interacting intraepithelial T cells. The cause of this condition is diverse and seems to involve T cell–mediated immunologic changes, neuropeptides, genetic disposition to autoimmunity, and distress [30,173,174]. As the infundibulum of the hair follicle is surrounded by an extensive capillary network and the permeability of its epithelium allows the transport of molecules or particles to the blood circulatory system. There is a high density of immune cells in and around the infundibulum epithelium which could be targets also for hair follicle immune system and topical vaccination.

The sebaceous glands associated with hair follicles provide another potential target for delivering drugs against acne, androgenetic alopecia and other sebaceous gland dysfunctions. Different nanoparticles formulations have been prepared in order to treat acne vulgaris, which is an inflammatory disease of the pilosebaceous units, most densely concentrated on the face and torso [175].Pathogenesis is multifactorial, but *Propionibacterium acnes* a Gram-positive bacterium plays a central role in the promotion of inflammation in acne. The most commonly formulations are prepared with different topical antimicrobials, either alone or in combination with other drugs. It is expected that an agent able to inhibit *P. acnes* growth and to suppress the inflammatory response will provide significant benefits to patients with acne vulgaris [30,176,177]. For that reason triclosan has been used in several systems. It was reported the characterization of triclosan loaded polymeric nanoparticles. They showed a good encapsulation efficiency and also a good physical stability representing an alternative as a treatment of acne [33]. Triclosan loaded nanoparticles made of chitosan and cyclodextrins were prepared using a very simple ionic gelation technique. This new approach permits to enhance the entrapment of hydrophobic drugs by forming molecular inclusion complexes with cyclodextrins in aqueous media. Such a device could be of interest for conferring protection to some specific drug molecules through the complexation followed by entrapment in the polymer matrix [109]. Another drug used to treat this disorder is tretinoin (all-*trans*-retinoic acid) which is the active form of a metabolic product of Vitamin A, also called retinoic acid. Tretinoin-loaded nanocapsules improved tretinoin photostability, independently on the type of oily phase used (capric/caprylic triglycerides and sunflower seed oil) in this study, and represent a potential system to be incorporated in novel topical or systemic dosage forms containing tretinoin [178].

Formulations of nanoparticles are often used in combination with penetration chemical and physical enhancers to modify the physical state of the stratum corneum, affecting the degree of transdermal drug penetration. DNA has been entrapped in nanoparticle of polysaccharide such as chitosan/poly-γ-glutamic acid and in a multifunctional core-shell polymeric nanoparticle of PLGA core and a positively-charged glycol chitosan (GC) shell. Another drugs used in the preparation of nanoparticles made of propyl-starch derivatives are flufenamic acid, testosterone and caffeine [179,180]. Insulin is a protein which has also been introduced in chitosan nanoparticles [83]. Poly (lactide-co-glycolide) polymer has been used to prepare biodegradables nanoparticles containing dexamethasone phosphate and 5-Fluorouracil [181,182]. Chlorhexidine loaded polymeric nanoparticle are used to treat cutaneous infections [183].

Inflammatory skin diseases account for a large proportion of all skin disorders and constitute a major health problem worldwide. Psoriasis, atopic dermatitis, poison ivy, and eczema are another skin disorders. Contact dermatitis, atopic dermatitis, and psoriasis represent the most prevalent inflammatory skin disorders and share a common efferent T-lymphocyte mediated response. Oxidative stress and inflammation have recently been linked to cutaneous damage in T-lymphocyte mediated skin diseases, particularly in contact dermatitis [184]. Poison ivy and atopic dermatitis may also present with bullous and vesicular changes [185]. Lipid nanoparticles have been investigated to improve the treatments of skin diseases such as atopic eczema, psoriasis, skin mycosis and inflammations. Apart from the treatment of skin diseases by topical application, e.g. gastrointestinal side effects of non-steroidal anti-inflammatory drugs can be decreased by topical antirheumatic therapy. Drugs under investigations for dermal application using lipid nanoparticles at the present are for instance glucocorticoids, retinoids, non-steroidal anti-inflammatory drugs, COX-2 inhibitors and antimycotics. It was showed that it is possible to enhance the percutaneous absorption with lipid nanoparticles. These carriers may even allow drug targeting to the skin or even to its substructures. Thus they might have the potential to improve the benefit/risk ratio of topical drug therapy [186].

Perioral dermatitis is commonly seen in women aged 20–35 years. It presents as red papules that form superficial plaques around the perioral area, nasolabial folds and/or lower eyelids. It is minimally itchy [187,188]. Topical corticosteroids are the first-line therapy of acute exacerbations of atopic dermatitis and contact dermatitis. Prednicarbate is superior to the halogenated glucocorticoids because of an improved benefit/risk ratio. However, at the present the separation of desired anti-inflammatory effects and undesired antiproliferative effects is still not satisfying. Therefore, lipid nanoparticles were investigated as a delivery system for prednicarbate. The report show an improved extent of prednicarbate uptake by human skin *in vitro*, if applied as SLN dispersion or cream containing prednicarbate-loaded SLN. The authors found that a prednicarbate targeting to the epidermis [189]. This is particular relevant because prednicarbate in the dermis is responsible for the induction of irreversible skin atrophy while the inflammatory process is most pronounced within the epidermis [186]. Therefore, a better benefit/risk ratio is expected for the application of pretnicarbate in SLN containing topical formulations.

Tretinoin, a metabolite of vitamin A is used for topical treatment of various proliferateive and inflammatory skin diseases such as psoriasis, acne (as mentioned before), photo aging, epidermotropic T-cell lymphomas and epithelial skin cancer. One of the major disadvantages associated with the topical application of tretinoin is local skin irritation such as erythrema, peeling and burning as well as increased sensitivity to sunlight. To overcome these problems tretinoin was incorporated into SLN [190]. *In vitro* permeation studies through rat skin indicated that SLN-based tretinoin gel has a permeation profile comparable to that of the market tretinoin cream. Furthermore, Draize patch test showed that SLN-based tretinoin gel resulted in remarkably less erythremic episodes compared to the currently marketed tretinoin cream (Figure 10). Therefore, also for formulations containing tretinoin-loaded SLN a better benefit/risk ratio is expected.

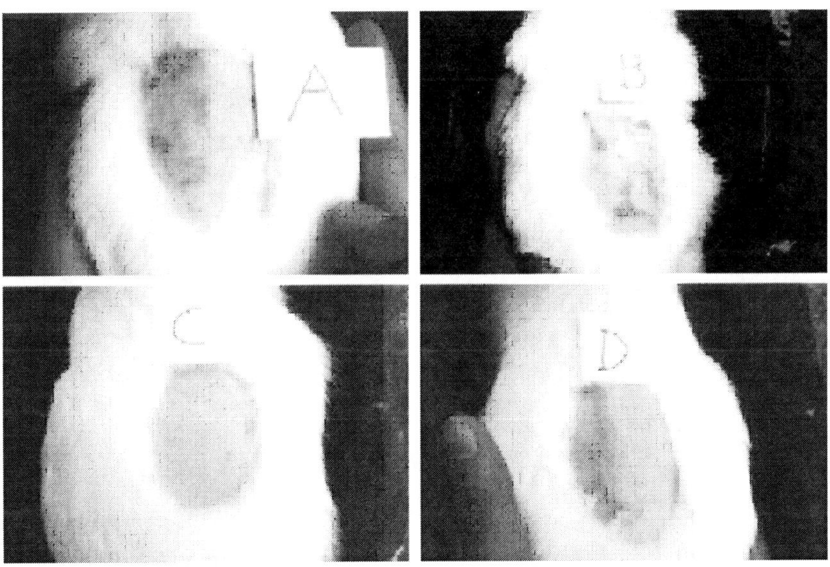

Figure 10. Pictures of Draize skin irritation studies carried out on New Zealand rabbits 24 h after application of (A) control (no application); (B) marketed formulation (Retino-A® cream); (C) SLN-based gel without tretinoin; (D) SLN-based gel containing tretinoin (0.05%, w/w). The Marketed tretinoin cream clearly shows erythemal lesions, which are not visible in SLN based tretinoin gel [190]. "Reprinted from International Journal of Pharmaceutics, 345/1-2, Kumar A. Shah, Abhijit A. Date,Medha D. Joshi,Vandana B. Patravale, Solid lipid nanoparticles (SLN) of tretinoin: Potential in topical delivery, 163-171., Copyright (2007), with permission from Elsevier"

Liposomes which were one of the first strategies for transdermal delivery are being successfully used in cancer therapy [139,141]. However to date, many liquid-type nanocosmetics carriers, such as liposomes, are structurally unstable. Specifically, when passing through the skin, they adhere to the inside walls of the skin cells causing the collapse of phospholipid association bodies and the leak of their encapsulated ingredients. As a result, their ability to transport active ingredients to deep skin is not likely good. Some authors report the use of flexible vesicles called transfersomes or transformable liposomes in comparison with rigid vesicles to enhance penetration [148,191]. The application of transformable liposomes more flexible, which are prepared using surfactants or alcohol (ethosomes) in the lipid bilayer, to be able to deform them when a pressure is applied in the transdermal route has been increased.

In some researches, dendrimers are used for transdermal drug delivery. They show promising results in the delivery of drugs such as tamsulosin [192], indomethacin [193], ketoprofen and diflunisal [194] and 5-fluorouracil [195]. The main problems with this kind of transdermal carrier are poor biodegradation and inherent cytotoxicity [158]. In order to get down the toxicity dendrimers have been linked to peptides (dendrimers-peptides) from amino acids linked via peptide-amide bonds to the branches of dendrimers in the core or on the surface [159,160].

For 5-fluorouracil (5FU) (log $P = -0.89$) which is one hydrophilic model drug used to treat skin diseases, has been reporte to have very poor penetration in skin [196-198]. Many strategies to increase skin permeation of this drug have been tested: prodrugs, terpenes, fatty acids, iontophoresis, sonophoresis, laser ablation and dendrimers which increased 5FU permeation across the skin by altering the skin structure [198-203].

Nowadays transdermal delivery using nanoemulsions it is not so used as nanoparticles or liposomes delivery due to the stability problems inherent to this dosage form. Nevertheless, Gamma Tocopherol, Caffeine, Plasmid DNA, Aspirin, Methyl Salicylate, Insulin, Nimesulide have been included in nanoemulsion. The use of these nanocarriers to deliver analgesics, corticosteroids, anti cancer agents, etc. is very important since these drugs are able to act immediately because they do not need to cross extra barriers. The drug is bioavailable easily and faster [204-210].

8. CONCLUSIONS

Nanocarriers have shown many advantages for topical and trasdermal delivery of drugs. It could be shown already for various drugs that topical/trandermal formulations containing nanoparticles can enhance the penetration into the skin increasing treatment efficiency, target the epidermis or follicles, reducing side effects. Furthermore, an increased activities as well as prolonged activities have been reported. These delivery systems can deliver both hydrophilic and lipophilic molecules. Advances with regard to materials, fabrication methods and techniques facilitate the development of new and better nanocarriers. Nonetheless, future researches must ensure the benefit and evaluate the risk ratio for many drugs included in nanocarriers.

ACKNOWLEDGEMENT

Dr. José Juan Escobar-Chávez wishes to acknowledge PAPIIT TA200-312, Cátedra PACIVE NCONS-17 and PAPIME 203612. We are also grateful with Dr. Müller andDr. Pardeike for permission of their figures. The authors report no conflict of interests.

REFERENCES

1. V Brower, Is nanotechnology ready for primetime? Journal of National Cancer Institute 200698911

2. J. J Escobar-chávez, Rodriguez Cruz IM, Dominguez-Delgado CL. Chemical and Physical enhancers for transdermal drug delivery. In: Gallelli L (ed.) Pharmacology. Rijeka: InTech; 2012397433

3. R Díaz-torres, Transdermal nanocarriers. In: José Juan Escobar-Chávez/Virginia Merino (Eds.) Current Technologies to Increase the Transdermal Delivery of Drugs. The Netherlands: Bentham Science Publishers; 2010120140

4. Hadgraft, , Lane ME. Skin permeation: The years of enlightenment. International Journal of Pharmaceutics 2005, 305 (1-2):2-12.

5. B. W Barry, Novel mechanisms and devices to enable successful transdermal drug delivery. European Journal of Pharmaceutical Sciences 20011421014

6. M Rizwan, M Aqil, S Talegaonkar, A Azeem, Y Sultana, A Ali, Enhanced transdermal drug delivery techniques: an extensive review of patents. Recent Patents on Drug Delivery & Formulations 2009105124

7. Escobar-ChávezJJ& Merino V edsCurrent Technologies to increase the Transdermal Delivery of Drugs. The Netherlands: Bentham Science Publishers; 2010

8. J. J Escobar-chávez, D Bonilla-martínez, A Villegas-gonzález, I. M Rodríguez-cruz, C. L Domínguez-delgado, The use of sonophoresis in the administration of drugs through the skin. Journal of Pharmacy and Pharmaceutical Sciences 200912188115

9. Escobar-Chávez JJ, Merino V, López-Cervantes M, Rodríguez-Cruz IM, et al.The use of iontophoresis in the administration of nicotine and new non-nicotine drugs through the skin for smoking cessation. Current Drug Discovery Technologies 2009; 6(3):171-185.

10. J. J Escobar-chávez, D Bonilla-martínez, A Villegas-gonzález, A. L Revilla-vazquez, The electroporation as an efficientphysical enhancer for transdermal drug delivery. Journal of Clinical Pharmacology 20094911126283

11. D Papakostas, F Rancan, W Sterry, U Blume-peytavi, A Vogt, Nanoparticles in dermatology. Archives of Dermatology Research 2011303853350

12. J Kristl, K Teskac, P. A Grabnar, Current view on nanosized solid lipid carriers for drug delivery to the skin. Journal of Biomedical Nanotechnology 20106552942

13. A Schroeter, T Engelbrecht, R. H Neubert, A. S Goebel, New nanosized technologies for dermal and transdermal drug delivery: A review.Journal of Biomedical Nanotechnology 20106551128

14. R. H Neubert, Potentials of new nanocarriers for dermal and transdermal drug delivery. European Journal of Pharmaceutics and Biopharmaceutics 201177112

15. P Desai, R. R Patlolla, M Singh, Interaction of nanoparticles and cell-penetrating peptides with skin for transdermal drug delivery. Molecular Membrane Biology 201027724759

16. Chourasia R Jain SKDrug targeting through pilosebaceous route. Current Drug Targets 2009101095067

17. M. R Gasco, P Gasco, Nanovector. Nanomedicine (London) 20072695560

18. S Magdassi, Delivery systems in cosmetics. Colloids and Surfaces A 1997671 EOF679 EOF

19. B Luppi, T Cerchiara, F Bigucci, R Basile, V Zecchi, Polymeric nanoparticles composed of fatty acids and polyvinylalcohol for topical

application of sunscreens. Journal of Pharmacy and Pharmacology 200456407411

20. I. P Kaur, R Agrawal, Nanotechnology: a new paradigm in cosmeceuticals. Recent Patents in Drug Delivery Formulations 20071171182

21. R Alvarez-roman, G Barre, R. H Guy, H Fessi, Biodegradable polymer nanocapsules containing a sunscreen agent: preparation and photoprotection. European Journal of Pharmaceutics and Biopharmaceutics 200152191195

22. Escobar Chávez JJMerino-Sanjuan V, López-Cervantes M, et al. The tape stripping technique as a method for drug quantification in skin. Journal of Pharmacy and Pharmaceutical Sciences 2008111104130

23. R. O Potts, M. L Francoeur, The influence of stratum corneum morphology on water permeability. Journal of Investigative Dermatology 199196495499

24. R. O Potts, R. H Guy, Predicting skin permeability. Pharmaceutical Research 199295663669

25. P. M Ellias, Epidermal barrier function: intercellular lamellar lipid structures, origin, composition and metabolism. Journal of Controlled Release 199115199208

26. Escobar-Chávez JJ, Melgoza-Contreras LM, López-Cervantes M, et al. The tape stripping technique as a valuable tool for evaluating topical applied compounds. In: Gary W. Caldwell /Atta-ur-Rahman / Z.Yan / M. Iqbal Choudhary (Eds.) Frontiers in Drug Design & Discovery, The Netherlands: Bentham Science Publishers; 2009. Vol. 4, p189-227.

27. J Hadgraft, Skin, the final frontier. International Journal of Pharmaceutics 2001

28. R. H Guy, and J Hadgraft, Transdermal drug delivery. New York: Marcel Dekker, Inc., 1232003

29. A. M Nevill, The need to scale for differences in body size and mass: and explanation of Klieber´s 0.75 mass exponent. Journal of Applied Physiology199428702873

30. C. L Domínguez-delgado, I. M Rodríguez-cruz, M López-cervantes, The skin a valuable route for administration of drugs. In: José Juan Escobar-Chávez/Virginia Merino (Eds). Current Technologies To

Increase The Transdermal Delivery Of Drugs. The Netherlands: Bentham Science Publishers Ltd; 2010122

31. Escobar-Chávez, , Merino V, Díez-Sales O, Nácher-Alonso A,Ganem-Quintanar A, Herráez M, Merino-Sanjuán M. Transdermal nortriptyline hydrochloride patch formulated within a chitosan matrix intended to be used for smoking cessation. Pharmaceutical Development Technology 2011, 16 (2):162-9.

32. J. J Escobar-chávez, M López-cervantes, A Naïk, Y. N Kalia, D Quintanar-guerrero, Ganem Quintanar A. Applications of the thermoreversible Pluronic F-127 gels in pharmaceutical formulations, Journal of Pharmacy andPharmaceutical Sciences 200693339358

33. C. L Domínguez-delgado, I. M Rodríguez-cruz, J. J Escobar-chávez, I. O Calderón-lojero, D Quintanar-guerrero, A Ganem, Preparation and characterization of triclosan nanoparticles intended to be used for the treatment of acne. European Journal of Pharmaceutics and Biopharmaceutics 2011102 EOF107 EOF

34. J. Y Fang, Y. L Leu, Y. Y Wang, and Y. H Tsai, In vitro topical application and in vivo phramacodynamic evaluation of nonivamidehydrogels using Wistar rat as an animal model,European Journal of Pharmaceutical Sciences 2002155417423

35. S. C Shin, C. W Cho, I. J Oh, Effects of non ionic surfactants as permeation enhancers towards piroxicam from the poloxamer gel through rat skins, International Journal of Pharmaceutics 2001

36. J Liaw, Y-C. h Lin, Evaluation of poly(ethylene oxide)-poly(propylene oxide)- poly(ethylene oxide) (PEO-PPO-PEO) gels as a release vehicle for percutaneous fentanyl. Journal ofControlled Release 200068273282

37. Y. Y Wang, C. T Hong, W. T Chiu, J. Y Fang, In vitro and in vivo evaluations of topically applied capsaicin and nonivamide from hydrogels, International Journal of Pharmaceutics 200022412

38. D. A Mattorano, L. L Kupper, L. A Nylander-french, Estimating dermal exposure to jet fuel (naphthalene) using adhesive tape strip samples. The Annals of Occupational Hygiene 2004482139146

39. Y-C. h Chao, L. A Nylander-french, Determination of Keratin Protein in a Tape stripped Skin Sample from Jet Fuel Exposed Skin.The Annals of Occupational Hygiene 20044816573

40. A. W Kenneth, Dermatological and transdermal formulations, New York, USA: Marcel Dekker Inc; 2002

41. G. M Gray, R. J White, R. H Williams, H. J Yardley, Lipid composition of the superficial stratum corneum cells of the epidermis. British of Journal Dermatology 19821065963

42. P. W Wertz, D. T Downing, Ceramides of pig epidermis: structure determination. Journal of Lipid Research 19832475965

43. S. A Long, P. W Wertz, S. J Strauss, D. T Downing, Human stratum corneum polar lipids and desquamation. Archives in Dermatology Research 198527728487

44. P. W Wertz, M. C Miethke, S. A Long, J. S Strauss, D. T Downing, The composition of ceramides from human stratum corneum and from comedones. Journal of Investigative Dermatology 19858441012

45. B. C Melnik, J Hollmann, E Erler, B Verhoeven, G Plewig, Microanalytical screening of all major stratum corneum lipids by sequential high-performance thin-layer chromatography. Journal of Investigative Dermatology 19899223134

46. R. M Lavker, T Sun, Heterogeneity in epidermal basal keratinocytes: morphological and functional correlations. Science 1982215123941

47. L Borradori, A Sonnenberg, Structure and function of hemidesmosomes: more than simple adhesion complexes. Journal of Investigative Dermatology 199911241118

48. D Sawamura, K Li, M-L Chu, J Uitto, Human bullous pemphigoid antigen (BPAG1): amino acid sequences deduced from cloned cDNAs predict biologically important peptide segments and protein domains. Journal of Biological Chemistry 19912661778490

49. K Li, K Tamai, Tan EML, Uitto J. Cloning of type XVII collagen. Complementary and genomic DNA sequences of mouse 180-kDa bullous pemphigoid antigen (BPAG2) predict an interrupted collagenous domain, a transmembrane segment, and unusual features in the 5'-end of the gene and the 3'-untranslated region of mRNA. Journal of Biological Chemistry 1993268882534

50. M. A Stepp, S Spurr-michaud, A Tisdale, J Elwell, I. K. α Gipson, Integrin heterodimer is a component of hemidesmosomes. Proceedings of the National Academy of Sciences 199087897074

51. R. E Burgeson, A. M Christiano, The dermal-epidermal junction. Current Opinion in Cell Biology 1997965158

52. S. E Cross, M. S Roberts, Subcutaneous absorption kinetics of interferon and other solutes. Journal of Pharmacy andPharmacology 19934560609

53. J. W Melski, The anatomy and physiology of the skin. In:Principles and Practice of Skin Excisions. Amsterdam: IEJEMEC & GBEJEMEC (eds), 1996114

54. I Steinstrasser, H. P Merkle, Dermal metabolism of topically applied drugs: Pathways and models reconsidered. Pharmaceutica Acta Helvetiae 199570324

55. A Szuba, S. G Rockson, Lymphedema: anatomy, physiology and pathogenesis. Vascular Medicine 1997232126

56. N Cameli, M Picardo, C Perrin, Expression of integrins in human nail matrix. British Journal of Dermatology 199413058388

57. M. P Philpott, Defensins and acne. Molecular Immunology 200340457462

58. B Illel, Formulation for transfollicular drug administration: some recent advances. Critical Reviews in Therapeutic Drug Carrier Systems 19971420719

59. T. X Xiang, B. D Anderson, Influence of chain ordering on the selectivity of dipalmitoylphosphatidylcholine bilayer membranes for permeant size and shape. Biophysical Journal 199875265871

60. S Geinoz, R. H Guy, B Testa, P. A Carrupt, Quantitative structure-permeation relationships (QSPeRs) to predict skin permeation: a critical evaluation. Pharmaceutical Research 2004218392

61. G Cevc, Drug delivery across the skin. Expert Opinion on Investigational Drugs 19976188737

62. G Cevc, U Vierl, Nanotechnology and the transdermal route. A state of the art review and critical appraisal. Journal of Controlled Release 2010141327799

63. A Schätzlein, G Cevc, Non-uniform cellular packing of the stratum corneum and permeability barrier function of intact skin: a high-resolution confocal laser scanning microscopy study using highly

deformable vesicles (Transfersomes). British Journal of Dermatology 199813858392

64. S Mitragotri, Modeling skin permeability to hydrophilic and hydrophobic solutes based on four permeation pathways. Journal of Controlled Release 2003866992

65. V Aguilella, K Kontturi, L Murtomiiki, P Ramírez, Estimation of the pore size and charge density in human cadaver skin. Journal of Controlled Release 19943224957

66. M. E Johnson, D Blankschtein, R Langer, Evaluation of solute permeation through the stratumcorneum: lateral bilayer diffusion as the primary transport mechanism. Journal of Pharmaceutical Science 199786116272

67. R Guy, Transdermal drug delivery. 2a ed; Marcel Dekker. New York, USA, 2003

68. V. M Meidan, M Docker, A. D Walmsley, W. J Irwin, Low intensity ultrasound as a probe to elucidate the relative follicular contribution to total transdermal absorption. Pharmeceutical Research 1998158592

69. Ogiso T, Shiraki T, Okajima K, et al. Transfollicular drug delivery: penetration of drugs through human scalp skin and comparison of penetration between scalp and abdominal skins in vitro. Journal of Drug Targeting 2002; 10:369-78.

70. S Dokka, S. R Cooper, S Kelly, G. E Hardee, J. G Karras, Dermal delivery of topically applied oligonucleotides via follicular transport in mouse skin. Journal of Investigative Dermatology 200512497175

71. Y. Y Grams, L Whitehead, G Lamers, N Sturman, J. A Bouwstra, Online diffusion profile of a lipophilic dye in different depths of a hair follicle in human scalp skin. Journal of Investigative Dermatology 200512577582

72. U Jacobi, R Toll, W Sterry, J Lademann, Follicles play a role as penetration pathways in in vitro studies on porcine skin? An optical study. Laser Physics 200515159498

73. A Teichmann, M Ossadnik, H Richter, W Sterry, J Lademann, Semiquantitative determination of the penetration of a fluorescent hydrogel formulation into the hair follicle with and without follicular closure by microparticles by means of differential stripping. Skin Pharmacology and Physiology 2006191015

74. R Alvarez-román, A Naik, Y. N Kalia, R. H Guy, H Fessi, Skin penetration and distribution of polymeric nanoparticles. Journal of Controlled Release 2004995362

75. E. A Essa, M. C Bonner, B. W Barry, Human skin sandwich for assessing shunt route penetration during passive and iontophoretic drug and liposome delivery. Journal of Pharmacy and Pharmacology 200254148190

76. J. J Escobar-chávez, M López-cervantes, Ganem Rondero A. Conventional Methods for Cutaneous Drug Sampling. In: Narasimha Murthy (Ed.). Dermatokinetics of therapeutic agents. Boca Raton: CRC Press Taylor and Francis Group; 201181130

77. Teichmann A, Jacobi U, Ossadnik M, et al. Differential stripping: determination of the amount of topically applied substances penetrated into the hair follicles. Journal of Investigative Dermatology 2005; 125: 264-69.

78. A. V Prochazka, New developments in smoking cessation. Chest 2000Suppl 1:169S EOF175S EOF

79. M. F Wilkosz, R. H Bogner, Trandermal drug delivery. Part 1: Current status. US pharmacist 2003

80. http://wwwezinearticles.com/?Transdermal-Drug-Delivery,-Transdermal-Patches&id=155961 (accessed: 1May 2012

81. X Huang, Y Du, Yuan H & Hu F. Preparation and pharmacodynamics of low-molecular-weight chitosan nanoparticles containing insulin. Carbohydrate Polymers 200976368373

82. S Goswami, Bajpai J & Bajpai AK. Designing Gelatin Nanocarriers as a Swellable System for Controlled Release of Insulin: An In-Vitro Kinetic Study. Journal of Macromolecular Science 201047119130

83. I. M Rodríguez-cruz, C. L Domínguez-delgado, J. J Escobar-chávez, G Leyva-gómez, A Ganem-quintanar, D Quintanar-guerrero, Nanoparticle infiltration to prepare solvent-free controlled drug delivery system. International Journal of Pharmaceutics 2009371177181

84. Sonneville-Aubrun, , SimonnetJT&AlloretFL. Nanoemulsions: a new vehicle for skincare products. Advances in Colloid and Interface Science 2004; 108-109:145-149.

85. A. D Bangham, Liposomes: the Babraham connection. Chemistry and Physics of Lipids 199364275285

86. Rodriguez-Justo O & Moraes ÂMAnalysis of process parameters on the characteristics of liposomes prepared by ethanol injection with a view to process scale-up: Effect of temperature and batch volume.Chemical Engineering Research and Design 2011896785792

87. G Cevc, G Blume, Lipid vesicles penetrate into intact skin owing to the transdermal osmotic gradients and hydration force. Biochimica et Biophysica Acta 19921104226232

88. S Svenson, D. A Tomalia, Dendrimers in biomedical applications-reflections on the field. Advanced Drug Delivery Reviews 20055721062129

89. V. A Kabanov, A. B Zezin, V. B Rogacheva, Z. G Gulyaeva, M. F Zansochova, Joosten JGH & Brackman J. Polyelectrolyte behavior of astramol poly(propyleneimine) dendrimers. Macromolecules 1998311425144

90. A. J Almeida, E Souto, Solid lipid nanoparticles as a drug delivery system for peptides and proteins. Advanced Drug Delivery Reviews 200759478490

91. C. A García-gonzález, Sampaio da Sousa AR, Argemí A, López Periago A, Saurina J, Duarte CM&Domingo C. Production of hybrid lipid-based particles loaded with inorganic nanoparticles and active compounds for prolonged topical release. International Journal of Pharmaceutics 2009296304

92. G. M El Maghraby, Barry BW & Williams AC. Liposomes and skin: From drug delivery to model membranes. European Journal of Pharmaceutical Sciences 200834203222

93. A. R Menjoge, Kannan RM & Tomalia DA. Dendrimer-based drug and imaging conjugates: design considerations for nanomedical applications. Drug Discovery Today 201015171185

94. B. A Rzigalinski, J. S Strobl, Cadmium-containing nanoparticles: Perspectives on pharmacology and toxicology of quantum dots. Toxicology and Applied Pharmacology 2009238280288

95. Y Dan, H Liu, Gao W & Chen S. Activities of essential oils from Asarum heterotropoides var. mandshuricum against five phytopathogens. Crop Protection 201029295-299

96. M. C Dodla, R. V Bellamkonda, Differences between the effect of anisotropic and isotropic laminin and nerve growth factor presenting scaffolds on nerve regeneration across long peripheral nerve gaps. Biomaterials 2008293346

97. Elsayed MMAAbdallah OY, Naggar VF & Khalafallah NM. Deformable liposomes and ethosomes: Mechanism of enhanced skin delivery. International Journal of Pharmaceutics 20063226066

98. I Rojas-oviedo, R. A Salazar-lópez, Reyes-Gasga J & Quirino-Barreda CT. Elaboration and structural analysis of aquasomes loaded with Indomethacin. European Journal of Pharmaceutical Sciences 200732223230

99. Y Jin, L Tong, P Ai, Li M & Hou X. Self-assembled drug delivery systems: 1. Properties and in vitro/in vivo behavior of acyclovir self-assembled nanoparticles (SAN). International Journal of Pharmaceutics 2006309199207

100. F. J Rossier-miranda, Schroën CGPH & Boom RM. Colloidosomes: Versatile microcapsules in perspective. Colloids and Surfaces A: Physicochemical and Engineering Aspects 20093434349

101. M Hong, S Zhu, Y Jiang, Tang G & Pei Y. Efficient tumor targeting of hydroxycamptothecin loaded PEGylated niosomes modified with transferrin. Journal of Controlled Release 200913396102

102. Elnaggar YSREl-Massik MA & Abdallah OY. Self-nanoemulsifying drug delivery systems of tamoxifen citrate: Design and optimization. International Journal of Pharmaceutics 2009

103. J. C Kim, M. E Song, M. J Kim, E. J Lee, S. K Park, M. J Rang, H. J Ahn, Preparation and characterization of triclosan-containing vesicles. Colloids and Surfaces B 200226235

104. J Pardeike, A Hommoss, R. H Müller, Lipid nanoparticles (SLN, NLC) in cosmetic and pharmaceutical dermal products. International Journal of Pharmaceutics 2009366170

105. G Cevc, Lipid vesicles and other colloids as drug carriers on the skin, Advanced Drug Delivery Reviews 200456675

106. F Maestrelli, M Garcia-fuentes, P Mura, M. J Alonso, A new drug nanocarrier consisting of chitosan and hydoxypropylcyclodextrin,European Journal of Pharmaceuitics andBiopharmaceutics 20066379

107. R Alvarez-román, A Naik, Y. N Kalia, H Fessi, R. H Guy, Visualization of skin penetration using confocal laser scanning microscopy. European Journal of Pharmaceutics and Biopharmaceutics 200458301

108. T Ogiso, T Shiraki, K Okajima, T. T Tanino, M. M Iwaki, and T. T Wada, Transfollicular drug delivery: penetration of drugs through human scalp skin and comparison of penetration between scalp and abdominal skins in vitro. Journal of Drug Targeting 20021036978

109. U Jacobi, R Toll, W Sterry, J Lademann, Follicles play a role as penetration pathways in in vitro studies on porcine skin?. An optical study. Laser Physics Letters 200515159498

110. T Wilson, 1990Confocal Microscopy. Academic Press, London.

111. H. C Gerritsen, and C. J De Grauw, Imaging of optically thick specimen using two-photon excitation microscopy. Microscopy Research andTechnique 199947206209

112. W Denk, K. R Delaney, A Gelperin, D Kleinfeld, B. W Strowbridge, D. W Tank, and R Yuste, Anatomical and functional imaging of neurons using 2-photon laser scanning microscopy. Journal of Neuroscience Methods 199454151162

113. K Svoboda, W Denk, D Kleinfeld, and D. W Tank, In vivo dendritic calcium dynamics in neocortical pyramidal neurons. Nature 1997385161165

114. C Xu, and W. W Webb, Nonlinear and two-photon induced fluorescence. In:Topics in Fluorescence Spectroscopy. New York, USA: J. R. Lakowicz, editor, 1997

115. W. J Denk, J. H Strickler, and W. W Webb, Two-photon laser scanning fluorescence microscopy. Science 19902487376

116. T. H Tsai, S. H Jee, C. Y Dong, S. J Lin, Multiphoton microscopy in dermatological imaging.Journal Dermatological Science 20095618

117. J Paoli, M Smedh, M. B Ericson, Multiphoton laser scanning microscopy-a novel diagnostic method for superficial skin cancers. Seminars in Cutaneous Medicine and Surgery 200928190195

118. T Chen, R Langer, J. C Weaver, 1998Skin electroporation causes molecular transport across the stratum corneum through localized transport regions. Journal of Investigative Dermatology Symposium Proceedings 1998; 3215965

119. C. L Domínguez- Delgado, Estudio sobre el transporte a través de piel in vitro de triclosán, formulado en una dispersión de nanopartículas poliméricas, como alternativa para el tratamiento del acné. Efecto del laureato de sacarosa como promotor de absorción. Informe de investigación de Maestría. Posgrado en Ciencias Químicas, UNAM, 2008

120. D Thiboutot, A Zaenglein, J Weiss, G Webster, B Calvarese, D Chen, An aqueous gel fixed combination of clindamycin phosphate 1.2% and benzoyl peroxide 2.5% for the once-daily treatment of moderate to severe acne vulgaris: assessment of efficacy and safety in 2813 patients. Journal of American Academy of Dermatology 200859792800

121. Y. W Chien, 1987Transdermal therapeutic systems. In: Controlled Drug Delivery Fundamentals and Applications. Robinson J. R., Lee V. H. L, editors. New York, USA: Marcel Dekker, Inc. 523549

122. U Schroeder, P Sommerfeld, S Ulrich, B. A Sabel, Nanoparticle technology for delivery of drugs across the blood-brain barrier. Journal of Pharmaceutical Science 19988713051307

123. R. S Raghuvanshi, Y. K Katare, K Lalwani, M. M Ali, O Singh, and A. K Panda, Improved immune response from biodegradable polymer particles entrapping tetanus toxoid by use of different immunization protocol and adjuvants. International Journal of Pharmaceutics 2002245109121

124. J. C Leroux, E Allemann, F De Jaeghere, E Doelker, R Gurny, Biodegradable nanoparticles-from sustained release formulations to improved site specific drug delivery. Journal of Controlled Release 1996

125. L Brannon-peppas, J. O Blanchette, Nanoparticle and targeted systems for cancer therapy, Advanced Drug Delivery Reviews 20045616491659

126. R Haag, Supramolecular drug-delivery systems based on polymeric core-shell architectures. Angewandte Chemie International Edition 200443278282

127. M. R Radowski, A Shukla, H Von Berlepsch, C Bottcher, G Pickaert, H Rehage, R Haag, Supramolecular aggregates of dendritic multishell architectures as universal nanocarriers. Angewandte Chemie International Edition 20074612651269

128. S. S Tinkle, J. M Antonini, B. A Rich, J. R Roberts, R Salmen, K Depree, E. J Adkins, Skin as a route of exposure and sensitization in chronic

beryllium disease. Environmental Health Perspectives 20031112021208

129. S Kim, Y. T Lim, E. G Soltesz, A. M De Grand, J Lee, A Nakayama, Near-Infrared fluorescent type II quantum dots for sentinel lymph node mapping. Nature Biotechnology 2004229397

130. Z Symon, A Peyser, D Tzemach, O Lyass, E Sucher, E Shezen, A Gabizon, Selective delivery of doxorubicin to patients with breast carcinoma metastases by stealth liposomes. Cancer 1999867278

131. A Gonçalves, A. C Braud, F Viret, D Genre, G Gravis, C Tarpin, Phase I study of pegylated liposomal doxorubicin (Caelyx) in combination with carboplatin in patients with advanced solid tumors. Anticancer Research 20032335433548

132. M. V Seiden, F Muggia, A Astrow, U Matulonis, S Campos, M Roche, A phase II study of liposomal lurtotecan (OSI-211) in patients with topotecan resistant ovarian cancer. Gynecologic Oncology 200493229232

133. El Maghraby GMMWilliams AC, Barry BW. Can drug-bearing liposomes penetrate intact skin? Journal of Pharmacy and Pharmacology 200658415429

134. N Weiner, N Williams, G Birch, C Ramachandran, C Shipman, G Flynn, Topical delivery of liposomally encapsulated interferon evaluated in a cutaneous herpes guinea pig model. Antimicrobial Agents and Chemotherapy 19893312171221

135. M. E Planas, P Gonzalez, L Rodriguez, S Sanchez, G Cevc, Noninvasive percutaneous induction of topical analgesia by a new type of drug carrier, and prolongation of local pain insensitivity by anesthetic liposomes. Anesthesia and Analgesia 199275615621

136. M Sentjurc, V Gabrijelcic, Transport of liposome-entrapped molecules into the skin as studied by electron paramagnetic resonance imaging methods. In: Non-Medical Application of liposomes. Lasic B, editor. New York, USA; CRC Press.199591114

137. G Cevc, D Gebauer, J Stieber, A Schatzlein, G Blume, 1998Ultraflexible vesicles, Transfersomes, have an extremely low pore penetration resistance and transport therapeutic amounts of insulin across the intact mammalian skin. Biochimica et Biophysica Acta 1998; 1368201215

138. A Paul, G Cevc, B. K Bachhawat, Transdermal immunisation with an integral membrane component, gap junction protein, by means of ultradeformable drug carriers, transfersomes. Vaccine 199816188195

139. Van den Bergh BABouwstra JA, Junginger HE, Wertz PW. Elasticity of vesicles affects hairless mouse skin structure and permeability. Journal of Controlled Release 199962367379

140. J Guo, Q Ping, G Sun, C Jiao, Lecithin vesicular carriers for transdermal delivery of cyclosporin A. International Journal of Pharmaceutics 2000194201207

141. J Guo, Q Ping, L Zhang, Transdermal delivery of insulin in mice by using lecithin vesicles as a carrier. Drug Delivery 20007113116

142. K Vrhovnik, J Kristl, M Sentjurc, J Smid-korbar, Influence of liposome bilayer fluidity on the transport of encapsulated substances into the skin, studied by EPR. Pharmaceutical Research 199815525530

143. K Egbaria, C Ramachandran, N Weiner, Topical application of liposomally entrapped cyclosporin evaluated by in vitro diffusion studies with human skin. Skin Pharmacology 199142128

144. G Cevc, A Schatzlein, H Richardsen, Ultradeformable lipid vesicles can penetrate the skin and other semi-permeable barriers unfragmented. Evidence from double label CLSM experiments and direct size measurements. Biochimica et Biophysica Acta 200215642130

145. P. L Honeywell-nguyen, A De Graaff, H. E Junginger, J. A Bouwstra, Interaction between elastic and rigid vesicles with human skin in vivo. Proceedings International Symposium of Controlled Release Bioactive Materials 200027237238

146. R Esfand, D. A Tomalia, Poly(amidoamine) (PAMAM) dendrimer: from biomimicry to drug delivery and biomedical applications. Drug Discovery Today 20016427436

147. D Emanuele, A Attwood, D. Dendrimerdrug interactions. Advanced Drug Delivery Reviews 20055721472162

148. H. S Parekh, The Advance of Dendrimers- A Versatile Targeting Platform for Gene/Drug Delivery. Current Pharmaceutical Design 20071328372850

149. P Niederhafner, J Šebestík, J Ježek, Peptide dendrimers. Journal of Peptide Science 200511757788

150. M. J Cloninger, Biological applications of dendrimers. Current Opinion in Chemical Biology 20026742748

151. I. L Medintz, H. T Uyeda, E. R Goldman, H Mattoussi, Quantum dot bioconjugates for imaging, labeling and sensing. Nature Materials 20054435446

152. S. D Caruthers, S. A Wickline, G. M Lanza, Nanotechnological applications in medicine. Current Opinion in Biotechnology 2007182630

153. R Arshady, 1999Microspheres, microcapsules, & liposomes. London: Citus books.

154. J Lademann, , H Richter, , A Teichmann, , N Otberg, , U Blume-Peytavi, , J Luengo, , B Weiss, , UF Schaefer, , CM Lehr, , Wepf, , Sterry W. Nanoparticles-an efficient carrier for drug delivery into the hair follicles. European Journal of Pharmaceutics and Biopharmaceutics 2007; 66:159-64.

155. M. A Bolzinger, S Briançon, J Pelletier, Y Chevalier, Penetration of drugs through skin, a complex rate-controlling membrane. Current Opinion in Colloid and Interface Science 2012;

156. R. H Müller, M Radtke, S. A Wissing, Solid lipid nanoparticles (SLN) and nanostructured lipid carriers (NLC) in cosmetic and dermatological preparations. Advance Drug Delivery Reviews 2002S131S155.

157. Z Mei, H Chen, T Weng, Y Yang, X Yang, Solid lipid nanoparticle and microemulsion for topical delivery of triptolide. European Journal of Pharmaceutics andBiopharmaceutics 200356189196

158. W Liu, M Hu, W Liu, C Xue, H Xụ, X Yang, Investigation of the carbopol gel of solid lipid nanoparticles for the transdermal iontophoretic delivery of triamcinolone acetonide acetate. International Journal of Pharmaceutics 2008364135141

159. E Ugazio, R Cavalli, M. R Gasco, Incorporation of cyclosporin A in solid lipid nanoparticles (SLN). International Journal of Pharmaceutics 2002241341344

160. N. P Aditya, S Patankar, B Madhusudhan, Murthy RSR, Souto EB. Arthemeter-loaded lipid nanoparticles produced by modified thin-film hydration: Pharmacokinetics, toxicological and in vivo anti-malarial activity. European Journal ofPharmaceutical Science 201040448455

161. V Sanna, G Caria, A Mariani, Effect of lipid nanoparticles containing fatty alcohols having different chain length on the ex vivo skin permeability of Econazole nitrate. Powder Technology 20102013236

162. M Joshi, V Patravale, Nanostructured lipid carrier (NLC) based gel of celecoxib. International Journal of Pharmaceutics 2008346124132

163. J Pardeike, R. H Müller, Q Coenzyme, loaded NLCs: preparation, occlusion properties and penetration enhancement (Cutanova Cream NanoRepair Q10). Pharmaceutical Technology Europe 2007194649

164. V Teeranachaideekul, E. B Souto, V B Junyaprasert, R. H Müller, Cetyl palmitate-based NLC for topical delivery of Coenzyme Q10-Development, physicochemical characterization and in vitro release studies. European Journal of Pharmaceutical Science 200767141148

165. A Dingler, Feste Lipid-Nanopartikel als kolloidale Wirstoffträgersysteme zur dermalen Applikation. Institut für Pharmazie. Freie Universität, Berlin, 1998

166. C Puglia, R Filosa, A Peduto, P De Caprariis, L Rizza, F Bonina, P Blasi, Evaluation of alternative strategies to optimize ketorolac transdermal delivery. AAPS Pharmaceutical Science Technology 2006Article 64).

167. V Teeranachaideekul, R. H Mülle, V. B Junyaprasert, Encapsulation of ascorbyl palmitate in nanostructured lipid carriers (NLC)-effects of formulation parameters on physicochemical stability. International Journal of Pharmaceutics 2007340198206

168. V Jenning, Solid Lipid Nanoparticles (SLN) as a Carrier System for the Dermal Application of Retinol. Free University of Berlin, Berlin, 1999

169. V Jenning, S. H Gohla, Encapsulation of retinoids in solid lipid nanoparticles (SLN). Journal of Microencapsulation 200118149158

170. J. P Jee, S. J Lim, J. S Park, C. K Kim, Stabilization of all-trans retinol by loading lipophilic antioxidants in solid lipid nanoparticles. European Journal of Pharmaceutics and Biopharmaceutics 200663134139

171. H Tsujimoto, and K Hara, Application 21-Development of functional skincare cosmetics using biodegradable PLGA nanospheres. Nanoparticle Technology Handbook (Second Edition), 2012501506

172. A Patzelt, H Richter, F Knorr, U Schäfer, C. M Lehr, L Dähne, W Sterry, J Lademann, Selective folicular targeting by modification of the particle sizes. Journal of Controlled Release 2011150458

173. J Shim, K. H Seok, W Park, S Han, J Kim, I Chang, Transdermal delivery of mixnoxidil with block copolymer nanoparticles. Journal of Controlled Release 200497477484

174. E. D Cetin, E Savk, M Uslu, M Eskin, A Karul, Investigation of the inflammatory mechanisms in alopecia areata. The American Journal of Dermatopathology 2009315360

175. A. L Zaenglein, E. M Graber, D. M Thiboutot, J. S Strauss, K Wolff, L. A Goldsmith, S. I Katz, B Gilchrest, A. S Paller, D. J Leffell, editors., Chapter 78. Acne vulgaris and acneiform eruptions. 2009Fitzpatrick's Dermatology in General Medicine. 7e: http://www.accessmedicine.com/content.aspx?aID=2963025

176. H Gollnick, W Cunliffe, D Berson, B Dreno, A Finlay, J. J Leyden, A. R Shalita, D Thiboutot, Management of acne: a report from the global alliance to improve outcomes in acne. Journal of the American Academy of Dermatology 2003S1S37.

177. R. A Bojar, K. T Holland, Acne and Propionibacterium acnes. Clinical Dermatology 2004223759

178. A. F Ourique, A. R Pohlmann, S. S Guterres, Beck RCR. Tretinoin-loaded nanocapsules: Preparation, physicochemical characterization, and photostability study. International Journal of Pharmaceutics 200835214

179. P Lee, S Peng, C Su, F Mi, H Chen, M Wei, H. J Lin, H. W Sung, The use of biodegradable polymeric nanoparticles in combination with a low-pressure gene gun for transdermal DNA delivery. Biomaterials 200829742751

180. M. J Santander-ortega, T Stauner, B Loretz, J. L Ortega-vinuesa, D Bastos-gonzález, G Wenz, U. F Schaefer, C. M Lehr, Nanoparticles made from novel starch derivatives for transdermal drug delivery. Journal of Controlled Release 20101418592

181. A. J Thote, R. B Gupta, Formation of nanoparticles of a hydrophilic drug using supercritical carbon dioxide and microencapsulation for sustained release. Nanomedicine:Nanotechnology 200518590

182. P. A Mccarron, M Hall, Incorporation of novel 1-alkylcarbonyloxymethyl prodrugs of 5-fluorouracil into poly(lactide-co-glycolide) nanoparticles. International Journal of Pharmaceutics 2008348115124

183. H Lboutounne, J Chaulet, C Ploton, F Falson, F Pirot, Sustained ex vivo skin antiseptic activity of chlorhexidine in poly(ε-caprolactone) nanocapsule encapsulated form and as a digluconate. Journal of Controlled Release 200282319334

184. J Fuchs, T. M Zollner, R Kaufmann, M Podda, Redox-modulated pathways in inflammatory skin diseases. Free Radical Bioogy and Medicine 20013033753

185. A. M Sanfilippo, V Barrio, C Kulp-shorten, J. P Callen, Common pediatric and adolescent skin conditions. Journal of Pediatric and Adolescent Gynecology 20031626983

186. M Schäfer-korting, W Mehnert, H. C Korting, Lipid nanoparticles for improved topical application of drugs for skin diseases. Advanced Drug Delivery Reviews 200759427443

187. A. M Layton, Acne vulgaris and similar eruptions. Medicine 2005334448

188. Zouboulis ChCSebaceous glands, acne and related disorders: basic and clinical research, clinical entities and treatment. Journal of Investigative Dermatology 199710837198

189. M. C Santos, W Mehnert, M Schaller, H. C Korting, A Gysler, A Haberland, M Schafer-korting, Drug targeting by solid lipid nanoparticles for dermal use. Journal of Drug Targeting 200210489495

190. K. A Shah, A. A Date, M. D Joshi, V. B Patravale, Solid lipid nanoparticles (SLN) of tretinoin: potential in topical delivery. International Journal of Pharmaceutics 2007345163171

191. G Cevc, D Gebauer, J Stieber, A Schatzlein, G Blume, Ultraflexible vesicles, Transfersomes, have an extremely low pore penetration resistance and transport therapeutic amounts of insulin across the intact mammalian skin. Biochimica et Biophysica Acta 19981368201215

192. Z Wang, Z Itoh, Y Hosaka, I Kobayashi, Y Nakano, I Maeda, F Umeda, J Yamakawa, M Kawase, K Yag, Novel Transdermal Drug Delivery System with Polyhydroxyalkanoate and Starburst Polyamidoamine Dendrimer. Journal of Bioscience and Bioengineering 200395541543

193. A. S Chauhan, S Sridevi, K. B Chalasani, A. K Jain, S. K Jain, N. K Jain, P. V Diwan, Dendrimer-mediated transdermal delivery: Enhanced bioavailability of indomethacin. Journal of Controlled Release 200390335343

194. C Yiyun, M Na, X Tongwen, F Rongqiang, W Xueyuan, W Xiaomin, W Longping, Transdermal delivery of nonsteroidal anti-inflammatory drugs mediated by polyamidoamine (PAMAM) dendrimer. Journal ofPharmaceutical Sciences 200796595602

195. Venuganti VVKPerumal OP. Effect of poly(amidoamine) (PAMAM) dendrimer on skin permeation of 5-fluorouracil. International Journal of Pharmaceutics 2008361230238

196. P. A Cornwell, B. W Barry, The routes of penetration of ions and 5-fluorouracil across human skin and the mechanisms of action of terpene skin penetration enhancers. InternationalJournal ofPharmaceutics 199394189194

197. T Tsuji, T Sugai, Topically administered fluorouracil in psoriasis. Archives in Dermatology 1975105208212

198. D. K Goette, Topical chemotherapy with 5-fluorouracil. A review. Journal of American Academy of Dermatology 19814633649

199. H. D Beall, K. B Sloan, Topical delivery of 5-fluorouracil (5-FU) by 1, 3-bisalkylcarbonyl-5-FU prodrugs. International Journal of Pharmaceutics 20022314349

200. S Gao, J Singh, Effect of oleic acid/ethanol and oleic acid/propylene glycol on the in vitro percutaneous absorption of 5-fluorouracil and tamoxifen and the macroscopic barrier property of porcine epidermis. International Journal of Pharmaceutics 19981654555

201. V. M Meidan, A. D Walmsley, M. F Docker, W. J Irwin, Ultrasound-enhanced diffusion into coupling gel during phonophoresis of 5-fluorouracil. International Journal of Pharmaceutics1999185205213

202. V Merino, A Lopez, Y. N Kalia, R. H Guy, Electrorepulsion versus electroosmosis: effect of pH on the iontophoretic flux of 5-fluorouracil. Pharmaceutical Research 199916758761

203. W. R Lee, S. C Shen, K. H Wang, C. H Hu, J. Y Fang, The effect of laser treatment on skin to enhance and control transdermal delivery of 5-fluorouracil. Journal of Pharmaceutical Science 20029116131626

204. F Kuo, B Subramanian, T Kotyla, T. A Wilson, S Yoganathan, R. J Nicolosi, Nanoemulsions of an anti-oxidant synergy formulation containing gamma tocopherol have enhanced bioavailability and anti-inflammatory properties. International Journal of Pharmaceutics 2008363206213

205. F Shakeel, W Ramadan, Transdermal delivery of anticancer drug caffeine from water-in-oil nanoemulsions. Colloid Surface B 201075356362

206. H Wu, C Ramachandran, A. U Bielinska, K Kingzett, R Sun, N. D Weiner, B. J Roessler, Topical transfection using plasmid DNA in a water-in-oil nanoemulsion. International Journal of Pharmaceutics 20012212334

207. B Subramanian, F Kuo, E Ada, T Kotyla, T Wilson, S Yoganathan, R Nicolosi, Enhancement of anti-inflammatory property of aspirin in mice by a nano-emulsion preparation. International Immunopharmacology 2008815331539

208. D Mou, , H Chen, , D Du, , C Mao, , J Wan, , H Xu, , Yang, . Hydrogel-thickened nanoemulsion system for topical delivery of lipophilic drugs. International Journal of Pharmaceutics 2008; 353:270-276.

209. H Wu, C Ramachandran, N. D Weiner, B. J Roessler, Topical transport of hydrophilic compounds using water-in-oil nanoemulsions. International Journal of Pharmaceutics 20012206375

210. M. P Alves, A. L Scarrone, M Santos, A. R Pohlmann, S. S Guterres, Human skin penetration and distribution of nimesulide from hydrophilic gels containing nanocarriers. International Journal of Pharmaceutics 2007341215220

Chapter 7

Planar Bioadhesive Microdevices: A New Technology For Oral Drug Delivery

Cade B. Fox,[1] Hariharasudhan D. Chirra,[1] and Tejal A. Desai[1,*]

[1]*Department of Bioengineering and Therapeutic Sciences, University of California, San Francisco, CA 94158, U.S.A*

ABSTRACT

The oral route is the most convenient and least expensive route of drug administration. Yet, it is accompanied by many physiological barriers to drug uptake including low stomach pH, intestinal enzymes and transporters, mucosal barriers, and high intestinal fluid shear. While many drug delivery systems have been developed for oral drug administration, the physiological components of the gastro intestinal tract remain formidable barriers to drug uptake. Recently, microfabrication techniques have been applied to create micron-scale devices for oral drug delivery with a high degree of control over microdevice size, shape, chemical composition, drug release profile, and targeting ability. With precise control over device properties, microdevices can be fabricated with characteristics that provide increased adhesion for prolonged drug exposure, unidirectional release which serves to avoid luminal drug loss and enhance drug permeation, and protection of a drug payload from the harsh environment of the intestinal tract. Here we review the recent developments in microdevice technology and discuss the potential of these devices to overcome unsolved challenges in oral drug delivery.

Keywords: Bio-MEMS; Microdevices; Microfabrication; Micropatches; Nanofabrication; Oral drug delivery

INTRODUCTION

Among the various routes of drug administration, including intravenous, intraperitoneal, intramuscular, and transdermal administration, oral administration is most preferred for its numerous advantages. Oral drug formulations are self-administrable and less invasive, leading to higher patient compliance and decreased cost of care. While oral drug administration is ideal in terms of cost and convenience, the oral bioavailability of many drugs is limited by a unique set of barriers to oral drug uptake, including low pH of the stomach, intestinal enzymes, transporter proteins expressed in intestinal epithelial cells, and the motile mucosal lining of the gastrointestinal (GI) tract. Many small molecule drugs suffer from poor oral bioavailability as a result of factors such as low permeation through the thick and hydrophobic mucosal layer and cell membrane, low drug solubility, and metabolic and transporter protein activity [1–2]. In addition, many recently developed biotherapeutics, including peptides, proteins, DNA, RNA, and macromolecules, have particularly low oral bioavailabilities [3]. This is due to their relatively large size and high complexity, causing decreased permeability and a tendency to denature under harsh conditions [4].

While there are many barriers to oral drug uptake, these barriers can be categorized into issues of 1) low drug permeation, 2) drug degradation, and 3) low drug solubility [5]. Current approaches to overcoming these barriers include enteric coating, permeation enhancers, drug modification, metabolic and transporter protein inhibitors, and micro-/nanoparticulate systems, which have been reviewed in detail before [6–12]. While these systems have made advances in improving oral bioavailability of select therapeutics, no reliable method for the oral delivery of biomolecules has been developed, and no single system has been shown to comprehensively address the three categories of barriers to drug uptake previously mentioned. Thus, more advanced approaches are required to provide sufficient systemic uptake of orally administered drugs ranging from small molecules to macromolecular structures.

In the past decade, micro-electro-mechanical systems (MEMS) technology originally developed by the semiconductor industry has been applied to many biomedical applications. Microfabrication techniques have facilitated the development of novel devices with precise control over device shape and size, allowing for the creation of devices specifically designed to address issues affecting drug uptake. Specifically, microfabricated devices for oral drug delivery, termed microdevices, are designed to maximize residence time in the GI tract, provide unidirectional drug release toward the intestinal epithelium, and release drug in a sustained manner. Together, these characteristics allow microdevices to simultaneously address issues with drug permeation, solubility, and degradation in a manner not provided by other drug delivery systems. This mini-review will highlight the unique mechanisms by which microdevices address the barriers to oral drug delivery, describe recent innovations in the design and fabrication of these devices, and review the efficacy of these devices *in vitro* and *in vivo*.

Barriers to oral drug delivery

The comprehensive set of barriers to oral drug uptake must be considered when examining the rationale behind microdevice design. Orally administered drugs face a sequential set of barriers to systemic drug uptake as outlined in Figure 1A. Following oral administration, drugs encounter pH values ranging from 1.5 to 3.5 and digestive enzymes in the stomach and are subsequently exposed to pH values of 5 to 7 and additional proteolytic and metabolic enzymes in the small intestine [13]. After entering the small intestine, the primary site of drug and nutrient uptake, drugs must then pass through a hydrophobic mucous membrane composed of a motile layer moving in contact with an underlying firmly adherent layer [14]. The motile mucus layer ranges between 100 and 500 μm in thickness, and the adherent mucus layer ranges from 0 to 20 μm in thickness [15–16]. After penetrating the mucus layer, drugs must pass through the glycocalyx, an extracellular matrix approximately 0.5 to 1 μm thick composed of negatively charged glycoproteins, proteoglycans, glycosaminoglycans, and glycolipids [17–18]. Drugs must then pass through the polarized enterocyte monolayer by either paracytosis directly through enterocytes or transcytosis through junctions between enterocytes. Paracytosis involves permeation through the apical cell membrane into enterocytes and subsequent permeation through the basal cell membrane into the interstitium. Paracytosis occurs through both passive diffusion through the cell membrane and facilitated diffusion involving transporter proteins or endocytosis [19–21]. Within the cytosol of enterocytes, drugs are exposed to influx and efflux protein transporters differentially expressed on the apical and basal cell membranes as well as metabolic enzymes [1]. The alternate pathway of transcytosis involves travel between cells through tight junctions, structures between closely associated cells composed of multiprotein complexes with pores approximately 1 to 3 nm in diameter [22–23]. This small pore size presents a significant obstacle to drug uptake, particularly for high molecular weight therapeutics. These barriers present a unique set of challenges not encountered in other routes of drug administration.

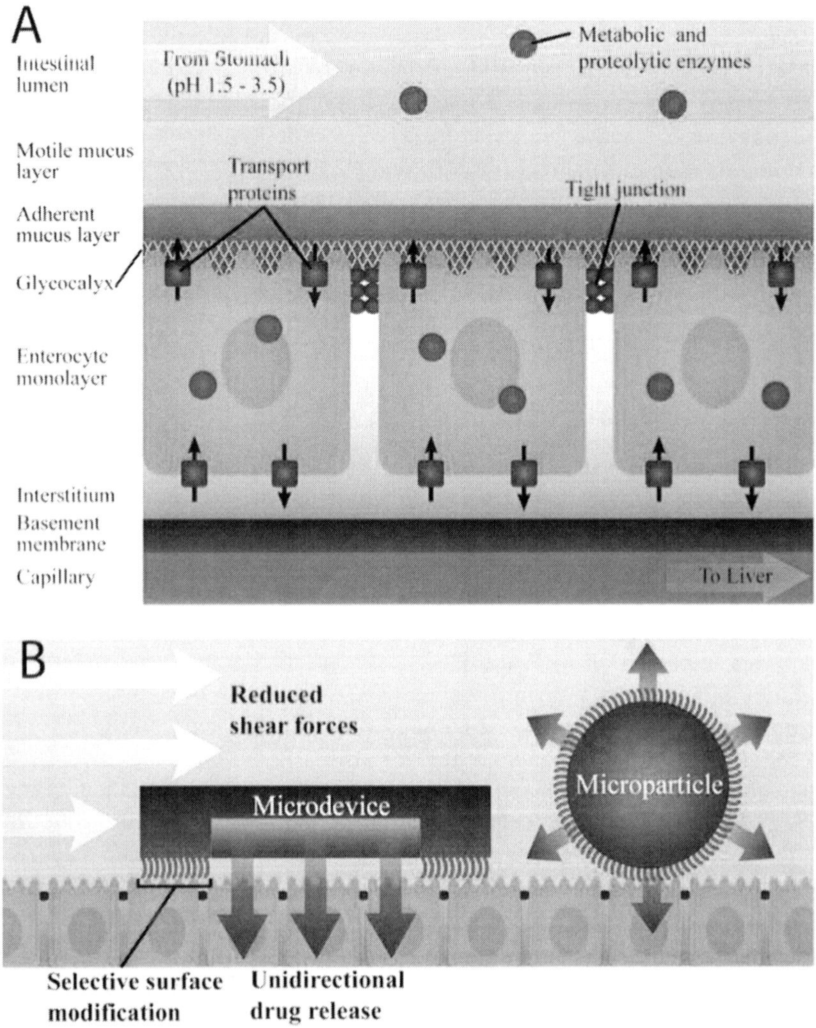

Figure 1. A. Physiological barriers to oral drug delivery. After encountering digestive enzymes and the low pH of the stomach, drugs enter the small intestine, the primary site of drug uptake where drugs encounter additional metabolic and proteolytic enzymes. Drugs **...**

Rationale for designing oral drug delivery microdevices

Like most previously developed oral drug delivery microparticulate systems made via precipitation methods, microdevices are designed on a scale small enough to fit within the features of the intestinal wall, which is made up of micron-sized folds and pits of the intestinal villi [24]. Also, microdevices are designed to be large enough to prevent device uptake into cells through endocytosis. While microdevices are similar in size to many oral drug delivery particulate systems, conventional methods of microfabrication deliver precise and consistent dimensions of microdevices, resulting in much higher monodispersity in size and shape [25]. In addition to providing monodispersity, microfabrication also provides the ability to create devices with custom shapes and dimensions. This abililty has been utilized to design devices with a planar design that simultaneously address drug permeability barriers, drug degradation, and low drug solubility (Figure 1B).

To address issues with poor drug solubility and increase overall drug exposure, microdevices are designed to be relatively flat, providing multiple advantages for drug transfer. A flat shape increases the surface area in contact with the GI wall, improving adhesive properties of the device [26]. In addition, a flat microdevice shape decreases the shear force per mass on the devices as shown in Figure 1B, preventing detachment of the device from the intestinal epithelium and further increasing residence time [27]. To overcome a second major barrier to oral drug uptake - issues with drug permeability, microdevices are designed with reservoirs on only one side of the device, allowing drug to be released in only one direction. In addition to asymmetric shape, devices can be asymmetrically modified with targeting moieties, mucoadhesive materials, and micro- and nanotopography, providing selective binding of the device side from which drug will be selectively released [26]. Thus, microdevices are designed to adhere to the mucosal or GI epithelial layer and release drug to enterocytes in a proximal, unidirectional manner as shown in Figure 1B. Releasing drug directly toward the epithelial barrier rather than into the lumen provides a more efficient mechanism of drug release by decreasing the loss of drug downstream through the lumen and also increasing the exposure of the enterocytes to the drug. Furthermore, the unidirectional release of drug in a localized, high concentration at the device-intestinal wall interface creates a strong concentration gradient, thereby enhancing drug permeation across the intestinal enterocytes [27]. This localized release of drug in high concentrations may also increase drug uptake through a second mechanism, as high drug concentrations may saturate metabolic enzymes and protein transporters, in turn increasing the bioavailability of the drug [28].

Numerous treatments have been studied for the disruption of drug transporters, metabolic enzymes, and endothelial tight junctions to allow for increased passage of drug [10–12]. However, each of these mechanisms is accompanied by severe clinical side effects [1, 23, 29]. If the total surface area of administered microdevices is significantly smaller than the surface area of the GI tract, microdevices have the potential to deliver drug to select regions of intestinal tissue. While drugs delivered in standard tablet form will diffuse to reach enterocytes with a relatively even distribution within a region of the GI tract, larger microdevices with high residence times have potential to only deliver drug to a small subset of intestinal epithelial cells. This could be advantageous in combination therapy for increased drug bioavailability. Given their potential to target a small subset of cells with high drug concentrations, microdevices could be designed to release permeation enhancers in localized regions in which the drug of interest is also released while leaving the majority of the GI tract unaffected. Finally, to address the third major barrier to oral drug uptake - drug degradation, microdevices are fabricated to include drug reservoirs that allow for sustained release of drug, thereby decreasing the exposure of drugs to harsh conditions of the GI relative to a bolus dose [26]. With a variety of sustained drug release systems developed in the recent past for oral delivery including pH-sensitive hydrogels [30–32], enteric coating [33], and degradable polymers, well established microfabrication techniques can be effectively used to incorporate microdevice reservoirs with these drug systems [27, 34–36].

In addition to delivery of drugs for systemic uptake, oral microdevices have the potential to treat diseases local to the GI tract including Crohn's disease, inflammatory bowel disease (IBD), and irritable bowel syndrome (IBS). While the site of action of most therapeutics for GI disorders is in intestinal tissue, many of these drugs in conventional large doses lead to severe systemic side effects [37–38]. For more efficient treatment of these diseases, microdevices could be modified to adhere to only the diseased GI tissue for localized delivery of drugs directly to the therapeutic target [39]. Direct targeting of sites of inflammation could improve drug efficacy while reducing severity of side effects associated with therapies for diseases of the GI tract.

Materials utilized for microdevice structure

To prevent toxicity and inflammation, microdevices must be made from biocompatible material. The first systems of oral microdevices used standard semiconductor materials, including silicon oxide and porous silicon as the device material [40–41]. While silicon and silicon oxide have been found to be relatively non-toxic in some studies [42–44], they have also been associated with inflammation [45–46]. In order to overcome this issue, microdevice fabrication shifted towards the use of relatively non-toxic polymers, including hydrogels and biodegradable materials. Poly(methyl methacrylate) (PMMA), an FDA-approved polymeric material [47] used in contact lenses and bone cement and also known to be stable at low pH values [26], has been utilized in numerous oral microdevice designs [34–35, 48–51]. Microdevices have also been fabricated from SU-8, an epoxy-based negative photoresist originally developed as an ultra-thick photoresist [36, 52]. While SU-8 is not currently FDA approved, studies have shown that SU-8 is non-toxic as an implantable material [53–55]. Other biocompatible polymers utilized for microdevice fabrication include chitosan [56], gelatin [57], poly(lactic-co-glycolic) acid (PLGA) [56–58], polypropylene (PP) [59], and poly(ethylene glycol) (PEG) [27, 52, 56]. The intrinsic biocompatibility, biodegradation, hydrophobicity, and structural properties of individual polymers can be tuned by adjusting the chemical structure of the monomer(s) used in polymer synthesis, the molecular weight of the polymer, and/or the crosslinking density [56, 60–62].

While materials utilized in microfabrication are typically chosen for biocompatibility and structural properties, there is a strong interest in utilizing bioresponsive or bioadhesive materials for additional properties to aid in drug delivery. The Peppas Lab has developed pH-sensitive hydrogel systems for selective insulin release in the small intestine [30–32]. These hydrogels retain insulin in low pH environments similar to that of the stomach but swell and release insulin as pH values increase to a simulated intestinal environment with a pH of 6.5 [31]. Similar pH-responsive materials could be incorporated into planar microdevices for selective drug release upon entry into the small intestine. In addition, microdevices could be surface coated with or encapsulated within enteric materials insoluble at low stomach pH values but soluble in the nearly neutral pH of the small intestine to prevent drug release and degradation [33].

Techniques utilized in the micro- and nanofabrication of oral drug delivery devices

A variety of fabrication methods, including photolithography, electron beam lithography, x-ray lithography, and soft lithography techniques are available for the fabrication of microdevices. The use of micro- and nanofabrication techniques for biological applications has been reviewed in detail elsewhere [63–65], but this review will highlight a selection of techniques that are particularly useful for oral microdevice fabrication. Many studies to this date have utilized conventional photolithography techniques originally developed by the microchip industry for the fabrication of microdevices [27, 34, 36, 40, 50–52, 62, 64–67]. Photolithography involves selective UV exposure of a photosensitive material, termed a photoresist, and is often followed by an etching step to transfer the photoresist pattern to a substrate. Typically, the substrate is spin-coated to form a thin film deposition, and a photoresist layer is spin-coated onto the substrate. The photoresist is then exposed to UV light through a mask created with custom patterns by computer-aided design (CAD), transferring the mask pattern through selective polymerization or cleavage of the photoresist in regions exposed to UV light, and non-polymerized or cleaved resist is removed by chemical development. The resist pattern can then be transferred to the substrate through either wet or dry etching processes with the resist acting to selectively protect regions of the substrate. In a straightforward fabrication technique, Tao et al. applied photolithography to fabricate microdevices from SU-8 in a two-step process [30]. A layer of SU-8 was exposed to UV light to form a device base, and then a second layer of SU-8 was UV-exposed to form the walls of drug reservoirs. The use of SU-8 photoresist as the structural component of microdevices eliminates the need for etching of a substrate material following UV exposure. To create microdevices from non-photoreactive materials, photolithography followed by reactive ion etching (RIE), a dry etching technique involving directional destruction of material by bombardment with chemically reactive plasma, has been employed [34]. As shown in Figure 2, Chirra and Desai used two series of steps each composed of photolithography followed by RIE with oxygen plasma to create microdevices with three reservoirs partially etched through the microdevice structure.

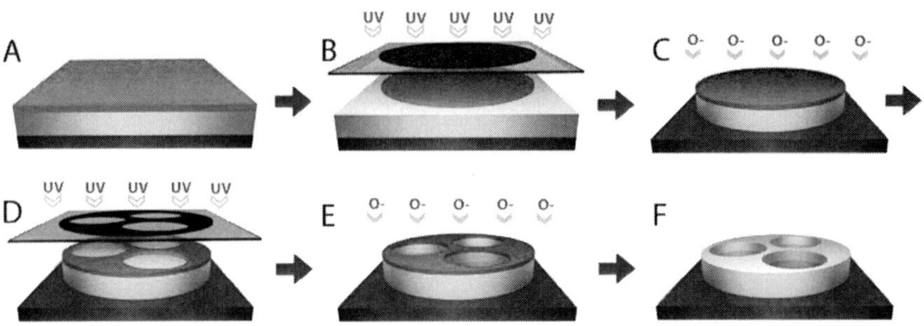

Figure 2. Photolithography-based techniques for microfabrication of multi-reservoir PMMA devices. **A**. PMMA and, subsequently, photoresist are spin-cast onto a silicon wafer. **B**. A circular pattern is transferred from a UV-blocking photomask to the photoresist through **...**

While photolithography techniques are often expensive and require access to cleanroom facilities, soft lithography allows for replication of a hard patterned substrate to create an inverse pattern with a soft elastomer such as polydimethylsiloxane (PDMS). The patterned elastomer can then be used as a master mold to repeatedly pattern a wide range of materials under standard laboratory settings. These patterned elastomers can be used as either a mold, to create devices from recessed regions, or as a stamp, which can be coated with material to create devices or patterned surface modifications in regions of contact. Guan et al. have demonstrated a variety of soft lithography techniques that can be utilized to fabricate microdevices (Figure 3) [56, 59–60, 68]. In one study, a micropillar PDMS stamp was coated with PPMA before bringing the stamp into contact with a glass slide coated with polyvinyl alcohol (PVA), creating PPMA microdevices in regions of contact (Figure 3A) [59]. In contrast, Guan et al. also used a microwell stamp to collect PPMA within recessed regions before bringing the stamp into contact with PVA-coated glass, creating microdevices from the wells of the microstamp (Figure 3B) [59]. In later studies, a mixture of PEGMA and PEGDMA was applied to a PDMS microwell stamp, allowing for microdevice formation through discontinuous dewetting (Figure 3C) [56]. As a result of the interactions at the interface of the polymer solution and the PDMS, the PEGMA/PEGDMA resin selectively collected in the microwells before UV exposure induced polymerization via a photoinitiator [56, 69]. A number of other studies have utilized similar soft lithography techniques in the fabrication of microdevices [65, 70–74].

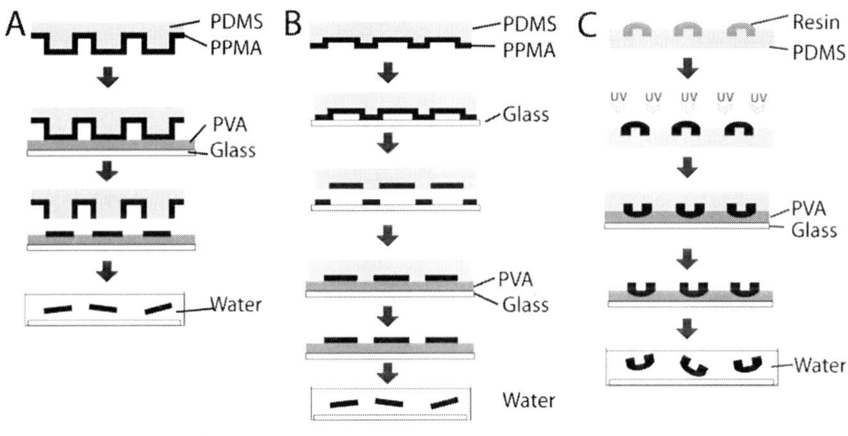

Figure 3. Soft lithography-based techniques for microdevice fabrication. **A.** Microcontact printing can be utililized for fabrication of microdevices in regions of contact of PVA with micropillar stamp with subsequent dissolution of PVA in water for device release **...**

Currently, only a select number of microparticulate drug delivery systems have incorporated nanostructures to enhance oral drug delivery [75–77]. As microdevice design advances, future studies may apply a wide range of nanofabrication techniques to microdevice design as nanotopography has been shown to enhance muco- and cytoadhesion and interact with epithelial cells to enhance drug permeability [77–78]. However, resolutions below 100 nm are difficult to achieve with UV-based photolithography due to the diffraction limit of light [79]. For fabrication of devices with nanoscale features, nanofabrication techniques are required. One such technique, X-ray lithography, uses electromagnetic radiation with wavelengths ranging from 0.5 to 4 nm and is capable of achieving resolutions approaching 20 nm [80–82]. Similar to UV lithography, X-ray lithography uses an X-ray source such as a synchrotron or laser-induced plasma generator to irradiate X-ray-sensitive material through an X-ray absorbing mask [80, 82]. Maskless lithography techniques, including electron beam, ion beam, and dip-pen lithography are also available for nanofabrication. Electron beam lithography directs a beam of electrons to create a pattern on a material sensitive to electron irradiation, which is later developed or etched to form features on the irradiated material or an underlying substrate [83]. Similarly, ion beam lithography utilizes a focused beam of ions to either remove a substrate material or deposit a dissociated precursor material onto the substrate [84]. Dip-pen lithography adapts a scanning atomic force microscopy probe to direct inorganic or biological ink molecules across a substrate where they subsequently adsorb [85]. Because maskless techniques require low throughput *de novo* pattern creation, they are often used to create a master mold, which is then used to transfer the inverse pattern to other materials through nanoimprinting [86–89]. In addition, templating of polymeric material with nanoporous membranes provides high-throughput fabrication of nanowire arrays [90–91]. In template synthesis, a polymer is exposed to a nanoporous membrane at a temperature greater than the polymer's glass transition temperature. The polymer is incorporated into the membrane, and the membrane is then selectively dissolved, leaving the polymer with the inverse nanowire array. Membrane selection provides control over the nanowire diameter and spacing, and templating time and temperature are adjusted to control nanowire length.

Structural and chemical modifications to improve microdevice muco- and cytoadhesion

Because microdevices rely on adhesion to increase residence time and prolong drug exposure in the GI tract, many designs to increase device adhesion have been developed, including the utilization of mucoadhesive materials, targeted molecular interactions, gecko-inspired micro/nano topography, and mechanical interactions as shown in Figure 4. Conjugation of lectins to microdevices and other drug delivery systems has proven effective in increasing both adhesion to Caco-2 cell monolayers and intestinal mucosa (Figure 4A) [34–35, 50, 92–93]. Conjugation of tomato lectin to PMMA microdevices resulted in a two- to six-fold increase binding of microdevices to a Caco-2 cell monolayer relative to unmodified microparticles [50]. Furthermore, upon binding to a cell monolayer under physiologically relevant shear stress levels, only a small fraction of lectin-conjugated devices detach, and the majority of the detached devices reattach

[94]. However, given the sensitivity of lectins and other peptide targeting molecules to the low pH values and enzymes of the GI tract, incorporation of bioadhesive materials and topographies that are less prone to degradation may provide more stable mechanisms for bioadhesion. Features on the micron-scale such as microposts and microneedles have also been developed to aid in transdermal drug delivery [18, 95–96]. These features have also been incorporated into planar oral drug delivery microdevices (Figure 4B) to penetrate the mucous membrane in order to increase both drug permeation and device residence time [57]. Utilizing topographical features at a smaller scale, gecko-inspired nanotopography can be incorporated into microdevices to promote adhesion by providing dramatically increased surface area, resulting in increased vaan der Waals interactions. The Desai Lab has created hierarchical nanoengineered microparticles (NEMPs) composed of silicon oxide microspheres coated with silicon nanowires (Figure 4C) that demonstrated a 100-fold increase in lift-off force*in vitro* and a ten-fold increase in residence time *in vivo* relative to uncoated silica microspheres [34, 90]. Fabrication conditions have been altered to produce planar NEMPs, resulting in significant increases in drug uptake both *in vitro* and *in vivo* [76]. Nanostructured surfaces are also capable of loosening the epithelial barrier and providing increased permeability to protein therapeutics [78]. Therefore, incorporation of nanotopography into microdevices has potential to simultaneously increase device adhesion to the epithelial barrier and interact with epithelial cells to increase drug permeability, and studies regarding this concept are currently underway in our laboratory. Given the mucoadhesive properties of chitosan [94, 97–99], chitosan-based hydrogels show great promise as materials for oral drug delivery microdevices. Hydrogels also have dynamic interactions with water, providing swelling properties that can be utilized to incorporate folding properties into microdevices [60]. Guan et al. created self-folding bilayered microdevices composed of layers of crosslinked chitosan and poly(PEGMA-co-PEGDMA) [56]. These devices were fabricated with arms that folded upon contact with water as a result of differential swelling of the two polymer layers (Figure 4D), resulting in mechanical attachment of the devices to excised pig intestinal mucosa [56].

Approaches to drug loading of microdevices

Drug loading of microdevices is a difficult challenge with the ultimate goal of loading drug in a high-throughput, low-waste, and precise manner. Initial studies utilized microinjection of drug into the device reservoir with subsequent drying and crystallization of the drug [40]. For higher throughput drug loading and custom release kinetics, drug-laden PEGDMA hydrogels were spin-coated onto devices and selectively polymerized by photolithography

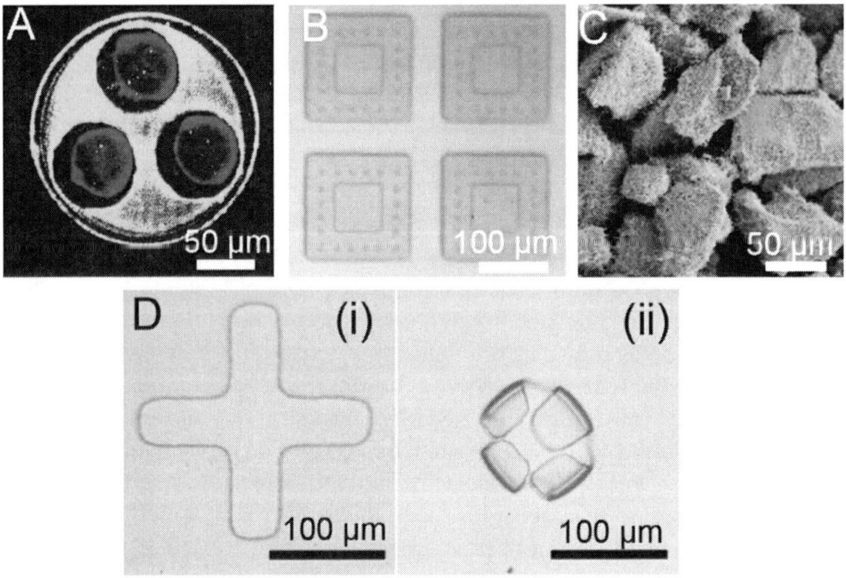

Figure 4. Chemical and structural modifications to enhance microdevice adhesion and increase residence time. **A**. Fluorescence micrograph of FITC-lectin (green) asymmetrically coated onto the drug-releasing side of microdevices for targeted bioadhesion to the intestinal **...**

(Figure 5A) [52]. The ability to lithographically pattern drug-loaded hydrogels was utilized by Ainslie et al. to layer drugs into reservoirs for sequential drug release kinetics [52]. In addition, Chirra and Desai loaded PMMA hydrogels crosslinked with PEGDMA each containing a differing model drug into three device reservoirs [34]. While photolithographically based drug loading allows for high precision and incorporation of hydrogels with custom release kinetics, drug is lost in the process of spin-coating and developing the drug-hydrogel solution, which is a major drawback when utilizing expensive biotherapeutics. To load drugs in a low-waste manner, discontinuous dewetting properties have been utilized. In one study, microdevices were brushed with a solution of sodium chloride, and surface interactions with drug solvent resulted in collection and crystallization of the model drug in the device reservoirs to provide a high-throughput method of drug loading with minimal drug waste (Figure 5B) [56]. In a more precise drug loading method, Marizza et al. recently developed inkjet printing for loading of drug solutions into microdevice reservoirs (Figure 5C) [100]. While this method requires sequential loading of each microdevice in a semi-automated manner and is not currently as high-throughput as discontinuous dewetting or photolithographic drug loading techniques, it is capable of precise, quasi-zero-waste performance. Marizza et al. went on to combine this method with supercritical fluid impregnation, allowing for the loading of hydrophobic drugs without the use of an organic solvent (Figure 5D) [101]. Other techniques for loading drug over the entire microdevice surface rather than into reservoirs have been utilized. NEMPs, with a large surface area as a result of nanowire coating, were loaded with insulin via a solvent-evaporation-induced capillary effect, in which drug crystallized over the microparticle surface at the base of the nanowires [76,102].

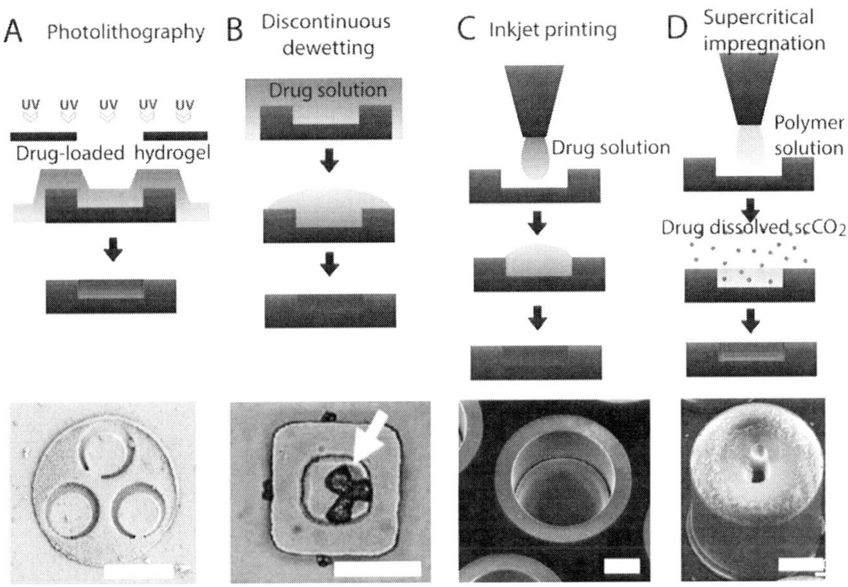

Figure 5. Current techniques for drug loading of microdevices with images of drug-loaded devices for each approach. **A**. Photolithography can be utilized to selectively induce crosslinking of drug-laden hydrogels in device reservoirs [34]. **B**. Discontinuous dewetting **...**

Efficacy of microdevices *in vitro* **and** *in vivo*

Within the last few years, the utilization of semi-conductor industry principles to fabricate oral microdevices has advanced leaps and bounds. Yet, the testing of these devices in improving the overall efficacy of most therapeutics is still at its relative infancy. Recent studies related to *in vitro* drug release and permeation have been done over monolayers of Caco-2 epithelial cells using Transwell® inserts. Ainslie et al. showed that the localized high concentration of drug at the device-cell interface resulted in an enhancement of drug permeation across the Caco-2 monolayer under physiological fluid flow, with a ten-fold increase in fluorescein permeation when released from microdevices relative to fluorescein free in solution [27]. Also, a sequential release of different sized drugs, insulin and camptothecin was achieved with the use of a dual layered hydrogel system that was present in microdevices made up of a single reservoir [103]. While sequential release can be harnessed to improve drug bioavailability by first releasing a permeation enhancer followed by the drug of interest, the release kinetics of the drug are co-dependent on the release kinetics of the permeation enhancer from its respective top hydrogel layer. To overcome this co-dependence issue, Chirra et al. used multiple reservoirs that can be filled with different drugs using different hydrogel/biodegradable polymeric systems [34]. Figure 6B shows the independent release of multiple model fluorophore-tagged BSAs from respective reservoirs as shown in Figure 6A. Such a device system can be used to release permeation enhancers, proteolytic enzyme inhibitors, and drugs of interest at independent rates and release times, thereby making oral microdevices effective for increasing drug efficacy as well as for combinatorial therapy. The Desai Lab also used Caco-2 monolayer coated parallel plate flow chambers to study the extent of oral microdevice retention under GI flow conditions. They have shown that 93% of tomato lectin microdevices remain attached to the cell surface under one hour of physiological shear conditions after initial binding, indicating that microdevices are capable of remaining attached to GI tissue for extended periods of time under physiological conditions [103].

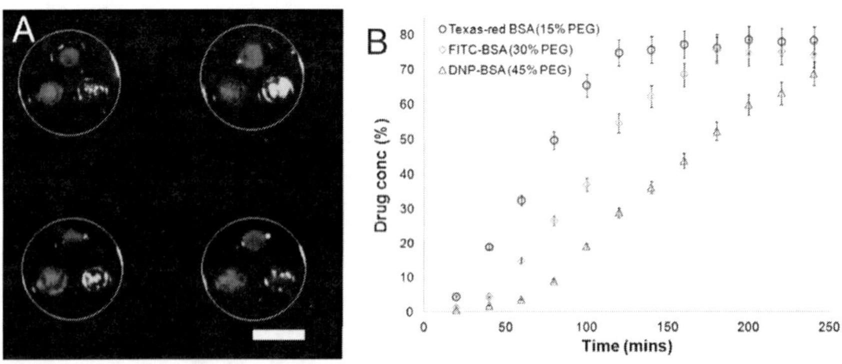

Figure 6. Microdevices loaded with multiple drugs with separate release profiles.**A**. Fluorescent image demonstrating separate drug loading of each microdevice reservoir with device shape outlined in white. Scale bar is 100 μm. **B**. Custom release profiles ...

The several advantages of using asymmetric planar oral microdevices including unidirectional release to avoid luminal drug loss, increased contact surface area and reduced shear stress with a planar design, and selective modification of reservoir side of device to introduce muco- or cytoadhesive properties were recently tested *in vivo*. Our lab observed that upon oral administration to mice, PMMA microdevices having the same contact surface area as that of symmetric PMMA microparticles have a 27% retention in the proximal small intestine after 2 hours due to the relatively low shear stress experienced by the thin device side walls, as compared to a retention of 12% for the curvilinear microparticles [66]. Also, a further enhancement of microdevice retention to 41% was observed after conjugation of the bioadhesive protein lectin, which targeted the intestinal epithelial cell wall. We have shown that with the help of microdevices, even the poorly permeable drug Acyclovir had a five-fold increase in oral bioavailability in mice as compared to that of a conventional solution of Acyclovir of same dosage [66]. This enhancement of oral bioavailability drastically reduces the overall dosage needed for effective therapy. Such a reduction in dosage with improved bioavailability proves vital in significantly alleviating issues of systemic side effects, thereby opening up oral administration to an array of toxic and expensive therapeutics. While most of the recent *in vivo* work was done using small molecule drugs, microdevices can be applied to the oral delivery of macromolecules and high-efficacy low-dosage drugs (e.g. Leuprolide, human growth hormone, etc.). Detailed studies on improving drug loading, dosage optimization, improved protection against GI environment, sustained release for systemic delivery, targeted attachment, and GI pathology oriented localized delivery are currently underway and are of much interest for future work. Therefore, the use of microfabricated planar oral devices holds promise in augmenting the range of oral therapeutics used, while solving pharmacokinetic issues associated with low permeability and avoiding systemic side effects.

CONCLUSION

Microfabrication techniques allow for the creation of devices with properties not easily attainable through other methods of conventional particle synthesis. This advanced control over device properties allows microdevices to be designed to simultaneously address many barriers to oral drug uptake. Specifically, planar device shape, asymmetric modifications for cyto- and mucoadhesion, and unidirectional drug release are utilized to overcome barriers to oral drug delivery. These devices can be modified with targeting moieties and tunable drug release profiles to address specific drug delivery applications. Future developments in microdevice fabrication may involve the incorporation of additional biocompatible and bioresponsive materials for smart drug delivery and mucoadhesion. Also, new micro- and nanofabrication techniques may be developed to provide device topography in order to improve cellular permeability and enhance device adhesion. Challenges in the application of microdevices to oral drug delivery include a limited drug loading capacity and the high cost of some microfabrication techniques. These potential drawbacks can be mitigated by selecting highly potent drugs to reduce the number of devices required per dosage and by optimizing fabrication and drug loading techniques to minimize cost and maximize drug capacity. Microdevices have demonstrated promising results for oral drug delivery both *in vitro* and *in vivo*. However, the exact mechanisms by which these devices interact with the mucous and epithelial layers are yet to be fully determined. Further trials will determine whether the devices function primarily through either mucoadhesion or cytoadhesion and which drugs and drug combinations are best suited for use in planar microdevices. Multi-drug carrying microdevices may allow for combination therapy involving a drug of interest and a disruptive drug for increasing bioavailabilty through localized and transient disruption of digestive enzymes, transporter proteins, and epithelial tight junctions. In addition to applications in increasing systemic drug bioavailability, microdevices could be modified for targeted and localized drug delivery to diseased tissue in the GI tract. As new materials, fabrication methods, and drug combinations are developed, microdevice technology has the potential to significantly improve the oral delivery of a wide range of therapeutics including small molecule drugs, mid-size peptide therapeutics including insulin and antigens for immunization, and larger protein-based therapeutics.

ACKNOWLEDGEMENTS

This work was supported by Z Cube Zambon Research Venture. CBF was supported by an NIH Training Grant (5T32GM007175-37).

REFERENCES

1. International Transporter C, Giacomini KM, Huang SM, Tweedie DJ, Benet LZ, Brouwer KL, Chu X, Dahlin A, Evers R, Fischer V, Hillgren KM, Hoffmaster KA, Ishikawa T, Keppler D, Kim RB, Lee CA, Niemi M, Polli JW, Sugiyama Y, Swaan PW, Ware JA, Wright SH, Yee SW, Zamek-Gliszczynski MJ, Zhang L. Membrane transporters in drug development. Nature reviews. Drug discovery. 2010;9(3):215–236.

2. Thanki K, Gangwal RP, Sangamwar AT, Jain S. Oral delivery of anticancer drugs: challenges and opportunities. Journal of

controlled release : official journal of the Controlled Release Society.2013;170(1):15–40.

3. Goldberg M, Gomez-Orellana I. Challenges for the oral delivery of macromolecules. Nature reviews. Drug discovery. 2003;2(4):289–295.

4. Gupta S, Jain A, Chakraborty M, Sahni JK, Ali J, Dang S. Oral delivery of therapeutic proteins and peptides: a review on recent developments. Drug delivery. 2013;20(6):237–246.

5. Agrawal U, Sharma R, Gupta M, Vyas SP. Is nanotechnology a boon for oral drug delivery? Drug discovery today. 2014

6. Felton LA, Porter SC. An update on pharmaceutical film coating for drug delivery. Expert opinion on drug delivery. 2013;10(4):421–435.

7. Kriegel C, Attarwala H, Amiji M. Multi-compartmental oral delivery systems for nucleic acid therapy in the gastrointestinal tract. Advanced drug delivery reviews. 2013;65(6):891–901.

8. Hunter AC, Elsom J, Wibroe PP, Moghimi SM. Polymeric particulate technologies for oral drug delivery and targeting: a pathophysiological perspective. Nanomedicine : nanotechnology, biology, and medicine.2012;8(Suppl 1):S5–S20.

9. Dahan A, Khamis M, Agbaria R, Karaman R. Targeted prodrugs in oral drug delivery: the modern molecular biopharmaceutical approach. Expert opinion on drug delivery. 2012;9(8):1001–1013.

10. Zee YK, Goh BC, Lee SC. Pharmacologic modulation strategies to reduce dose requirements of anticancer therapy while preserving clinical efficacy. Future oncology. 2012;8(6):731–749.

11. Kuppens IE, Breedveld P, Beijnen JH, Schellens JH. Modulation of oral drug bioavailability: from preclinical mechanism to therapeutic application. Cancer investigation. 2005;23(5):443–464.

12. Lemmer HJ, Hamman JH. Paracellular drug absorption enhancement through tight junction modulation.Expert opinion on drug delivery. 2013;10(1):103–114.

13. Kararli TT. Comparison of the gastrointestinal anatomy, physiology, and biochemistry of humans and commonly used

laboratory animals. Biopharmaceutics & drug disposition. 1995;16(5):351–380.

14. Allen A, Flemstrom G, Garner A, Kivilaakso E. Gastroduodenal mucosal protection. Physiological reviews. 1993;73(4):823–857.

15. Atuma C, Strugala V, Allen A, Holm L. The adherent gastrointestinal mucus gel layer: thickness and physical state in vivo. American journal of physiology. Gastrointestinal and liver physiology.2001;280(5):G922–G929.

16. Gustafsson JK, Ermund A, Johansson ME, Schutte A, Hansson GC, Sjovall H. An ex vivo method for studying mucus formation, properties, and thickness in human colonic biopsies and mouse small and large intestinal explants. American journal of physiology. Gastrointestinal and liver physiology.2012;302(4):G430–G438.

17. Pries AR, Secomb TW, Gaehtgens P. The endothelial surface layer. Pflugers Archiv : European journal of physiology. 2000;440(5):653–666.

18. van den Berg BM, Nieuwdorp M, Stroes ES, Vink H. Glycocalyx and endothelial (dys) function: from mice to men. Pharmacological reports : PR. 2006;58(Suppl):75–80.

19. Lipinski CA, Lombardo F, Dominy BW, Feeney PJ. Experimental and computational approaches to estimate solubility and permeability in drug discovery and development settings. Advanced drug delivery reviews. 2001;46(1–3):3–26.

20. Mayor S, Pagano RE. Pathways of clathrin-independent endocytosis. Nature reviews. Molecular cell biology. 2007;8(8):603–612.

21. Le Roy C, Wrana JL. Clathrin- and non-clathrin-mediated endocytic regulation of cell signalling. Nature reviews. Molecular cell biology. 2005;6(2):112–126.

22. Pantzar N, Lundin S, Wester L, Westrom BR. Bidirectional small-intestinal permeability in the rat to some common marker molecules in vitro. Scandinavian journal of gastroenterology. 1994;29(8):703–709.

23. Salama NN, Eddington ND, Fasano A. Tight junction modulation and its relationship to drug delivery.Advanced drug delivery reviews. 2006;58(1):15–28.

24. Tang R, Li WX, Huang W, Yan F, Chai WM, Yang GY, Chen KM. CO2-based in-line phase contrast imaging of small intestine in mice. Scientific reports. 2013;3:2313.

25. Sant S, Tao SL, Fisher OZ, Xu Q, Peppas NA, Khademhosseini A. Microfabrication technologies for oral drug delivery. Advanced drug delivery reviews. 2012;64(6):496–507.

26. Chirra HD, Desai TA. Emerging microtechnologies for the development of oral drug delivery devices.Advanced drug delivery reviews. 2012;64(14):1569–1578.

27. Ainslie KM, Lowe RD, Beaudette TT, Petty L, Bachelder EM, Desai TA. Microfabricated devices for enhanced bioadhesive drug delivery: attachment to and small-molecule release through a cell monolayer under flow. Small. 2009;5(24):2857–2863.

28. Tachibana T, Kato M, Takano J, Sugiyama Y. Predicting drug-drug interactions involving the inhibition of intestinal CYP3A4 and P-glycoprotein. Current drug metabolism. 2010;11(9):762–777.

29. Konig J, Muller F, Fromm MF. Transporters and drug-drug interactions: important determinants of drug disposition and effects. Pharmacological reviews. 2013;65(3):944–966.

30. Besheer A, Wood KM, Peppas NA, Mader K. Loading and mobility of spin-labeled insulin in physiologically responsive complexation hydrogels intended for oral administration. Journal of controlled release : official journal of the Controlled Release Society. 2006;111(1–2):73–80.

31. Kim B, Peppas NA. In vitro release behavior and stability of insulin in complexation hydrogels as oral drug delivery carriers. International journal of pharmaceutics. 2003;266(1–2):29–37.

32. Lowman AM, Morishita M, Kajita M, Nagai T, Peppas NA. Oral delivery of insulin using pH-responsive complexation gels. Journal of pharmaceutical sciences. 1999;88(9):933–937.

33. Al-Hilal TA, Alam F, Byun Y. Oral drug delivery systems using chemical conjugates or physical complexes. Advanced drug delivery reviews. 2013;65(6):845–864.

34. Chirra HD, Desai TA. Multi-reservoir bioadhesive microdevices for independent rate-controlled delivery of multiple drugs. Small. 2012;8(24):3839–3846.

35. Tao SL, Desai TA. Micromachined devices: the impact of controlled geometry from cell-targeting to bioavailability. Journal of controlled release : official journal of the Controlled Release Society. 2005;109(1–3):127–138.

36. Tao SL, Popat K, Desai TA. Off-wafer fabrication and surface modification of asymmetric 3D SU-8 microparticles. Nature protocols. 2006;1(6):3153–3158.

37. Harris LA, Hansel S, DiBaise J, Crowell MD. Irritable bowel syndrome and chronic constipation: emerging drugs, devices, and surgical treatments. Current gastroenterology reports. 2006;8(4):282–290.

38. Wald A. Irritable bowel syndrome--diarrhoea. Best practice & research. Clinical gastroenterology.2012;26(5):573–580.

39. Wilson DS, Dalmasso G, Wang L, Sitaraman SV, Merlin D, Murthy N. Orally delivered thioketal nanoparticles loaded with TNF-alpha-siRNA target inflammation and inhibit gene expression in the intestines. Nature materials. 2010;9(11):923–928.

40. Ahmed A, Bonner C, Desai TA. Bioadhesive microdevices with multiple reservoirs: a new platform for oral drug delivery. Journal of controlled release : official journal of the Controlled Release Society.2002;81(3):291–306.

41. Foraker AB, Walczak RJ, Cohen MH, Boiarski TA, Grove CF, Swaan PW. Microfabricated porous silicon particles enhance paracellular delivery of insulin across intestinal Caco-2 cell monolayers.Pharmaceutical research. 2003;20(1):110–116.

42. Hiebl B, Hopperdietzel C, Hunigen H, Jung F, Scharnagl N. Influence of a silicon (Si14)-based coating substrate for biomaterials on fibroblast growth and human C5a. Clinical hemorheology and microcirculation.2013;55(4):491–499.

43. Peng F, Su Y, Zhong Y, Fan C, Lee ST, He Y. Silicon nanomaterials platform for bioimaging, biosensing, and cancer therapy. Accounts of chemical research. 2014;47(2):612–623.

44. Santos HA, Makila E, Airaksinen AJ, Bimbo LM, Hirvonen J. Porous silicon nanoparticles for nanomedicine: preparation and biomedical applications. Nanomedicine. 2014;9(4):535–554.

45. Fubini B, Hubbard A. Reactive oxygen species (ROS) and reactive nitrogen species (RNS) generation by silica in inflammation and fibrosis. Free radical biology & medicine. 2003;34(12):1507–1516.

46. Tungjai M, Whorton EB, Rithidech KN. Persistence of apoptosis and inflammatory responses in the heart and bone marrow of mice following whole-body exposure to (2)(8)Silicon ((2)(8)Si) ions. Radiation and environmental biophysics. 2013;52(3):339–350.

47. Food; Drug Administration, H. H. S. Medical devices; reclassification of polymethylmethacrylate (PMMA) bone cement. Final rule. Federal register. 2002;67(137):46852–46855.

48. Xu Y, Xie F, Qiu T, Xie L, Xing W, Cheng J. Rapid fabrication of a microdevice with concave microwells and its application in embryoid body formation. Biomicrofluidics. 2012;6(1):16504–1650411.

49. Reedy CR, Price CW, Sniegowski J, Ferrance JP, Begley M, Landers JP. Solid phase extraction of DNA from biological samples in a post-based, high surface area poly(methyl methacrylate) (PMMA) microdevice.Lab on a chip. 2011;11(9):1603–1611.

50. Tao SL, Lubeley MW, Desai TA. Bioadhesive poly(methyl methacrylate) microdevices for controlled drug delivery. Journal of controlled release : official journal of the Controlled Release Society.2003;88(2):215–228.

51. Wei S, Vaidya B, Patel AB, Soper SA, McCarley RL. Photochemically patterned poly(methyl methacrylate) surfaces used in the fabrication of microanalytical devices. The journal of physical chemistry. B. 2005;109(35):16988–16996.

52. Ainslie KM, Kraning CM, Desai TA. Microfabrication of an asymmetric, multi-layered microdevice for controlled release of orally delivered therapeutics. Lab on a chip. 2008;8(7):1042–1047. [PMC free article]

53. Kotzar G, Freas M, Abel P, Fleischman A, Roy S, Zorman C, Moran JM, Melzak J. Evaluation of MEMS materials of construction for implantable medical devices. Biomaterials. 2002;23(13):2737–2750.

54. Voskerician G, Shive MS, Shawgo RS, von Recum H, Anderson JM, Cima MJ, Langer R. Biocompatibility and biofouling of MEMS drug delivery devices. Biomaterials. 2003;24(11):1959–1967.

55. Vernekar VN, Cullen DK, Fogleman N, Choi Y, Garcia AJ, Allen MG, Brewer GJ, LaPlaca MC. SU-8 2000 rendered cytocompatible for neuronal bioMEMS applications. Journal of biomedical materials research. Part A. 2009;89(1):138–151.

56. Guan J, He H, Lee LJ, Hansford DJ. Fabrication of particulate reservoir-containing, capsulelike, and self-folding polymer microstructures for drug delivery. Small. 2007;3(3):412–418.

57. Tao SL, Desai TA. Microfabrication of multilayer, asymmetric, polymeric devices for drug delivery. Adv Mater. 2005;17(13):1625-+.

58. Ryu WH, Vyakarnam M, Greco RS, Prinz FB, Fasching RJ. Fabrication of multi-layered biodegradable drug delivery device based on micro-structuring of PLGA polymers. Biomedical microdevices.2007;9(6):845–853.

59. Guan JJ, Chakrapani A, Hansford DJ. Polymer microparticles fabricated by soft lithography. Chem Mater. 2005;17(25):6227–6229.

60. Guan J, He H, Hansford DJ, Lee LJ. Self-folding of three-dimensional hydrogel microstructures. The journal of physical chemistry. B. 2005;109(49):23134–23137.

61. He H, Cao X, Lee LJ. Design of a novel hydrogel-based intelligent system for controlled drug release.Journal of controlled release : official journal of the Controlled Release Society. 2004;95(3):391–402.

62. Ito Y, Hasuda H, Morimatsu M, Takagi N, Hirai Y. A microfabrication method of a biodegradable polymer chip for a controlled release system. Journal of biomaterials science. Polymer edition.2005;16(8):949–955.

63. Norman JJ, Desai TA. Methods for fabrication of nanoscale topography for tissue engineering scaffolds.Annals of biomedical engineering. 2006;34(1):89–101.

64. Qian T, Wang Y. Micro/nano-fabrication technologies for cell biology. Medical & biological engineering & computing. 2010;48(10):1023–1032.

65. Ziaie B, Baldi A, Lei M, Gu Y, Siegel RA. Hard and soft micromachining for BioMEMS: review of techniques and examples

of applications in microfluidics and drug delivery. Advanced drug delivery reviews.2004;56(2):145–172.

66. Chirra HD, Shao L, Ciaccio N, Fox CB, Wade JM, Ma A, Desai TA. Planar Microdevices for Enhanced In Vivo Retention and Oral Bioavailability of Poorly Permeable Drugs. Advanced healthcare materials. 2014

67. Tao SL, Desai TA. Gastrointestinal patch systems for oral drug delivery. Drug discovery today.2005;10(13):909–915.

68. Guan J, Ferrell N, James Lee L, Hansford DJ. Fabrication of polymeric microparticles for drug delivery by soft lithography. Biomaterials. 2006;27(21):4034–4041.

69. Jackman RJ, Duffy DC, Ostuni E, Willmore ND, Whitesides GM. Fabricating large arrays of microwells with arbitrary dimensions and filling them using discontinuous dewetting. Analytical chemistry.1998;70(11):2280–2287.

70. Ferrell N, Woodard J, Hansford D. Fabrication of polymer microstructures for MEMS: sacrificial layer micromolding and patterned substrate micromolding. Biomedical microdevices. 2007;9(6):815–821.

71. Geipel A, Goldschmidtboeing F, Jantscheff P, Esser N, Massing U, Woias P. Design of an implantable active microport system for patient specific drug release. Biomedical microdevices. 2008;10(4):469–478.

72. Husler R, Schlittler FL, Kreutziger J, Streit M, Banic A, Schoni-Affolter F, Hunger RE, Constaninescu MA. Staged surgical therapy of basal cell carcinoma of the head and neck region: an evaluation of 500 procedures. Swiss medical weekly. 2008;138(49–50):746–751.

73. Lu X, Kim-Han JS, O'Malley KL, Sakiyama-Elbert SE. A microdevice platform for visualizing mitochondrial transport in aligned dopaminergic axons. Journal of neuroscience methods. 2012;209(1):35–39.

74. Park J, Li J, Han A. Micro-macro hybrid soft-lithography master (MMHSM) fabrication for lab-on-a-chip applications. Biomedical microdevices. 2010;12(2):345–351.

75. Fischer KE, Nagaraj G, Hugh Daniels R, Li E, Cowles VE, Miller JL, Bunger MD, Desai TA. Hierarchical nanoengineered surfaces for

enhanced cytoadhesion and drug delivery. Biomaterials.2011;32(13):3499–3506.

76. Uskokovic V, Lee K, Lee PP, Fischer KE, Desai TA. Shape effect in the design of nanowire-coated microparticles as transepithelial drug delivery devices. ACS nano. 2012;6(9):7832–7841. [PMC free article]

77. Uskokovic V, Lee PP, Walsh LA, Fischer KE, Desai TA. PEGylated silicon nanowire coated silica microparticles for drug delivery across intestinal epithelium. Biomaterials. 2012;33(5):1663–1672.

78. Kam KR, Walsh LA, Bock SM, Koval M, Fischer KE, Ross RF, Desai TA. Nanostructure-mediated transport of biologics across epithelial tissue: enhancing permeability via nanotopography. Nano letters.2013;13(1):164–171.

79. Lohmuller T, Aydin D, Schwieder M, Morhard C, Louban I, Pacholski C, Spatz JP. Nanopatterning by block copolymer micelle nanolithography and bioinspired applications. Biointerphases. 2011;6(1):MR1–MR12.

80. Marmiroli B, Amenitsch H. X-ray lithography and small-angle X-ray scattering: a combination of techniques merging biology and materials science. European biophysics journal : EBJ. 2012;41(10):851–861.

81. Betancourt T, Brannon-Peppas L. Micro- and nanofabrication methods in nanotechnological medical and pharmaceutical devices. International journal of nanomedicine. 2006;1(4):483–495.

82. Chen Y, Pepin A. Nanofabrication: conventional and nonconventional methods. Electrophoresis.2001;22(2):187–207.

83. Hu WC, Sarveswaran K, Lieberman M, Bernstein GH. Sub-10 nm electron beam lithography using cold development of poly(methylmethacrylate) J Vac Sci Technol B. 2004;22(4):1711–1716.

84. Pulsifer DP, Lakhtakia A. Background and survey of bioreplication techniques. Bioinspiration & biomimetics. 2011;6(3):031001.

85. Wu CC, Reinhoudt DN, Otto C, Subramaniam V, Velders AH. Strategies for patterning biomolecules with dip-pen nanolithography. Small. 2011;7(8):989–1002.

86. Chou SY, Krauss PR. Imprint lithography with sub-10 nm feature size and high throughput.Microelectron Eng. 1997;35(1–4):237–240.

87. Chou SY, Krauss PR, Renstrom PJ. Imprint lithography with 25-nanometer resolution. Science.1996;272(5258):85–87.

88. Lan H, Liu H. UV-nanoimprint lithography: structure, materials and fabrication of flexible molds. Journal of nanoscience and nanotechnology. 2013;13(5):3145–3172.

89. Truskett VN, Watts MP. Trends in imprint lithography for biological applications. Trends in biotechnology. 2006;24(7):312–317.

90. Fischer KE, Aleman BJ, Tao SL, Hugh Daniels R, Li EM, Bunger MD, Nagaraj G, Singh P, Zettl A, Desai TA. Biomimetic nanowire coatings for next generation adhesive drug delivery systems. Nano letters.2009;9(2):716–720.

91. Porter JR, Henson A, Popat KC. Biodegradable poly(epsilon-caprolactone) nanowires for bone tissue engineering applications. Biomaterials. 2009;30(5):780–788.

92. Wood KM, Stone GM, Peppas NA. The effect of complexation hydrogels on insulin transport in intestinal epithelial cell models. Acta biomaterialia. 2010;6(1):48–56.

93. Arbos P, Arangoa MA, Campanero MA, Irache JM. Quantification of the bioadhesive properties of protein-coated PVM/MA nanoparticles. International journal of pharmaceutics. 2002;242(1–2):129–136.

94. Amidi M, Mastrobattista E, Jiskoot W, Hennink WE. Chitosan-based delivery systems for protein therapeutics and antigens. Advanced drug delivery reviews. 2010;62(1):59–82.

95. Bariya SH, Gohel MC, Mehta TA, Sharma OP. Microneedles: an emerging transdermal drug delivery system. The Journal of pharmacy and pharmacology. 2012;64(1):11–29.

96. Escobar-Chavez JJ, Bonilla-Martinez D, Villegas-Gonzalez MA, Molina-Trinidad E, Casas-Alancaster N, Revilla-Vazquez AL. Microneedles: a valuable physical enhancer to increase transdermal drug delivery.Journal of clinical pharmacology. 2011;51(7):964–977.

97. Illum L. Chitosan and its use as a pharmaceutical excipient. Pharmaceutical research. 1998;15(9):1326–1331.

98. Kumar MN, Muzzarelli RA, Muzzarelli C, Sashiwa H, Domb AJ. Chitosan chemistry and pharmaceutical perspectives. Chemical reviews. 2004;104(12):6017–6084.

99. Patel MP, Patel RR, Patel JK. Chitosan mediated targeted drug delivery system: a review. Journal of pharmacy & pharmaceutical sciences : a publication of the Canadian Society for Pharmaceutical Sciences, Societe canadienne des sciences pharmaceutiques. 2010;13(4):536–557.

100. Marizza P, Keller SS, Boisen A. Inkjet printing as a technique for filling of micro-wells with biocompatible polymers. Microelectron Eng. 2013;111:391–395.

101. Marizza P, Keller SS, Mullertz A, Boisen A. Polymer-filled microcontainers for oral delivery loaded using supercritical impregnation. Journal of controlled release : official journal of the Controlled Release Society. 2014;173:1–9.

102. Fischer KE, Jayagopal A, Nagaraj G, Daniels RH, Li EM, Silvestrini MT, Desai TA. Nanoengineered surfaces enhance drug loading and adhesion. Nano letters. 2011;11(3):1076–1081.

103. Ainslie KM, Desai TA. Microfabricated implants for applications in therapeutic delivery, tissue engineering, and biosensing. Lab on a chip. 2008;8(11):1864–1878.

Chapter 8

Progress and Challenges in Developing Aptamer-Functionalized Targeted Drug Delivery Systems

Feng Jiang [1,2,†], Biao Liu [1,2], Jun Lu [1,2], Fangfei Li [1,2], Defang Li [1,2], Chao Liang [1,2], Lei Dang [1,2], Jin Liu [1,2], Bing He[1,2], Shaikh Atik Badshah [1,2], Cheng Lu [1,2], Xiaojuan He [1,2], Baosheng Guo [1,2], Xiao-Bing Zhang [3,4,*], Weihong Tan [3,4,*], Aiping Lu [1,2,*] and Ge Zhang [1,2,*]

[1]*Institute for Advancing Translational Medicine in Bone & Joint Diseases, School of Chinese Medicine, Hong Kong Baptist University, Hong Kong, China*

[2]*Hong Kong Baptist University Branch of State Key Laboratory of Chemo/Biosensing and Chemometrics of Hunan University, Hong Kong, China*

[3]*Molecular Science and Biomedicine Laboratory, State Key Laboratory of Chemo/Bio-Sensing and Chemometrics, College of Chemistry and Chemical Engineering, Hunan University, Changsha 410000, China*

[4]*College of Biology, Hunan University, Changsha 410000, China*

ABSTRACT

Aptamers, which can be screened via systematic evolution of ligands by exponential enrichment (SELEX), are superior ligands for molecular recognition due to their high selectivity and affinity. The interest in the use of aptamers as ligands for targeted drug delivery has been increasing due to their unique advantages. Based on their different compositions and preparation methods, aptamer-functionalized targeted drug delivery systems can be divided into two main categories: aptamer-small molecule conjugated systems and aptamer-nanomaterial conjugated systems. In this review, we not only summarize recent progress in aptamer selection and the application of aptamers in these targeted drug delivery systems but also discuss the advantages, challenges and new perspectives associated with these delivery systems.

Keywords: aptamer-functionalized targeted drug delivery systems; aptamer; SELEX; aptamer-small molecule conjugated systems; aptamer-nanomaterial conjugated systems

1. INTRODUCTION

Many drugs that are used in clinical treatment, particularly anticancer drugs, do not explicitly discriminate between pathological cells and normal cells. The indiscriminate exposure of normal cells to these drugs often leads to the use of a suboptimal dosage and is responsible for many of the toxicities associated with these drugs. Combining this issue with the emergence of drug-resistant pathological cells, it is clear that selective and efficacious therapies with minimal toxicity are urgently required [1]. It is known that targeted drug delivery systems can improve the specificity of delivered agents for an intended location, can increase the accumulation in the cellular of disease tissue and can potentially provide an alternative delivery mechanism [2]. Therefore, such systems are desirable and have become an exciting area of research in recent decades, and hundreds of approaches and strategies for selectively delivering potent drugs to pathological cells have been developed. Targeted drug delivery systems are especially of significance for drug delivery to tumors, as evidenced by the fact that certain drugs have now been approved for oncological indications by the Food and Drug Administration (FDA) [1,3].

Targeted drug delivery systems are simple but efficient and are composed of therapeutic agents and ligands that undergo noncovalent or covalent conjugation for targeted delivery. To achieve effective drug delivery, the first step is to develop a suitable ligand that can recognize the morphological and physiological differences between pathological and normal tissues. Early work in targeted drug delivery systems used polyunsaturated fatty acids, folic acid, hyaluronic acid or oligopeptides as tumor recognition moieties. However, research on these ligands is at a standstill for a variety of reasons, such as the need to clarify the detailed mechanism of tumor targeting by polyunsaturated fatty acids; the fact that the application of folic acid in small-size conjugates has only limited success [4]; and the observation that peptides are susceptible to enzymatic degradation in the systemic circulation, making them unsuitable for many *in vivo* applications [5].

Nucleic acid aptamers are identified through an *in vitro* selection process called systematic evolution of ligands by exponential enrichment (SELEX) [6]. Since their discovery in the 1980s, aptamers have attracted considerable interest for medical applications as therapeutic agents, diagnostic tools and moieties for targeted drug delivery [7]. In particular, aptamers are short, single-stranded DNA (ssDNA) or RNA oligonucleotides with specific secondary and tertiary structures, which exert their biological and physiological effects by binding to targeted proteins with high affinity and specificity [8]. Due to their specificity, low immunogenicity and toxicity, easily modified chemical structure and wide range of targets, aptamers are superior ligands encouraging the development of aptamer-targeted drug delivery systems.

Depending on their different compositions and preparation methods, aptamer-targeted drug delivery systems can be divided into two main categories: aptamer-small molecule conjugated systems (in which aptamers directly deliver drug molecules as both a carrier and a ligand) and aptamer-nanomaterial conjugated systems (in which aptamers function together with nanoparticles (NPs) for targeted delivery of drugs) [9]. This review is focused on the recent advances in the development of aptamer SELEX, aptamer-small molecule conjugated systems and aptamer-nanomaterial conjugated systems.

2. APTAMER SELEX

SELEX is a well-established and efficient technology for the screening of oligonucleotides with high affinities for their targets from random-sequence libraries [10]. This technique was introduced in 1990 by Andrew Ellington and Larry Gold and has been an important tool ever since for the identification and screening of aptamers. In fact, a wide variety of aptamers have been identified using the SELEX technique since the first report on SELEX 20 years ago [11]. After decades of development, this method has undergone dramatic changes and improvements. In addition to conventional SELEX [12,13,14], there are improved versions such as capillary electrophoresis-SELEX [15,16,17], magnetic bead-based SELEX [18,19,20], cell-SELEX [21,22,23,24,25,26,27], automated SELEX [28,29,30,31], complex-target SELEX [32,33,34,35], and so on.Table 1 shows some examples of nucleic acid aptamers that bind to targets of therapeutic interest. Since there are already many published reviews on aptamer SELEX [12,24,29,36], in this section, we highlight the cell-SELEX and complex-target SELEX strategy, which select aptamers able to bind to a specific cell type or a complex-target.

Table 1. Example of nucleic acid aptamers.

Aptamer	Molecular Target	Associated Disease	Aptamer Structure	Ref.
Macugen	Vascular endothelial growth factor (VEGF)	Age-related macular degeneration	5′-CGGAAUCAGUGAAUG CUUAUACAUCCG-3′	[37]
AS1411	Nucleoin	Cancer	5′-d(GGTGGTGGTGGT TGTGGTGGTGGTGG)-3′	[38]
Sgc8	Protein tyrosine kinase 7 (PTK-7)	Cancer	5′-ATCTAACTGCTGC GCCGCCGGGAAAATA CTGTACGGTTAGA-3′	[39]
TD05	Immunoglobulin μ heavy chains (IGHM)	Lymphoma	5′-AACACCGGGAGGAT AGTTCGGTGGCTGTTCA GGGTCTCCTCCCGGTG-3′	[40]
ARC1779	A1 Domain of von Willebrand factor (vWF)	Thrombotic microangiopathies and carotid artery disease	5′-GCGUGCAGUGCCU UCGGCCGUGCGGT GCCUCCGUCACGC-3′	[41]
TBA	α-Thrombin	Thrombosis	5′-GGTTGGTGTGGTTGG-3′	[42]

2.1. Cell-SELEX

In this method, to identify a cell-specific aptamer, cells of a certain type can be used as positive targets, and normal cells can be used as negative targets [21]. The screening process of Cell-SELEX is as follows. First, an oligonucleotide library with random sequences is constructed with constant primers flanking the 5′ and 3′ ends [22]. The total size of the library can be as large as 10^{14}, covering nearly all of the possible three-dimensional conformations that can be applied to target nearly all types of natural molecules [23]. The oligonucleotide library is then incubated with target cells at a certain temperature, and the aptamers that bind to target cells are isolated as a library for negative selection. Meanwhile, the aptamers that bind to both target cells and non-target cells are removed. Finally, the aptamers are washed and amplified by PCR or RT-PCR to generate a secondary library for the next round of screening [24,25]. The three major steps of cell-SELEX, including incubation, partitioning and amplification, are shown in Figure 1. Using this method, Lu and Zhang's group [26] specifically selected aptamers from a library composed of 10^{15} different ssDNA sequences. In this study, rat primary osteoblasts were used as target cells, whereas a rat liver cell line BRL-3A and rat PBMCs were used as negative targets. The researchers obtained an aptamer that specifically recognized and bound to the target cells after only 14 rounds of selection, and they successfully applied the aptamer to the development of an osteoblast-specific delivery system.

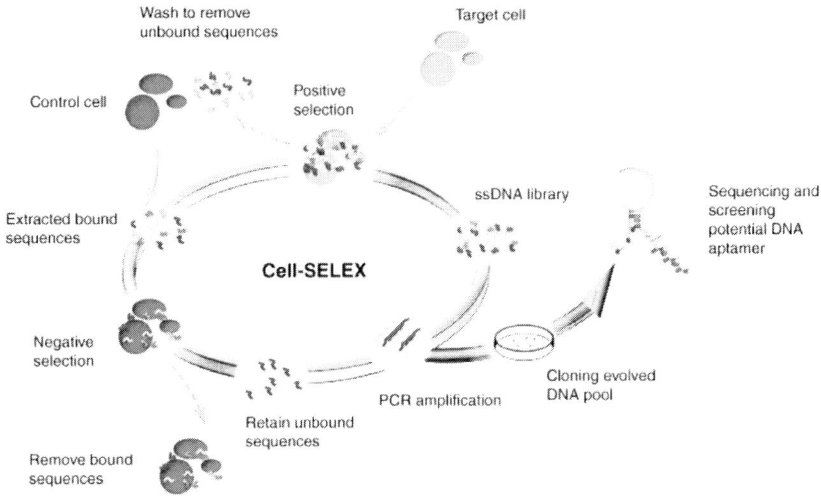

Figure 1. Schematic illustration of the cell-SELEX (systematic evolution of ligands by exponential enrichment) process. Reproduced with permission from Reference [43].

Compared with conventional SELEX technology, Cell-SELEX exhibits the following advantages. First, through this process, functional aptamers against various targets—ranging from small targets such as proteins and peptides to complex targets such as complete tumor cells, bacteria and other pathogens—can be isolated [27]. Second, this well-developed method uses whole cells as targets. Based on this principle, it is feasible to screen aptamers specifically recognizing target cells as long as there are different expression levels of target molecules between the positive and negative target cells, even without information about the specific target molecule. Third, when using whole cells as targets, the target protein on the cell surface can retain its native conformation. Aptamer selection in this context can facilitate the identification of biomarkers expressed on target cells.

2.2. Complex-Target SELEX

Early targets for SELEX were usually purified soluble proteins. However, it is difficult to screen aptamers with only one protein as the target, as a protein normally functions by forming a complex with other proteins, which is the case for transmembrane proteins or ion channels [32]. Thus, it was necessary to develop a novel approach to screen aptamers against complex targets. In particular, complex-target SELEX can be used to screen different aptamers against multiple molecules in parallel [33]. However, if the target protein is just one part of a composite target, other methods, such as subtractive SELEX, should be applied to eliminate interference factors. Morris's group used complex-target SELEX to identify an ssDNA aptamer that could bind to red blood cells after 25 rounds of selection, and they demonstrated that different aptamers could specifically recognize different molecular targets [34]. Similar results were reported for aptamer screening against human plasma and γ-carboxyglutamic acid protein complex targets. Shamah [35] summarized recent studies performing aptamer screening using complex-target SELEX and also discussed the process for selecting aptamers against composite targets, including cell debris, cell membranes, bacteria, parasites, viruses and animal cells. This approach can also be applied in a selection process using a membrane protein as the target: the membrane protein is expressed on the cell surface through genetic engineering, and complex-target SELEX can then be performed for aptamer selection.

The development of complex-target SELEX greatly enriched the pool of aptamer screening targets and expanded the potential applications of aptamers in disease diagnosis, drug development and other research fields.

Although aptamer is an excellent candidate for targeted drug delivery, the wild-type RNA and DNA molecules can be easily degraded by nucleases [44]. In order to increase the stability of the aptamer, several schemes have been proposed, such as modification of aptamer with 2′-aminopyrimidine [45], 2′-fluoropyrimidine [45], or 2′-*O*-methyl nucleotides [46]. Precise site-specific modifications facilitate the engineering of aptamers for improved specificity and stability. The mechanism of action for aptamers and the reasons for their marked tumor selectivity were also studied. Bates *et al.*[47,48,49,50] found that AS1411 is a quadruplex forming oligodeoxynucleotide that binds to nucleolin. The nucleolin is not only serving as the receptor for AS1411 and leading to selective uptake in cancer cells, but also essential for the stimulation of macropinocytosis of AS1411 which leads to further uptake of the aptamer. In cancer cells, uptake by macropinocytosis allows endosomal escape of AS1411, and AS1411 induces cell death due to hyperstimulation of macropinocytosis.

In summary, after years of improvement and development, aptamer-screening technology has achieved fruitful progress in many respects. Many new SELEX methods have been developed for an easier and faster selection of aptamers, and both the screening efficiency and the range of application have been improved. However, each method has its limitations, including a long screening time, high labor requirements and/or high cost. Additionally, there is still a lack of a common screening method. Therefore, it is necessary to develop new technologies to solve these problems. In the future, the core work of aptamer screening should focus on the following aspects: first, the mechanisms by which aptamers have affinity for their target molecules; second, new methods to shorten the screening cycle and to reduce screening costs; and third, the establishment of a more optimal aptamer-screening platform to improve the degree of automation.

3. APTAMER-SMALL MOLECULE CONJUGATED SYSTEMS

Because targeted drugs that undergo nonchemical conjugation are unstable due to the reversible nature of the noncovalent conjugation process [36], aptamer-small molecule conjugated systems mainly refer to aptamer-drug conjugates with covalent linkers. Currently, aptamer-small molecule conjugates are still in the early stages of development. One well-known example is sgc8c-doxorubicin (DOX), developed by Tan and colleagues [51]. Sgc8c is a DNA aptamer that specifically binds to protein tyrosine kinase 7 (PTK7) on the surface of CCRF-CEM (T-cell acute lymphoblastic leukemia, T-cell ALL) cells with high binding affinity. The study showed that sgc8c-DOX conjugate possess high binding affinity of the sgc8c aptamer, and could be efficiently internalized by target cells. The sgc8c-DOX conjugate not only demonstrates anticancer potency similar to unconjugated DOX, but also exhibits limited toxicity toward non-target cells. Another aptamer-small molecule conjugate was also developed by Tan and colleagues, who designed and synthesized a therapeutic module via solid-phase synthesis, consisting of a phosphoramidite containing an anticancer drug moiety and a photocleavable linker [52]. Such examples demonstrated that aptamer-small molecule conjugates could achieve targeted drug delivery similar to antibody-drug conjugates (ADCs). An aptamer-small molecule conjugate consists of three main components—an aptamer, a linker and a small molecule drug (also frequently referred as the payload)—and its development relies on two key research areas: the selection of highly potent but cytotoxic drugs and the application of appropriate linkers. In our opinion, the drugs and linkers applied in ADCs may also be suitable for aptamer-drug chemical conjugates. Therefore, in this section, we review recent research advances in these two areas related to ADCs and ApDCs.

3.1. Cytotoxic Drugs

The selection of cytotoxic drug is critical because it affects efficacy toward the targeted tumor. Early research focused on well-established and clinically approved cytotoxic drugs, such as DOX, methotrexate, mitomycin, 5-fluorouracil, and Vinca alkaloids. However, the low efficacies of such

ADCs in preclinical and clinical evaluations showed that these drugs had limited potency, which encouraged the use of much more potent drugs that were too toxic to be used in an untargeted manner in ADCs [53]. Today, two main categories of highly potent cytotoxic anticancer agents are used in ADCs: DNA-modifying agents and microtubule-disrupting agents, which demonstrate at least 100–1000-fold greater potency compared with conventional chemotherapeutic agents [54].

The calicheamicins, which were originally isolated from a broth extract of the soil microorganism *Micromonospora echinospora* ssp. calichensis, are DNA-modifying agents that represent a novel structural class with powerful biological properties that confer potent antitumor activity (Figure 2). For example, calichensis shows extraordinary potency against murine tumors and is approximately 4000-fold more active than adriamycin, with an optimal dose of 0.5–1.5 μg/kg [55]. Maytansine, isolated from *Maytenus* species, is a microtubule-disrupting payload that inhibits the microtubule assembly, induces the microtubule disassembly and disrupts mitosis (Figure 3) [56]. Additionally, dolastatin 10 is a microtubule-disrupting agent for cancer chemotherapy that was isolated from the sea hare *Dolabella auricularia* [57]. Since these agents were first studied, many research groups have engaged in structure activity relation (SAR) studies of their synthetic analogs, termed "auristatins", such as monomethyl auristatin F (MMAF) and monomethyl auristatin E (MMAE) (Figure 4). The auristatins exhibit picomolar GI_{50} values in most cancer cell proliferation assays, caused by their ability to strongly inhibit microtubule assembly and tubulin-dependent GTP hydrolysis, resulting in cell cycle arrest and apoptosis [58].

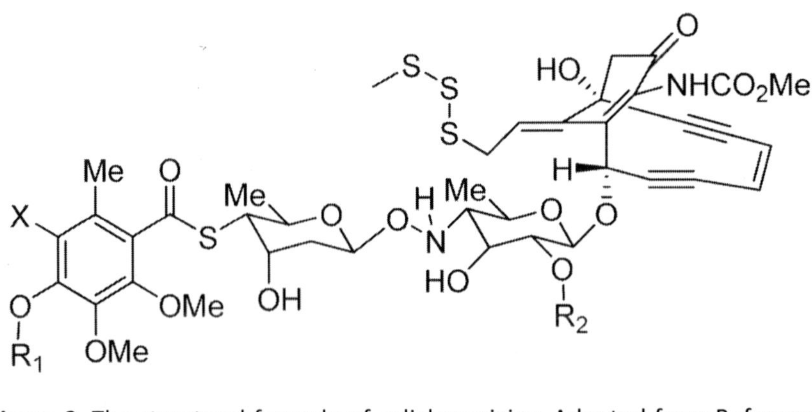

Figure 2. The structural formula of calicheamicins. Adapted from Reference [55].

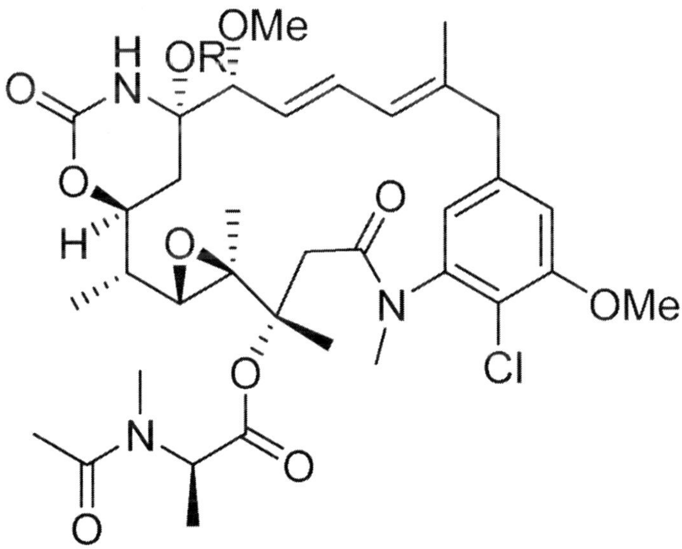

Figure 3. The structural formula of maytansines. Adapted from Reference [56].

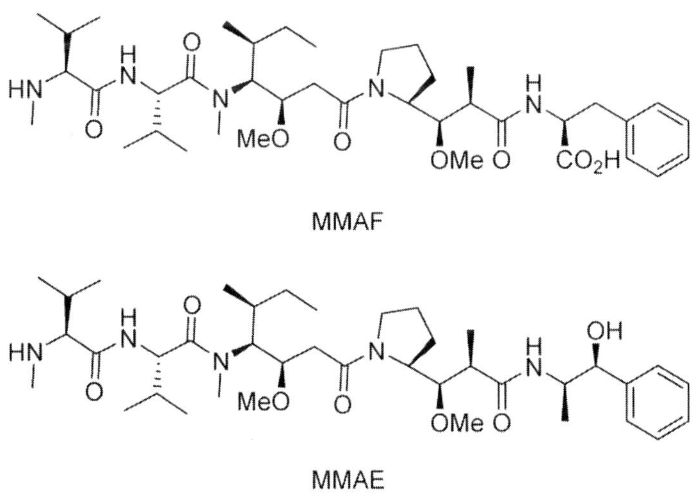

Figure 4. The structural formulas of monomethyl auristatin F (MMAF) and monomethyl auristatin E (MMAE). Adapted from Reference [57].

More recently, two promising cytotoxic agents were investigated: pyrrolobenzodiazepines (PBDs), which form dimers that bind to the minor groove of DNA in a sequence-selective manner [59,60], and the RNA polymerase II inhibitor α-amanitin [61]. Cytotoxic agents used in aptamer-drug chemical conjugates should possess the following properties: (1) high*in vitro* potency toward tumor cell lines, with IC_{50} values within the range of 0.01–0.1 nM; (2) a suitable functional group for linkage to a ligand (If a functional group is not already present, the desired substituent must be introduced at a suitable site to retain the potency of the parent drug); (3) reasonable solubility in aqueous solutions to enable reaction with the ligand; and (4) prolonged stability in the aqueous formulations commonly used for the ligand [62].

3.2. Linkers

One of the biggest challenges in the development of chemically conjugated targeted drugs is the application of suitable linkers for conjugating a ligand and a drug [63]. With the development of ADCs, a series of linkers had already been exploited. The linker that connects a cytotoxic drug to a ligand should be relatively stable in circulation to limit the damage caused to healthy tissues by highly active agents before they are delivered to the target site. Once the chemically conjugated targeted drug is internalized, however, the linker should facilitate efficient drug release. Based on the chemical structure and properties of the conjugated drug, three types of linkers can be selected: chemically labile linkers, enzyme-labile linkers and non-cleavable linkers.

3.2.1. Chemically Labile Linkers

Acid-Cleavable Linkers

Acid-cleavable linkers were the first type to be used. These linkers take advantage of the low pH in endosomes and lysosomal compartment to trigger hydrolysis of an acid-cleavable portion, such as a hydrazone bond, to release the payload.

A prominent example of an acid-cleavable linker-based ADC is BR96-DOX (Figure 5), which has curative efficacy in human tumor xenograft models. However, in clinical trials, significant toxicity was observed due to nonspecific cleavage of the relatively labile hydrazone linker and the expression of the antigen target in normal tissue [64]. The development of

DOX-based ADCs may still continue, and milatuzumab-DOX with a hydrazone bond linker, is currently in the advanced stages of clinical trials.

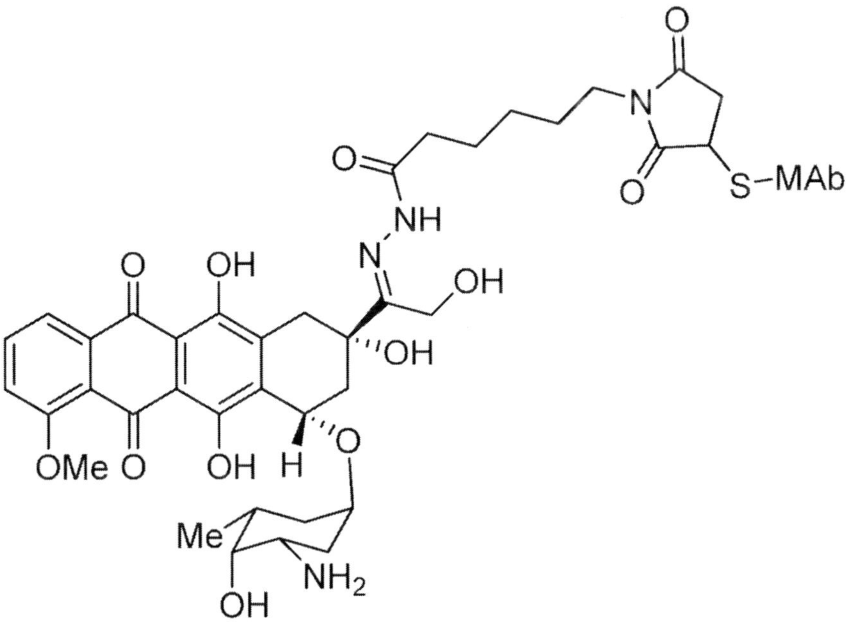

Figure 5. The structural formula of BR96-doxorubicin. Adapted from Reference [64].

Mylotarg was the first ADC ever approved by the FDA and is used as a second-line therapy for the treatment of relapsed acute myelocytic leukemia. This ADC consists of a humanized anti-CD33 monoclonal antibody (MAb) (hP67.6) attached to a calicheamicin drug through an acid-labile hydrazone linker. However, this linker is relatively unstable in plasma, with a half-life of approximately 48–72 h [65]. A hydrazone linker was also used in inotuzumab ozogamicin (an anti-CD22 calicheamicin conjugate). Despite using the same linker and payload as Mylotarg, this ADC appears to be more stable. This conjugate is also currently undergoing clinical evaluation [66].

The linker was also applied in ApDC to conjugate the aptamer Sgc8c and DOX, the resulting conjugate showed similar anticancer potency to unconjugated DOX [51].

Disulfide Linkers

The use of this type of linker is attractive because the concentration of glutathione in tumor cells is much higher than that in normal cells. In particular, irregular blood flow results in both a hypoxic environment in the tumor and enhanced activity of reductive enzymes, therefore leading to higher glutathione concentrations [67]. Thus, the linker, which is stable in circulation, can efficiently reduce the release of the active drug from the non-toxic prodrug at the tumor site, and elevate the stability by increasing steric hindrance [68].

Representative disulfide linker-based conjugates include maytansine derivatives conjugated to different MAbs. One maytansine derivative, DM1, has been conjugated to trastuzumab via a disulfide bond, allowing targeted delivery to tumor cells overexpressing human epidermal growth factor receptor 2 (HER2) [69]. The comparison between these antibody-maytansinoid conjugates with a "cleavable" linker containing a disulfide bond and a "non-cleavable" linker containing a thioether were shown in Figure 6. The conjugate huC242-SMCC-DM1, with a non-cleavable linker, was at least as potent *in vitro* as the conjugate huC242-SPDB-DM4, with a disulfide bond. However, huC242-SMCC-DM1 exhibited significantly lower *in vivo* activity in multiple xenograft tumor models [70,71]. The conjugates of huC242 with DM1 which linked via a disulfide bond could exert the bystander effect, whereas the conjugates with thioether bond could not [72]. Moreover, the clearance rate of the conjugate containing a reducible disulfide bond was approximately four times faster than that of the antibody component. The covalent linkage of 3–4 maytansinoid drug molecules to huC242 did not change the distribution of the antibody in mice [73].

Certain representative maytansine ADCs containing a reducible disulfide bond as a linker are currently undergoing clinical evaluation: SAR3419, lorvotuzumab mertansine (IMGN901), BT-062, BAY-94-9343 and SAR-566658.

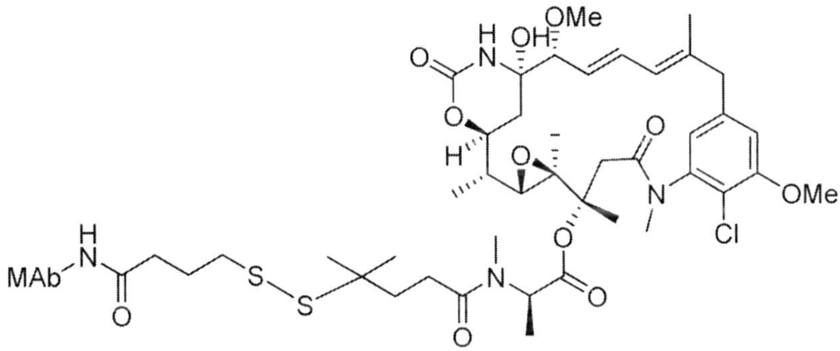

MAb-SPDB-DM4

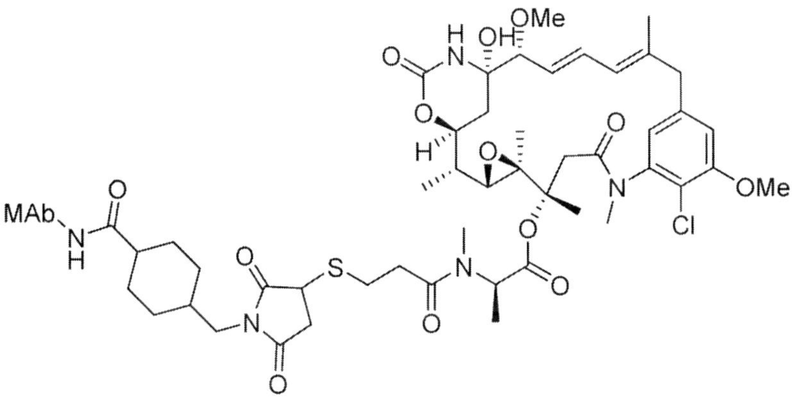

MAb-SMCC-DM1

Figure 6. The structural formulas of huC242-SPDB-DM4 and huC242-SMCC-DM1. Adapted from Reference [70].

3.2.2. Enzyme-Labile Linkers

Peptide Linkers

Peptide linkers, as the name suggests, attach a drug to an antibody via a peptide bond. Because proteases are normally not active outside cells due to unfavorable pH conditions and serum protease inhibitors, these peptide linkers are expected to have good serum stability. However, these linkers can be cleaved by lysosomal proteases, leading to elevated levels of free drug in certain tumor tissues. Many ADCs designed to utilize this linker strategy thus display an optimal balance between plasma stability and intracellular protease cleavage.

The most well-known conjugates involving a peptide linker are the MMAE and MMAF ADCs. Wahl described the *in vitro*and *in vivo* properties of MAb-drug conjugates consisting of auristatin E (AE) and MMAE linked to the chimeric MAbs cBR96 and cAC10 via an acid-labile hydrazone and protease-sensitive dipeptides. The researchers found that the peptide-linked MAb-valine-citrulline (Val-Cit)-MMAE and MAb-phenylalanine-lysine(Phe-Lys)-MMAE conjugates were much more stable in buffers and plasma compared with the MAb conjugates linked via the hydrazone of 5 benzoylvaleric acid-AE ester (AEVB) (Figure 7). The MAb-Val-Cit-MMAE conjugates released only 2% of MMAE in the following 10 days after being incubated in human plasma, whereas they could easily be cleaved by lysosomal protease after internalization (Scheme 1). Compared with the corresponding hydrazone conjugates, the MAb-Val-Cit-MMAE conjugates showed greater*in vitro* specificity and lower *in vivo* toxicity. The therapeutic indices were shown to be as high as 60-fold for the peptide-linked conjugates induced regression and cure of established tumor xenografts. The treatment with cAC10-Val-Cit-MMAE at 1 mg mAb component/kg/injection and at 0.5 mg/kg/injection resulted in 100% and 80% tumor inhibition, respectively [74,75,76].

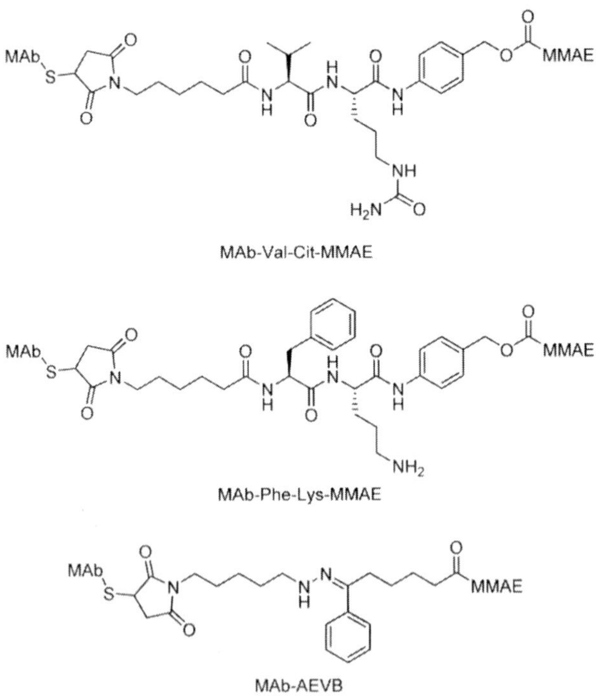

Figure 7. The structural formulas of MAb-Val-Cit-MMAE, MAb-Phe-Lys-MMAE and MAb-AVEB. Adapted from Reference [74,75,76].

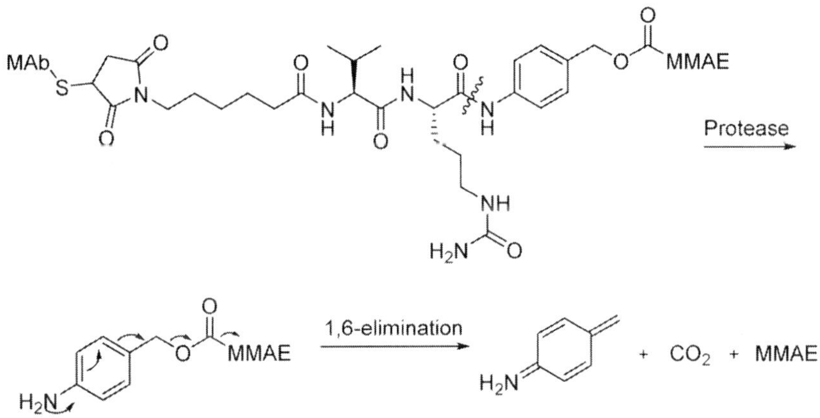

Scheme 1. Presumed release mechanism of MMAE from an ADC incorporating a peptide linker. Adapted from Reference [74,75].

In July 2011, Adcetris, which consists of an anti-CD30 MAb attached through a Val-Cit dipeptide to an average of four MMAE molecules, was approved by the FDA for the treatment of patients with Hodgkin's lymphoma after failed autologous stem cell transplantation, or patients with systemic anaplastic large-cell lymphoma after the failure of at least one prior multi-agent chemotherapy regimen [65]. There are also certain representative MMAE- or MMAF-containing ADCs containing a peptide bond linker that are undergoing clinical evaluation now: RG-7596, glembatumumab vedotin (CDX-011), PSMA-ADC and ASG-5ME.

β-Glucuronide Linkers

This type of linker is exploited such that drug release requires cleavage by β-glucuronidase, an enzyme present in lysosomes and the tumor interstitium [77]. The linker is hydrophilic, so it could reduce the ADC aggregation of the hydrophobic drugs [78].

Cyclopropyl indole minor-groove binders (MGBs) have been conjugated to the MAbs cAC10 (anti-CD30) and h1F6 (anti-CD70) using a β-glucuronide linker. The resulting ADCs were non-aggregated and monomeric, even when heavily loaded with hydrophobic anticancer drugs (up to eight drugs/MAb). The free drug was released through the cleavage of the β-glucuronide linker [79].

This linker was also used in ADCs bearing camptothecin analogs, and the ADCs showed high potency, immunologically specificity and efficacy at well-tolerated doses in a renal cell carcinoma xenograft model [80].

3.2.3. Non-Cleavable Linkers

Non-cleavable linkers have also proven to be very potent. The release mechanism is believed to occur via internalization of the ADC, followed by the degradation in the lysosome, resulting in the release of the drug while still attached to the linker. As such, when the payloads could maintain the antitumor effects even if being chemically modified, the non-cleavable linker is a good choice.

Early studies that described non-cleavable linkers were aimed at amine-to-thiol coupling. One advantage of these linkers is their greater stability in the circulation compared with cleavable linkers, which can potentially improve the therapeutic index because it may be better tolerated. Senter found that MMAF, conjugated to a MAb via a non-cleavable maleimidocaproyl linker (cAC10-L4-MMAF), was approximately as potent as when conjugated via a cleavable peptidic linker (cAC10-L1-MMAF) against a large panel of cell lines both *in vitro* and *in vivo*. Importantly, cAC10-L4-MMAF was tolerated at a dose more than three times higher than the MTD of cAC10-L1-MMAF. The half-lives of such non-cleavable linkers in circulation are much longer than those of earlier-generation cleavable linkers (Figure 8). Nearly all animals that received single ADC injections at 2 mg antibody component/kg body weight were cured. When lowering the dose to 1 mg/kg, there were two of six and three of six animals cured by cAC10-L1-MMAF4 and cAC10-L4-MMAF4 ADCs, respectively [81,82].

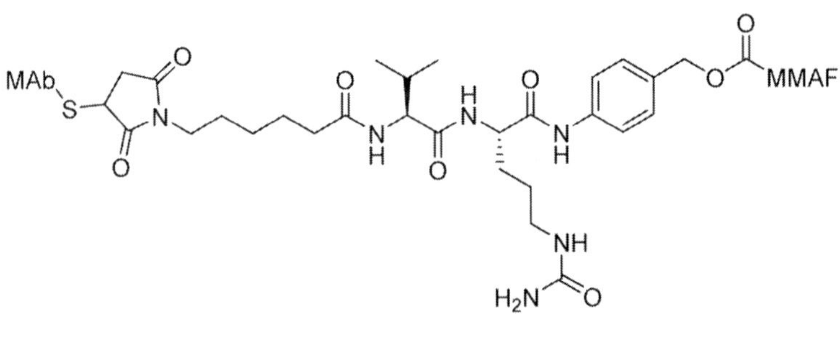

cAC10-L1-MMAF

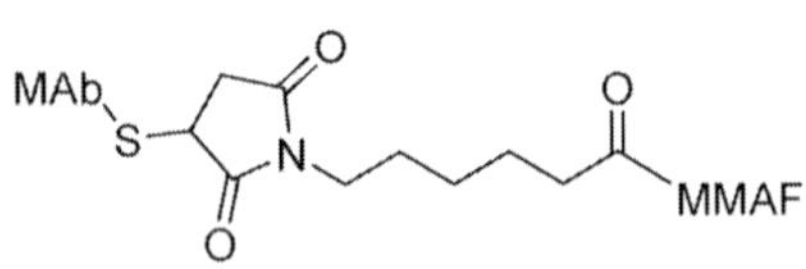

cAC10-L4-MMAF

Figure 8. The structural formulas of cAC10-L1-MMAF and cAC10-L4-MMAF. Adapted from Reference [81,82].

When examining the intracellular release of a series of disulfide-linked ADCs, it was unexpectedly found that a trastuzumab-DM1 control conjugate, in which the disulfide was replaced with a non-cleavable thioether bond, was the most effective ADC variant in *in vivo* therapy tests [83].

This type of linker is successfully used in Kadcyla, which was approved by FDA in 2013 and has been used to treat patients with HER2-positive metastatic breast cancer. In addition, IMGN529 containing a thioether bond linkage has entered phase I clinical trials. However, through a retro-Michael reaction, the thioether succinimide can revert to the starting materials thiol and maleimide. The reversibility of this connection between antibody and linker has been demonstrated in several ADCs and has been shown to result in a decrease in the efficacy of the corresponding drug conjugate [80]. In order to solve the problem, there were a series of recent attempts to design more stable linkers [84,85,86,87].

Recently, Zu *et al.* [88] identified a single-strand DNA aptamer specifically binding to the biomarker CD117, which is highly expressed on AML cells, then conjugated the aptamer to methotrexate (MTX) using a non-cleavable bond. The formed Apt-MTX conjugates specifically inhibited AML cell growth and had little effect on CD117-negative cells under the same treatment conditions.

In summary, a series of studies have demonstrated that the appearance of ADCs and ApDCs boosted the development of chemically conjugated targeted drugs. Regarding ADCs in particular, the development of aptamer-drug chemical conjugates has relied on two key areas of research: the choice of potent cytotoxic drugs and the application of appropriate linkers. The drugs selected for a chemical conjugation and targeting should have two characteristics in common: high potency and ease of modification. The exact linker between a cytotoxic agent and a ligand has profound effects on the selectivity, PK, therapeutic index, and overall success of a chemically conjugated targeted drug. Each linker has its advantages and disadvantages, and many factors must be evaluated when selecting an appropriate linker, including the existing groups presented in the carrier and the introduction of special reactive groups into agents. Certain cytotoxic payloads that are not active with substitutions require a cleavable linker. In contrast, other cytotoxic payloads that can accommodate substitutions while maintaining potency are good substrates for the development of non-cleavable linkers.

Although chemically conjugated targeted drugs exhibit certain advantages in terms of the targeted delivery of drugs to tumor cells, certain challenges remain, including low drug loading and costly multistep procedures involved in the design and synthesis. Future work will be necessary to increase drug loading, to reduce costs or to identify a better approach for achieving targeted drug delivery.

4. APTAMER-NANOMATERIAL CONJUGATED SYSTEMS

Nanomaterials have unique physicochemical properties, such as an ultra-small size, a large surface-area-to-mass ratio, and high reactivity, in contrast to bulk materials of the same composition [89]. These properties could be used to overcome several of the limitations of traditional therapeutic and diagnostic agents. These particles allow exquisite modification for binding to target cell membranes, components of the microenvironment, or cytoplasmic/nuclear receptor sites [89]. Combining the advantages of nanomaterials with the high affinity and specificity of aptamers, aptamer-functionalized nanomaterials have drawn great attention in drug delivery. Based on their different compositions, the nanomaterials applied in aptamer-targeted drug delivery systems can be divided into two categories: inorganic and organic nanomaterials. In this section, several representative aptamer-nanomaterial conjugated systems are discussed.

4.1. Inorganic Nanomaterials

4.1.1. Gold Nanoparticles

Gold NPs (AuNPs) have attracted considerable attention as drug delivery platforms because of their numerous advantageous properties. First, AuNPs have been shown to be essentially inert, non-toxic, and biocompatible. Second, AuNPs can be easily modified with aptamers due to their predictable and reliable surface modification chemistry, and the resulting AuNPs possess the affinity and specificity of the aptamer. Third, the optical and electronic properties of AuNPs are highly shape- and size-dependent, which has led to their use in many biomedical applications [90,91].

Huang *et al.* [92] devised an aptamer/hairpin DNA-AuNP (apt/hp-AuNP) conjugate for targeted drug delivery. The DNA aptamer sgc8c which possesses strong affinity for protein tyrosine kinase 7 (PTK7) was used as the ligand, and the hairpin DNA was used for loading of the anticancer drug DOX. The capability of the apt/hp-AuNP conjugate binding to target cells was investigated by flow cytometry and atomic absorption spectroscopy, which showed that the aptamer-functionalized nanoconjugates were selective for cancer cells. When illuminated with a continuous-wave laser, corresponding to the resonant wavelength of Au NPs, the conjugate could release DOX and more effectively kill targeted cancer cells. Most interestingly, the nanoconjugates could enhance the antitumor efficacy with few side effects when illuminated with plasmon-resonant light (532 nm) (Scheme 2). The inhibition concentration (IC_{50}) of Dox:sgc8c/hp AuNPs was quite similar to that of free Dox.

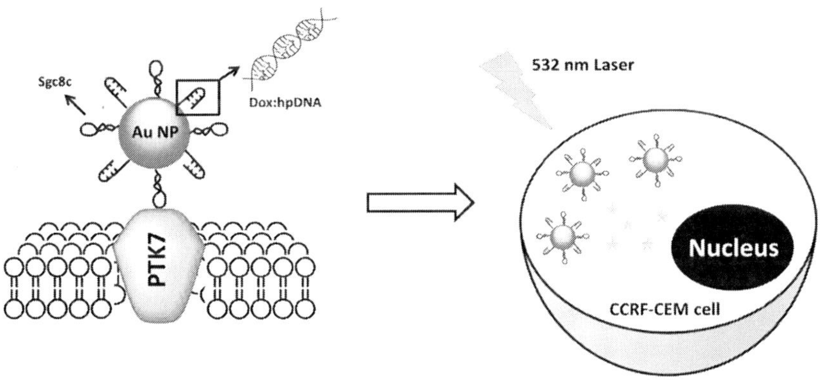

Scheme 2. Light-induced Dox release from Dox:apt/hp-Au NP nanocomplexes inside targeted cancer cells (schematic). Reproduced with permission from Reference [92].

Moreover, Qu *et al.* [93] constructed novel gold nanorods (GNRs) with a mesoporous silica framework (AuMPs). The surfaces of these nanorods were then modified with the AS1411 aptamer (Ap-AuMPs). One oligonucleotide containing the AS1411 aptamer with a 12-base extension was chosen as the capping agent (DNA-2). Another 12-mer oligonucleotide (DNA-1) was covalently attached to the exterior of the mesoporous particles. Thus, the DNA could form an assembly with two identical duplex regions connected by a dimeric G-quadruplex. The duplex portion could serve as a linker anchored to the mesoporous silica NP (MSN) surface, and the G-quadruplex was utilized as a cap for trapping the guest molecules. When exposed to a laser beam matching the absorption peak of the GNRs, the duplex DNA denatured at an increased temperature, allowing the release of the entrapped drug.

Aptamer-photosensitizer conjugates have been used for photodynamic therapy (PDT), with a high degree of selectivity [94,95]. In particular, Tan *et al.* [94] used the photosensitizer molecule chlorin e6 (Ce6) conjugated to aptamer-GNRs for photothermal therapy (PTT). In this case, aptamer sgc8 which binds the cell membrane protein tyrosine kinase-7 (PTK7) with high affinity and selectivity was used as the ligand. Due to the high expression on cell surface and internalization function of PTK7 in cancer cells, sgc8 was taken up by cells and delivered into the endosomes. When the targeted cancer cells were absent, Ce6 was quenched and demonstrated no PDT effect. However, when the targeted cancer cells were present, the aptamer changed its structure to release Ce6 and produced singlet oxygen for PDT upon light irradiation. PTT/PDT dual therapy presented a more effective therapeutic outcome than either therapeutic modality alone by combining photosensitizer and photothermal agents (Scheme 3). In another case, aptamers were conjugated to the surfaces of GNRs through thiol-gold covalent linkages, and the resulting conjugates were successfully used to target and kill both cancer cells and cancer stem cells by near-infrared (NIR) laser irradiation [95].

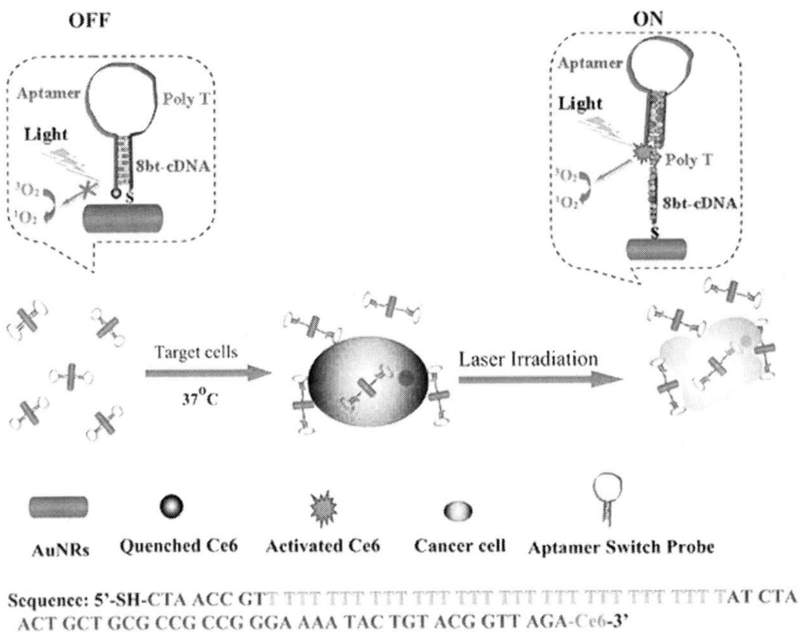

Scheme 3. Schematic representation of ASP_photosensitizer_AuNRs for PTT and PDT. Reproduced with permission from Reference [94].

However, *in vivo* experiments were not performed in none of the studies. Furthermore, the accumulation and potential long-term toxicity of AuNPs have not yet been investigated, and the high costs of production in large scale also remain a challenge [96].

4.1.2. Nano-Scale Iron Oxides

Nano-scale iron oxides have exhibited great potential for biological applications [97]. Recently, a smart multifunctional nanostructure (SMN) containing an iron porous hollow magnetite NP (PHMNP), a heterobifunctional polyethylene glycol (PEG) ligand, and an aptamer sgc8 was constructed by Tan *et al.* [98]. In this case, multiple aptamers on the surface of a single SMN led to enhanced binding and uptake by target cancer cells due to the multivalent effect. The pores of the PHMNPs were stable at physiological pH, limiting the damage caused to healthy tissues by highly active anticancer agents. However, once the SMN reached the lysosomes of the target cancer cells through receptor-mediated endocytosis, the relatively low lysosomal pH levels resulted in the corrosion of the PHMNP pores, facilitating the release of DOX to kill the target cancer cells. The results showed that enhanced killing efficacy and improved targeting specificity could be achieved by using SMNs.

Among nano-scale iron oxides, superparamagnetic iron oxide NPs (SPIONs) have attracted significant attention as drug delivery platforms due to their special structure. SPIONs typically have one of the following two structural configurations: (i) a magnetic particle core is coated with a biocompatible polymer or (ii) a porous biocompatible polymer in which magnetic particles are precipitated inside the pores. The coating can be functionalized by attaching other molecules to increase the targeting yield. These molecules then act as attachment points for the coupling of cytotoxic drugs or target antibodies to the carrier complex [99,100]. Furthermore, SPIONs have several advantages, including stronger enhancement of proton relaxation and lower detection limits [101,102].

Ferumoxtran-10 (Combidex) is currently in phase III clinical trials for prostate cancer (PCa) imaging in which the dextran is coated with SPION [103]. The major shortcoming of Combidex is its inability to detect PCa disease outside the lymph nodes. To address this shortcoming, Farokhzad *et al.* [104] reported a novel thermally cross-linked SPION aptamer (TCL-SPION-Apt) bioconjugate in which the TCL-SPION was functionalized with an A10 RNA aptamer. DOX was intercalated in the aptamer and complexed with the TCL-SPION through charge interactions. The TCL-SPION-Apt bioconjugates could be used as therapeutic carriers for the selective delivery to PSMA-expressing cells. The cytotoxicity of Dox-loaded TCL-SPION–Apt bioconjugates was nearly as potent as free Dox. Using a similar process, aptamer-conjugated TCL-SPION was reported to be PCa-specific nanotheranostic agents [105]. These agents were capable for prostate tumor detection *in vivo* by simultaneous magnetic resonance imaging (MRI) and selective delivery of drugs to the tumor tissue. However, the exposure of the iron oxide core would cause toxicity in neurological systems; therefore, the potential of TCL-SPION-Apt bioconjugates as a targeted delivery system requires further validation prior to *in vivo* use. In addition, despite of the considerable potential benefits of SPIONs, SPIONs might still cause cellular disturbances, such as modulation of the actin cytoskeleton, alterations in gene expression profiles, and so on [106,107]. There is thus a need to identify potential cellular damages associated with these NPs.

Nano-scale iron oxides have attracted considerable attention as drug delivery platforms for their numerous advantageous properties, such as superparamagnetic property, low cytotoxicity, colloidal stability, and bioactive molecule conjugation capability. The cost of production is acceptable for application in drug delivery systems [97].

4.1.3. Carbon Nanomaterials

Representative carbon nanomaterials that have been applied in aptamer-nanomaterial conjugated systems are single-walled carbon nanotubes (SWNTs). These materials are emerging as promising delivery vehicles for targeted drug delivery due to several important properties: (1) relatively easy internalization by cells; (2) a high capacity for drug loading and a large surface area for attaching specific targeting molecules either covalently or non-covalently; and (3) unique optical properties, such as NIR fluorescence and absorption, Raman scattering, and photoacoustic properties [108].

Tan *et al.* proposed a molecular complex AP-SWNT which was composed of a photosensitizer Ce6, an ssDNA aptamer (AP), and SWNTs for controllable singlet oxygen generation (SOG). Singlet oxygen is an alternative noninvasive cancer treatment generated during PDT [109]. The AP-SWNT design was based on the attachment of a DNA aptamer and photosensitizer (Ce6) to SWNTs, with the aptamer interacting noncovalently with the SWNTs by π-stacking between nucleotide bases and the SWNT side walls. When the target was absent, the SOG would be quenched by SWNT for the close proximity of the photosensitizer to the SWNT surface. However, in the presence of its target, the interaction between the aptamer and the target molecule could cause the DNA to fall off the SWNT surface, then lead to a restoration of SOG for PDT applications (Scheme 4). The resulting AP-SWNTs maintained the advantages of aptamers, presented an excellent specific response to thrombin and enhanced the efficiency and reliability of PDT. More specifically, when 2.0 μM thrombin was introduced, the singlet oxygen sensing of the AP-SWNTs exhibited a 13-fold enhancement.

Zhang's group [110] constructed a multifunctional tumor-targeting drug delivery system, in which the SWNTs were used as drug carriers, AS1411 as a targeting ligand and DOX as a model chemotherapy drug. The delivery system could retain the optical properties of SWNTs and the cytotoxicity of DOX. More importantly, it could accumulate in tumors and facilitate the combination of chemotherapy and PTT, effectively promoting DOX cellular uptake and intracellular accumulation.

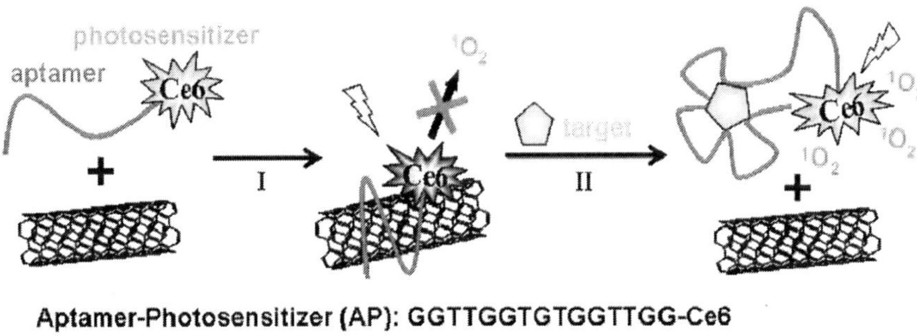

Scheme 4. Schematic of aptamer-photosensitizer-SWNT complex and the regulation of SOG upon target binding. Reproduced with permission from Reference [109].

Graphene oxide is another carbon nanomaterial that has been applied in aptamer-nanomaterial conjugated systems. For example, Cai *et al.* [111] used cell type-specific aptamer-functionalized graphene oxide to deliver decitabine. In particular, the researchers conjugated graphene oxide to an aptamer A1 (a 45-base oligonucleotide that binds to A549 cell with high specificity and affinity) and then loaded decitabine onto the surface of the graphene oxide. The resulting complexes could specifically distinguish A549 cells from other types of cancer cells and subtypes of lung cancer cells. The drug release rate is much slower at physiological conditions than at acidic conditions. Importantly, cell viability assay results revealed that the complex displayed much higher therapeutic efficacy compared with the free drug.

Although these materials provide the advantages of low cost and high biocompatibility, the toxicological effects of carbon nanomaterials should also be considered. The concentration, physical form, degree of functionalization, and agglomeration state might all affect the toxicity of carbon nanomaterials [112,113].

4.1.4. Mesoporous Silica Nanoparticles

Mesoporous silica nanoparticles (MSNs) are considered as the most promising candidate for targeted drug delivery because of the following unique properties: (1) a tunable particle size; (2) a stable and rigid framework; (3) a uniform and tunable pore size; (4) a high surface area and a large pore volume; (5) two functional surfaces; and (6) a unique porous structure [114].

Wang *et al.* [115] devised an efficient target drug delivery system for cancer therapy which was based on MSN–PEM–aptamer conjugate. In the system, MSNs were employed as drug carriers. Polyelectrolyte multilayers (PEMs) composed of thiolated poly(methacrylic acid) (PMASH) coated with the MSNs were used to prevent the premature leakage of drugs, and to controllably release drugs under reducing conditions. sgc8 was selected as the targeting recognition molecule. When this conjugate was exposed to a reducing environment, the deconstruction of PEM occurred for the cleavage of disulfide bonds and released the drug. The resulting drug delivery system combines the advantages of high drug-loading efficacy, stimulus responsiveness in the intracellular environment and high cell recognition ability.

Lu *et al.* [116] integrated a cancer-targeting DNA aptamer with mesoporous silica particles to construct a targeted drug delivery system. In this system, the nucleolin aptamer AS1411 was conjugated to the external MSN surface, and the drug molecules were encapsulated inside the nanopores. The functionalized MSN could selectively bind to nucleolin on the surface of MCF-7 breast cancer cells and was internalized via receptor-mediated endocytosis. This novel targeted drug delivery system showed pH-dependent controlled-release kinetics with specific binding to various target cells, thus ensuring the delivery of high doses of other therapeutics.

Similar to carbon nanomaterials, the targeted drug delivery systems based on MSNs is attractive for the numerous advantageous properties, such as low cost and high biocompatibility. However, many issues still need to be studied, such as the pharmacokinetics (PK) and pharmacodynamics (PD) of drugs loaded in MSNs, the acute and chronic toxicity, as well as the long-term *in vivo* degradation and compatibility of MSNs. In addition, the methods developed for synthesizing functional MSNs with a core-shell structure are limited to the production of a small amount of NPs [113,117].

4.1.5. Quantum Dots

QDs (Quantum Dots) are microscopic metal or semiconductor boxes that hold a certain number of electrons. Due to their unique optical properties, including broad absorption with narrow photoluminescence spectra, a high quantum yield, low photobleaching, and resistance to chemical degradation, this material has been extensively investigated as a potential drug delivery vehicle [118,119].

Farokhzad *et al.* [120] developed a novel QD-Apt-DOX conjugate. The conjugate was formed by the intercalation of DOX within the aptamer on the surface of QDs, and the QD and DOX fluorescence were quenched through a Bi-FRET mechanism (Scheme 5): the fluorescence of the QD was specifically quenched by DOX, while the fluorescence of DOX was simultaneously quenched by intercalation within the A10 PSMA aptamer. However, after selective uptake of the conjugate into the target cancer cell, the released DOX induced the recovery of fluorescence from both the QD and itself, thereby revealing the intracellular delivery of DOX and enabling the synchronous fluorescent localization and killing of cancer cells. The cytotoxicity of QD-Apt(Dox) conjugate was nearly equivalent potent to that of free Dox.

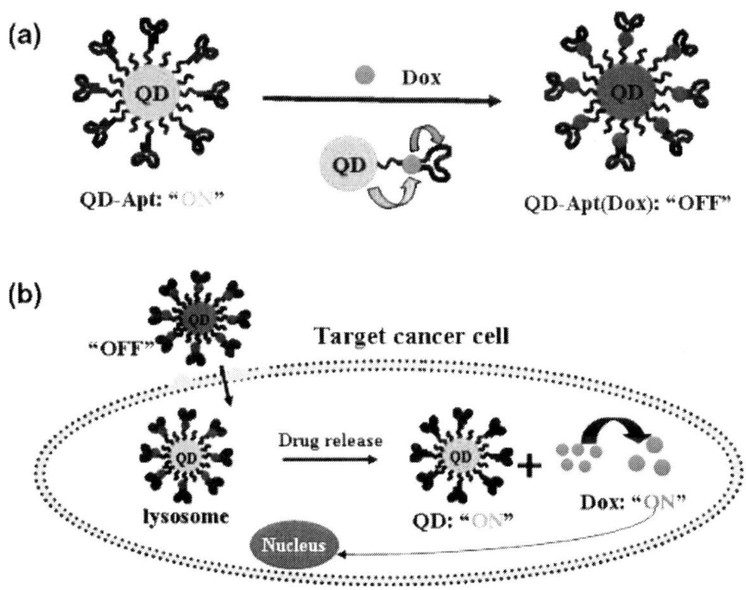

Scheme 5. (**a**) Schematic illustration of QD-Apt(Dox) Bi-FRET system; (**b**) Schematic illustration of specific uptake of QD-Apt(Dox) conjugates into target cancer cell through PSMA mediate endocytosis. Reproduced with permission from Reference [120].

Savla *et al.* [121] reported the design and delivery of a tumor-targeted, pH-responsive QD-mucin1 aptamer-DOX (QD-MUC1-DOX) conjugate for chemotherapy for ovarian cancer. To achieve active cancer targeting, QDs were conjugated to a DNA aptamer specific for mutated MUC1, which is overexpressed in many cancer cells, including ovarian carcinoma. DOX was attached to the QDs via a pH-sensitive hydrazone bond to stabilize the complex in the systemic circulation and to allow drug release in the acidic environment inside cancer cells. The conjugate had higher cytotoxicity than free DOX in multidrug-resistant cancer cells and was preferentially accumulated in ovarian tumors *in vivo*.

Although QDs offer potentially invaluable societal benefits for targeted drug delivery, they may also pose risks to human health and the environment under certain conditions. In fact, several studies have strongly suggested that minor concentrations of QDs may be sufficient to cause long-lasting, even transgenerational, effects [122,123,124].

4.2. Organic Nanomaterials

4.2.1. Liposomes

Due to their size and their dually hydrophobic and hydrophilic characteristics, liposomes are promising systems for drug delivery. Indeed, liposomes have been used successfully as drug delivery systems, and a number of liposome-based systems have been approved by FDA for disease treatment in the clinic [125,126]. Liposomes can both encapsulate hydrophilic therapeutic agents inside their aqueous core and load hydrophobic drugs within their lipid bilayer membranes. In certain approaches, the liposome surface can also be passivated with different ligands, such as PEG, extending liposomes' systemic circulation time and resulting in preferential accumulation at tumor sites. Liposomes can be further improved by incorporating molecular recognition moieties, which can lead to drug transport with better efficacy and fewer side effects [127].

Lu *et al.* [128] developed aptamer-liposome bioconjugates that could effectively deliver cisplatin in a cancer cell-specific manner, and the selectivity of cisplatin was largely improved. The liposomes with an aptamer derived from AS1411 as a ligand, which had high binding affinity to nucleolin, could be internalized into cells through nucleolin-mediated endocytosis. The cell viability was approximately 59.5% at day 4 when the MCF-7 cells were treated with the aptamer-liposome-cisplatin, however, it was only 88.6% and 88.9% at day 4 for two control tests involving cisplatin. This study provided further evidence that the aptamer-mediated cancer-targeting strategy was highly specific and could be modulated for desired drug delivery applications.

Tan *et al.* [129] used a well-established liposome preparation protocol to prepare aptamer-modified liposomes. To increase liposome stability and to sustain aptamer stability in the serum, the liposomes were coated with the polymer PEG, and aptamers were conjugated to the surface of the liposomes to facilitate target binding. The formed sgc8 aptamer-liposome conjugates were able to selectively bind to their target cells, with no binding to non-target NB4 cells. In each liposome, there were approximately several thousand fluorescein isothiocyanate-dextran molecules, indicating a very high loading capacity.

Lu *et al.* [130] reported AS1411 aptamer-functionalized liposomes as a drug delivery system for targeted anticancer drug doxorubicin. The research showed that the liposomes could increase cellular internalization and cytotoxicity to MCF-7 breast cancer cells compared with non-targeted liposomes. Furthermore, improved antitumor efficacy against xenograft MCF-7 breast tumors in athymic nude mice was attributed to the liposomes' enhanced tumor tissue penetration.

Kim *et al.* [131] designed novel anticancer drug-encapsulating liposomes which were conjugated to a PSMA-specific RNA aptamer. The resulting liposomes specifically bound to LNCaP prostate epithelial cells and, thus, significantly enhanced the*in vitro* cellular binding and uptake of the NPs compared with non-targeted NPs. Moreover, the liposomes were significantly more toxic to the targeted LNCaP cells than to non-targeted cancer cells. *In vivo* efficacy showed that the tumor was significantly reduced when was treated with the NPs as compared with the non-targeted NPs and free DOX.

Recently, Fattal *et al.* [132] successfully conjugated a 2'-F-pyrimidine-containing RNA aptamer to the surface of PEGylated liposomes using the thiol-maleimide click reaction. The functionalized liposomes exhibited significantly enhanced cellular uptake compared with blank liposomes as well as promising potency as a specific drug delivery system.

4.2.2. Poly(lactide-*co*-glycolic acid) Nanoparticles

Because poly(lactide-co-glycolic acid) (PLGA) is a biocompatible and biodegradable polymer that has been approved by FDA and has an established clinical safety record [133], drug delivery systems based on PLGA NPs play an important role in cancer therapy. PLGA NPs are usually functionalized with PEG because PEGylated polymeric NPs exhibit significantly reduced systemic clearance compared with similar particles without PEG [134,135].

Farokhzad *et al.* [136] described docetaxel (Dtxl)-encapsulating NPs formulated with PLGA-*b*-PEG copolymer, wherein the surface was functionalized with A10 2'-fluoropyrimidine RNA aptamers via an amide bond (Scheme 6). The resulting conjugates (Dtxl-NP-Apt) bound to the PSMA protein, which was expressed on the surface of LNCaP prostate epithelial cells, and were taken up by these cells, resulting in significantly enhanced *in vitro* cellular toxicity compared with non-targeted NPs lacking the PSMA aptamer. The Dtxl-NP-Apt conjugates also exhibited remarkable efficacy and reduced toxicity *in vivo*. Subsequently, A10 aptamer-functionalized PLGA-PEG NPs were used as a vehicle for transport of a platinum (Pt) (IV) prodrug. A Pt(IV) compound with alkyl chains at the axial positions was first prepared, after which the compound was encapsulated in PEGylated PLGA NP conjugates that bound to the A10 2'-fluoropyrimidine RNA aptamer. These resulting aptamer-derivatized Pt(IV)-encapsulating NPs were significantly superior to cisplatin or non-targeted NPs in terms of activity against LNCaP cells [137]. The conjugate was then investigated *in vivo*, and the results showed that the 10-days maximum tolerated doses following a single intravenous injection of the conjugate in rats and mice were 40 and 5 mg/kg, respectively. PK studies with the conjugate revealed prolonged Pt persistence in the systemic blood circulation and decreased accumulation of Pt in the kidney. Moreover, the conjugate exhibited significant dose-sparing characteristics with regard to the drug, with equal efficacy at 1/3 the dose of cisplatin administered in its conventional form [138].

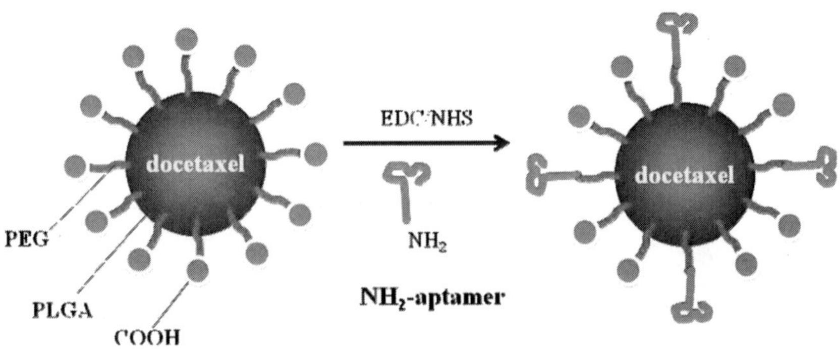

Scheme 6. Schematic representation of the synthesis of PLGA-PEG-COOH copolymer and strategy of encapsulation of Dtxl. Adapted from Reference [136].

Farokhzad *et al.* [139] prepared a biointegrated triblock copolymer (TCP) consisting of PLGA, PEG, and the A10 aptamer (Apt). This TCP could be used for the self-assembly of targeted NPs for PCa targeting. The researchers found that increasing the Apt density on the NP surface could increase the rate of NP uptake by LNCaP cells *in vitro* and that a high Apt surface density could increase NP accumulation in the liver and spleen. Therefore, it was necessary to balance the tumor-targeting ligand surface density and the antibiofouling surface properties. Using different ratios of functional biomaterials (PLGA-PEG conjugated to aptamer) to non-functional biomaterials, the surface aptamer density could be adjusted, and the optimal ratio for maximal PCa cell uptake both *in vitro* and *in vivo* was identified.

Tan *et al.* [140] synthesized aptamer-coated PLGA hybrid NPs with core-shell lipid-polymeric structures via simple nanoprecipitation and self-assembly. After self-assembly, hydrophobic PLGA with encapsulated hydrophobic paclitaxel (PTX) constituted the core structure, and lecithin, DSPE-PEG and lipid-PEG-aptamer loaded with DOX formed the hydrophilic shell. The cytotoxicity results showed that this targeted co-delivery system selectively enhanced antitumor efficacy.

4.2.3. Polymeric Micelles

Polymeric micelles (PMs) can be used for drug targeting due to their inherent properties, including stability in plasma and longevity *in vivo*. Additionally, the pathological characteristics of tumors allow PMs to be targeted to the tumor site via enhanced permeability and retention. The amphiphilic block copolymer of PMs is particularly suitable for the encapsulation of hydrophobic drugs, and PMs can be conjugated to many ligands on their outer shell to bring the micelles to cancer cells [141] The aptamer-micelles can simply interact with the cell membrane and quickly release the doped hydrophobic molecules into the cells [142].

Tan *et al.* [142] reported the design of a self-assembled aptamer-micelle nanostructure (Scheme 7). In the aptamer assembly, the aptamer TDO5 not only acted as the building block for the nanostructure but also performed a recognition function, identifying its specific target. Interestingly, this nanostructure showed an 80-fold enhancement in selective binding to Ramos cells at 37 °C compared with the free aptamer. Furthermore, the aptamer-micelles showed great dynamic specificity in flow channel systems used to mimic drug delivery in a flowing system.

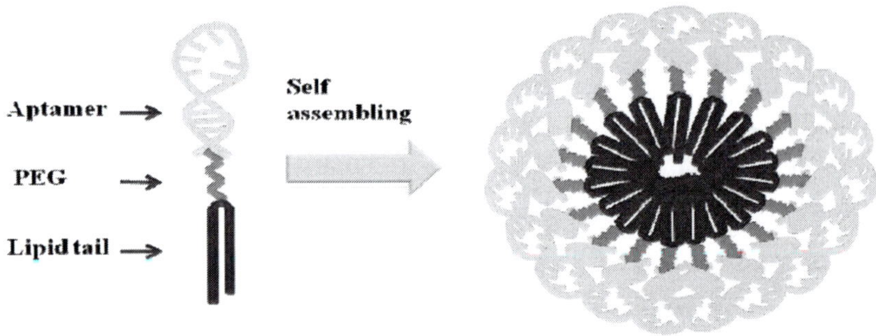

Scheme 7. Schematic illustration of aptamer–micelle formation. Adapted from Reference [142].

Hao *et al.* [143] developed novel aptamer-functionalized PEG-polylactic acid micelles (APPs) with the objective of targeting the transferrin receptor on brain endothelial cells. Flurbiprofen was loaded into the APPs using the co-solvent evaporation method. The experimental results indicated that the APPs significantly enhanced the surface association of micelles with bEND5 cells. Furthermore, the APPs significantly enhanced intracellular flurbiprofen delivery compared with unmodified micelles.

Gong *et al.* [144] prepared aptamer-conjugated unimolecular micelles that were formed by individual hyperbranched polymer molecules, consisting of a hyperbranched H40 polymer core and approximately 25 amphiphilic polylactide (PLA)-PEG block copolymer arms (H40-PLA-PEG-Apt). The micelles exhibited pH-sensitive, controlled drug release and much higher cellular uptake among PSMA-positive CWR22Rn1 prostate carcinoma cells than non-targeted unimolecular micelles. In CWR22Rn1 tumor-bearing mice, DOX-loaded H40-PLA-PEG-Apt micelles exhibited much higher levels of DOX accumulation in the tumor tissue than DOX-loaded H40-PLA-PEG micelles.

Qiu *et al.* [145] developed an AS1411-functionalized composite micelle (CM-Ap) composed of AS1411 (Ap)-modified Pluronic F127, β-cyclodextrin-linked PEG-b-PLA and DOX. The study revealed that the cellular uptake of CM-Ap was higher than the uptake of untargeted CM due to the nucleolin-mediated endocytosis effect. *In vivo* testing also demonstrated that the CM-Ap exhibited a prolonged circulation time in the blood, enhanced accumulation in tumors, improved antitumor activity, and decreased cardiotoxicity.

Chen *et al.* [146] developed Apt-mixed micelles consisting of a pH-responsive copolymer, D-α-tocopherol PEG 1000-block-poly-(β-amino ester)

(TPGS-b-PBAE, TP), and AS1411 aptamer (Apt)-decorated TPGS polymer (Apt-TPGS). PTX was encapsulated in the Apt-mixed micelles. The resulting PTX/Apt-mixed micelles were stable at pH 7 but dissolved and quickly released the encapsulated PTX in a weakly acidic environment (pH 5.5). Compared with non-Apt-modified mixed micelles, more Apt-modified mixed micelles were internalized by SKOV3 ovarian cancer cells, whereas no remarkable difference in cellular uptake was observed in normal cells (LO2 cells). Moreover, the micelles exhibited significantly increased cytotoxicity and G2/M phase arrest among SKOV3 cells compared with free PTX.

Although the past decade has witnessed the explosive development of biodegradable micelles for targeted and controlled anticancer drug delivery, certain challenges are yet to be addressed. In particular, the techniques are still far from optimal due to issues such as low *in vivo* stability, poor tumor penetration, inefficient cellular uptake, and slow intracellular drug release [147].

4.2.4. Dendrimers

Dendrimers have narrow polydispersity and a nanometer size range, allowing easier passage across biological barriers compared with other architectural forms of polymers that have been used in drug delivery systems [148,149]. Combination with the target specificity and sensitivity of an aptamer has resulted in aptamer-dendrimer bioconjugates, which have opened new vistas for targeted drug delivery [150]. Dendrimers are hyperbranched polymers characterized by a central inner core surrounded by layers of repeating units, with an outermost layer of multivalent functional groups. The cavities inside the dendritic structure can be modified to incorporate hydrophobic and hydrophilic drugs, and the outermost layer of functional groups can be modified to attach aptamers for tumor recognition [151].

Jon *et al.* [152] used poly(amidoamine) (PAMAM) succinamic acid dendrimers as vehicles for targeted drug delivery. Firstly, an oligodeoxynucleotide was conjugated to PAMAM dendrimers, resulting in a single-stranded ONT-conjugated dendrimer (sONT-DEN). Secondly, an A9 RNA aptamer was hybridized with the sONT-DEN to create a double-stranded Apt-dONT-DEN conjugate. Finally, DOX was intercalated into the double-stranded dONT site to yield DOX-loaded Apt-dONT-DEN conjugates. The resulting conjugates showed excellent antitumor efficacy and target specificity in an *in vivo*prostate tumor model. The conjugates also exhibited high drug-loading capacity and enhanced the *in vivo* stability of the oligonucleotides.

However, dendrimers also have limitations, including the complex and costly multistep procedures involved in the synthesis and processing of dendrimer-based NPs. Future work will be necessary to identify cost-effective synthesis strategies and to determine the relationship between dendrimers and drug molecules for successful commercialization of this technology [149].

4.2.5. Serum Albumin Nanoparticles

Serum albumin NPs have been extensively studied for drug delivery because they are biodegradable, relatively easy to prepare and have a controllable size range [153]. The hydrophobic albumin core of these NPs displays a high affinity for hydrophobic drugs, and the resulting combinations may form exceptional slow-release formulations under physiological conditions. As the primary structure of proteins contains $-NH_2$, $-COOH$, $-SH$ and other functional groups, serum albumin NPs can offer various possibilities for conjugating target molecules to particulate surfaces [154].

Gao *et al.* [155] prepared bovine serum albumin (BSA) NPs using a desolvation technique and coated them with PAH/PSS multilayers to activate their surface, to which the AS1411 aptamer was covalently bound. The functional BSA NPs were then loaded with DOX. The resulting conjugates exhibited a faster release rate at pH 5.5 than at pH 7.4. With this pH-responsive release and targeting properties, the BSA NPs loaded with DOX could induce the death of liver cancer cells more effectively than the free drug could while maintaining the same low toxicity to normal liver cells.

4.2.6. DNA Nanomaterials

Although various nanostructures have been applied in targeted drug delivery, there is still demand for a simple, target-specific and economical drug delivery platform with high maximum tolerated doses. Therefore, a series of DNA nanomaterials have been proposed as drug delivery platforms by Tan *et al.* [156,157,158,159,160].

Tan *et al.* [156] firstly reported targeted drug delivery employing polymeric aptamers to induce selective cytotoxicity inside target cells. Conjugates were specifically assembled by polymerization using a one-step procedure (Scheme 8). A reporting element, or 5′-acrydite-T_{10}-dye-3′, was introduced to maintain the appropriate configuration of the individual aptamers and to provide a tracking signal for both targeting and internalization. The targeting elements, or 5′-acrydite-aptamers, on the polymer chain could facilitate cellular delivery by multivalent binding. The resulting polymeric conjugate exhibited target recognition capability, enhanced cellular internalization, and cellular disruption. Importantly, the polymer backbone built into the conjugate was cytotoxic only inside cells. As a result, selective cytotoxicity was achieved equally in both normal cancer cells and drug-resistant cells.

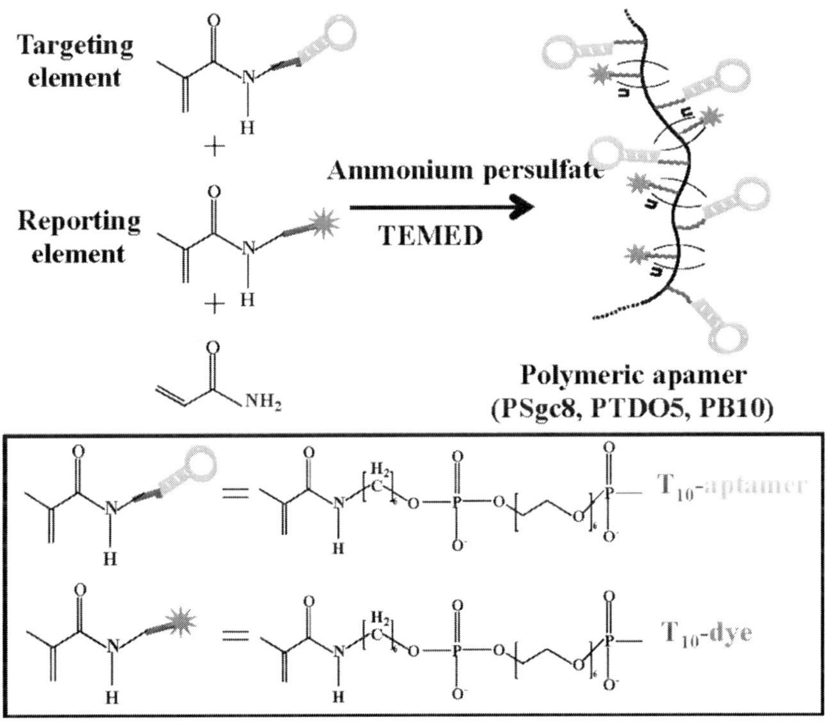

Scheme 8. Schematic of polymeric aptamer synthesis. Reproduced with permission from Reference [156].

As another example, Tan *et al.* [157] constructed a multifunctional aptamer-based DNA nanoassembly (AptNA) for targeted cancer therapy. The researchers first used base-pair hybridization to design the multifunctional AptNA with various Y-shaped functional DNA domains, including targeting aptamers, intercalated anticancer drugs, and therapeutic antisense oligonucleotides. These functional DNA domains were then linked to an X-shaped DNA core connector, termed a building unit. Finally, hundreds of these basic building units were further photo-cross-linked into a multifunctional aptamer-based nanoassembly structure (Scheme 9). The resulting AptNAs demonstrated excellent biostability in a physiological environment (pH 7.4), thus avoiding unnecessary leakage of intercalated drugs during delivery.

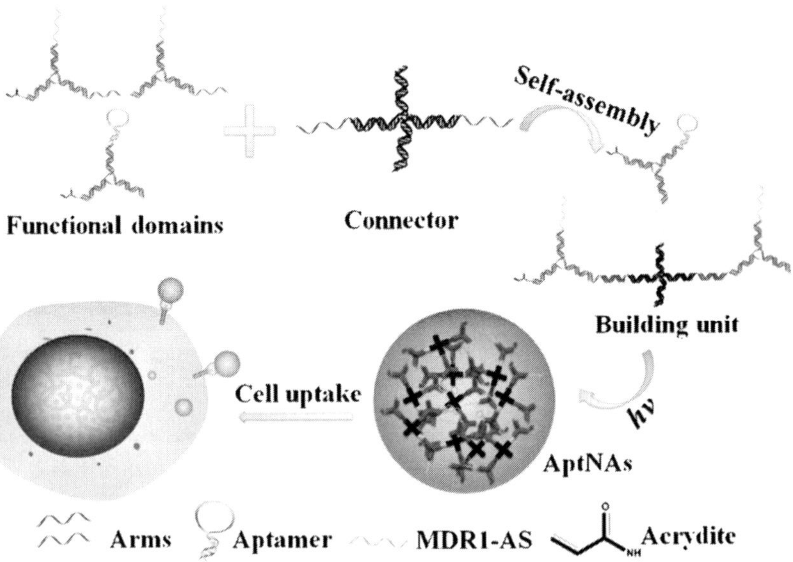

Scheme 9. Schematic illustration of the multifunctional self-assembled nanoassembly building units and photo-cross-linked nanoassembly structure. Reproduced with permission from Reference [157].

Recently, Tan *et al.* [158,159] presented monodisperse multifunctional DNA nanostructures assembled through DNA liquid crystallization, termed nanoflowers (NFs). These NFs were integrated with aptamers, bioimaging agents, and drug-loading sites. Compared with conventional DNA nanostructure assembly, NF assembly was independent of template sequences, avoiding the otherwise complicated design of DNA building blocks assembled into nanostructures by base pairing. The resulting NFs exhibited high fluorescence intensity and excellent photostability. Moreover, these NFs were demonstrated to exhibit selective cancer cell recognition and targeted anticancer drug delivery and to be of use in bioimaging.

Additionally, Tan *et al.* [160] designed and developed smart self-assembled, aptamer-tethered DNA nanotrains (aptNTrs) for the targeted transport of molecular drugs as cancer theranostics. In particular, long aptNTrs were self-assembled from only two short DNAs upon initiation by modified aptamers and worked similarly to locomotives, guiding nanotrains toward target cancer cells. Meanwhile, tandem "boxcars" served as carriers with high payload capacity, transporting drugs to target cells and inducing selective cytotoxicity. The aptNTrs enhanced the maximum tolerated dose in non-target cells, and the potent antitumor efficacy and reduced side effects of drugs delivered by biocompatible aptNTrs were demonstrated in a mouse xenograft tumor model.

Except the cytotoxic agent and photosensitizer, there was also a series of attempts to the delivery of siRNA. For example, Rossi *et al.* [161] used the novel dual inhibitory function aptamer-siRNA delivery system for HIV-1 therapy. Haam *et al.* [162] used the aptamer conjugated polyplexes (APs) to investigate the synergism between shRNAs against Bcl-xL and doxorubicin (DOX) in combination cancer therapy. Yu *et al.* [163] used CTLA4 aptamer delivering STAT3 siRNA to tumor-associated and malignant T cells.

In summary, nanomaterials possess many advantages over bulk structures. Due to the high selectivity and affinity of aptamers, aptamer-nanomaterial conjugates are ideal agents for drug delivery, with proven therapeutic effects and reduced toxicity to normal tissue. However, the unforeseen hazardous properties of the carrier nanomaterials themselves should be considered, and drug delivery systems based on biocompatible and biodegradable nanomaterials should be promising for either FDA approval or clinical trial.

5. CONCLUSIONS

Since their discovery in the 1990s, aptamers, with their high affinity and specific targeting abilities, have been widely used in targeted drug delivery systems. However, the conventional aptamer SELEX selection procedure is time consuming and labor intensive. Further refinements of CELL-SELEX techniques are needed to facilitate the identification of more disease targets and to improve targeting efficiency.

Due to the wide diversity of potential targets, no standard SELEX methods have been established so far, although different methods have been used for aptamer screening. After a decade of development, both the screening efficiency and applications have been improved. The targets have changed from single substances to animal cells and pathogens (bacteria). At the same time, complex screening systems such as multi-target screening and automated screening have emerged. In order to improve their efficiency, new outcomes from modern biophysical chemistry research and bio-separation technologies need to be integrated into SELEX so that high-throughput screening technologies can be developed. SELEX technologies will significantly push forward the studies related to gene regulation, proteomics, biochemical analysis, disease treatments and new drug exploration.

Furthermore, because aptamers are nucleic acids, they are highly susceptible to degradation by several nucleases *in vivo*, and the small size of aptamers also makes them liable to renal filtration. Consequently, aptamers are often modified with different types of agents, resulting in increased cost and side effects. Therefore, methods to reduce the cost of aptamer SELEX selection will be needed in the future.

Chemically conjugated targeted drugs exhibit good chemical stability and isotropic properties for targeted delivery to tumor cells (Table 2). However, low drug loading and the complex and costly multistep procedures involved in the design and synthesis have retarded the development of this approach. At the same time, the desirable properties of nanomaterials make them good candidate signal generation and transduction components as well as delivery vehicles. Compared with chemical conjugated targeted drugs, the aptamer-nanomaterial conjugated systems showed large drug loading. However, complex and costly preparation procedures, the PK and toxicology of nanomaterials require further investigation and development, and there are great difficulties in ascertaining the risks associated with this new technology. Chemically conjugated targeted drugs based on DNA nanotrains exhibit not only good chemical stability and isotropic properties but also a high loading capacity, biocompatibility and biodegradability, and may be a good choice for targeted drug delivery [52].

Table 2. The advantages and disadvantages of the aptamer-functionalized targeted drug delivery systems.

Aptamer-Functionalized Targeted Drug Delivery Systems	Advantage	Disadvantage
Aptamer-small molecule conjugated systems	Good chemical stability, isotropic properties	Low drug loading, complex and costly procedures
Aptamer-nanomaterial conjugated systems	High loading capacity	Unpredictable risks, complex and costly procedures

In summary, aptamer-based delivery systems appear to have potential for use in cancer therapy. However, major challenges and issues are yet to be resolved. Therefore, the study and development of promising therapeutic strategies is ongoing.

ACKNOWLEDGMENTS

We thank the other academic staff members in Aiping Lu and Ge Zhang's group at Hong Kong Baptist University and Weihong Tan and Xiaobing Zhang's group at Hunan University. We also thank Hong Kong Baptist University for providing critical comments and technical support. This study was supported by the Hong Kong General Research Fund (HKBU12102914 to Ge Zhang) and the Faculty Research Grant of Hong Kong Baptist University (FRG2/12-13/027 to Ge Zhang).

AUTHOR CONTRIBUTIONS

Feng Jiang, Biao Liu, Jun Lu and Fangfei Li wrote the manuscript; Defang Li, Chao Liang, Lei Dang, Jin Liu, Bing He, Shaikh Atik Badshah, Cheng Lu, Xiaojuan He and Baosheng Guo contributed the manuscript for literature research; Xiao-Bing Zhang, Weihong Tan, Aiping Lu and Ge Zhang revised and approved the manuscript.

CONFLICTS OF INTEREST

The authors declare no conflict of interest.

REFERENCES

1. Dosio, F.; Brusa, P.; Cattel, L. Immunotoxins and anticancer drug conjugate assemblies: The role of the linkage between components. Toxins 2011, 3, 848–883.

2. Muro, S. Challenges in design and characterization of ligand-targeted drug delivery systems. J. Control. Release 2012,164, 125–137.

3. Kamaly, N.; Xiao, Z.; Valencia, P.M.; Radovic-Moreno, A.F.; Farokhzad, O.C. Targeted polymeric therapeutic nanoparticles:

Design, development and clinical translation. Chem. Soc. Rev. 2012, 41, 2971–3010.

4. Jaracz, S.; Chen, J.; Kuznetsova, L.V.; Ojima, I. Recent advances in tumor-targeting anticancer drug conjugates. Bioorg. Med. Chem. 2005, 13, 5043–5054.

5. Jin, E.L.; Zhang, B.; Sun, X.R.; Zhou, Z.X.; Ma, X.P.; Sun, Q.H.; Tang, J.B.; Shen, Y.Q.; Kirk, E.V.; Murdoch, W.J.; et al. Acid active cell-penetrating peptides for in vivo tumor-targeted drug delivery. J. Am. Chem. Soc. 2013, 135, 933–940.

6. Jarosch, F.; Buchner, K.; Klussmann, S. In vitro selection using a dual RNA library that allows primerless selection.Nucleic Acids Res. 2006, 34, e86.

7. Kong, H.Y.; Byun, J. Nucleic acid aptamers: New methods for selection, stabilization, and application in biomedical science. Biomol. Ther. 2013, 21, 423–434.

8. Liang, H.; Zhang, X.B.; Lv, Y.; Gong, L.; Wang, R.; Zhu, X.; Yang, R.; Tan, W. Functional DNA-containing nanomaterials: Cellular applications in biosensing, imaging, and targeted therapy. Acc. Chem. Res. 2014, 47, 1891–1901.

9. Li, X.; Zhao, Q.; Qiu, L. Smart ligand: Aptamer-mediated targeted delivery of chemotherapeutic drugs and siRNA for cancer therapy. J. Control. Release 2013, 171, 152–162.

10. Mosing, R.K.; Bowser, M.T. Isolating aptamers using capillary electrophoresis-SELEX (CE-SELEX). Methods Mol. Biol.2009, 535, 33–43.

11. Glökler, J.; Schütze, T.; Konthur, Z. Automation in the high-throughput selection of random combinatorial libraries—different approaches for select applications. Molecules 2010, 15, 2478–2490.

12. Blind, M.; Blank, M. Aptamer selection technology and recent advances. Mol. Ther. Nucleic Acids 2015, 4, e223.

13. Elle, I.C.; Karlsen, K.K.; Terp, M.G.; Larsen, N.; Nielsen, R.; Derbyshire, N.; Mandrup, S.; Ditzel, H.J.; Wengel, J. Selection of LNA-containing DNA aptamers against recombinant human CD73. Mol. Biosyst. 2015, 11, 1260–1270.

14. Murgha, Y.E.; Rouillard, J.M.; Gulari, E. Methods for the preparation of large quantities of complex single-stranded oligonucleotide libraries. PLoS ONE 2014, 9, e94752.

15. Yang, J.; Bowser, M.T. Capillary electrophoresis-SELEX selection of catalytic DNA aptamers for a small-molecule porphyrin target. Anal. Chem. 2013, 85, 1525–1530.

16. Yang, Y.; Yang, D.; Schluesener, H.J.; Zhang, Z. Advances in SELEX and application of aptamers in the central nervous system. Biomol. Eng. 2007, 24, 583–592.

17. Tok, J.; Lai, J.; Leung, T.; Li, S.F. Selection of aptamers for signal transduction proteins by capillary electrophoresis.Electrophoresis 2010, 31, 2055–2062.

18. Bruno, J.G.; Phillips, T.; Montez, T.; Garcia, A.; Sivils, J.C.; Mayo, M.W.; Greis, A. Development of a fluorescent enzyme-linked DNA aptamer-magnetic bead sandwich assay and portable fluorometer for sensitive and rapid listeria detection. J. Fluoresc. 2015, 25, 173--183.

19. Stoltenburg, R.; Reinemann, C.; Strehlitz, B. FluMag-SELEX as an advantageous method for DNA aptamer selection.Anal. Bioanal. Chem. 2005, 383, 83–91.

20. Liang, C.; Li, D.; Zhang, G.; Li, H.; Shao, N.; Liang, Z.; Zhang, L.; Lu, A.; Zhang, G. Comparison of the methods for generating single-stranded DNA in SELEX. Analyst 2015, 140, 3439–3444.

21. Zhou, J.; Rossi, J.J. Cell-type-specific, aptamer-functionalized agents for targeted disease. Mol. Ther. Nucleic Acids2014, 3.

22. Chen, W.X.; Zhang, K.H.; Zou, X.S.; Chen, Y.Q.; Li, J.G. Screening and identification of the nucleic acid aptamers in nasopharyngeal carcinoma. Genet. Mol. Res. 2013, 12, 6850–6857.

23. Dickinson, H.; Lukasser, M.; Mayer, G.; Hüttenhofer, A. Cell-SELEX: In vitro selection of synthetic small specific ligands. Methods Mol. Biol. 2015, 1296, 213–224.

24. Zhu, H.; Li, J.; Zhang, X.B.; Ye, M.; Tan, W. Nucleic acid aptamer-mediated drug delivery for targeted cancer therapy.ChemMedChem 2015, 10, 39–45.

25. Kim, J.W.; Kim, E.Y.; Kim, S.Y.; Byun, S.K.; Lee, D.; Oh, K.J.; Kim, W.K.; Han, B.S.; Chi, S.W.; Lee, S.C.; et al. Identification of DNA

aptamers toward epithelial cell adhesion molecule via cell-SELEX. Mol. Cells 2014, 37, 742–746.

26. Liang, C.; Guo, B.; Wu, H.; Shao, N.; Li, D.; Liu, J.; Dang, L.; Wang, C.; Li, H.; Li, S.; et al. Aptamer-functionalized lipid nanoparticles targeting osteoblasts as a novel RNA interference-based bone anabolic strategy. Nat. Med. 2015,21, 288–294.

27. Conidi, A.; van den Berghe, V.; Huylebroeck, D. Aptamers and their potential to selectively target aspects of EGF, Wnt/β-catenin and TGFβ-smad family signaling. Int. J. Mol. Sci. 2013, 14, 6690–6719.

28. Hoinka, J.; Zotenko, E.; Friedman, A.; Sauna, Z.E.; Przytycka, T.M. Identification of sequence-structure RNA binding motifs for SELEX-derived aptamers. Bioinformatics 2012, 28, i215–i223.

29. Stoltenburg, R.; Reinemann, C.; Strehlitz, B. SELEX—A (r)evolutionary method to generate high-affinity nucleic acid ligands. Biomol. Eng. 2007, 24, 381–403.

30. Cox, J.C.; Ellington, A.D. Automated selection of anti-protein aptamers. Bioorg. Med. Chem. 2001, 9, 2525–2531.

31. Eulberg, D.; Buchner, K.; Maasch, C.; Klussmann, S. Development of an automated in vitro selection protocol to obtain RNA-based aptamers: Identification of a biostable substance P antagonist. Nucleic Acids Res. 2005, 33, e45.

32. Fang, X.; Tan, W. Aptamers generated from cell-SELEX for molecular medicine: A chemical biology approach. Acc. Chem. Res. 2010, 43, 48–57.

33. Seo, Y.J.; Chen, S.; Nilsen-Hamilton, M.; Levine, H.A. A mathematical analysis of multiple-target SELEX. Bull. Math. Biol. 2010, 72, 1623–1665.

34. Morris, K.N.; Jensen, K.B.; Julin, C.M.; Weil, M.; Larry, G. High affinity ligands from in vitro selection: Complex targets. Proc. Natl. Acad. Sci. USA 1998, 95, 2902–2907.

35. Shamah, S.M.; Healy, J.M.; Cload, S.T. Complex target SELEX. Acc. Chem. Res. 2008, 41, 130–138.

36. Sun, H.; Zhu, X.; Lu, P.Y.; Rosato, R.R.; Tan, W.; Zu, Y. Oligonucleotide aptamers: New tools for targeted cancer therapy. Mol. Ther. Nucleic Acids 2014, 3.

37. Gragoudas, E.S.; Adamis, A.P.; Cunningham, E.T.; Feinsod, M.; Guyer, D.R. Pegaptanib for Neovascular Age-Related Macular Degeneration. N. Engl. J. Med. 2004, 351, 2805–2816.

38. Bates, P.J.; Laber, D.A.; Miller, D.M.; Thomas, S.D.; Trent, J.O. Discovery and Development of the G-rich Oligonucleotide AS1411 as a Novel Treatment for Cancer. Exp. Mol. Pathol. 2009, 86, 151–164.

39. Shangguan, D.; Cao, Z.; Meng, L.; Mallikaratchy, P.; Sefah, K.; Wang, H.; Li, Y.; Tan, W. Cell-Specific Aptamer Probes for Membrane Protein Elucidation in Cancer Cells. J. Prot. Res. 2008, 7, 2133–2139.

40. Mallikaratchy, P.; Tang, Z.; Kwame, S.; Meng, L.; Shangguan, D.; Tan, W. Aptamer Directly Evolved from Live Cells Recognizes Membrane Bound Immunoglobin Heavy Mu Chain in Burkitt's Lymphoma Cells. Mol. Cell. Proteom.2007, 6, 2230–2238.

41. Tope, S.; Maske, S.; Nagulwar, V.; Sufi, J.; Welankiwar, A. Aptamers as therapeutics. Indo Am. J. Pharm. Res. 2013, 3, 2718–2743.

42. Bock, L.C.; Griffin, L.C.; Latham, J.A.; Vermaas, E.H.; Toole, J.J. Selection of single-stranded DNA molecules that bind and inhibit human thrombin. Nature 1992, 355, 564–566.

43. Tan, W.; Wang, H.; Chen, Y.; Zhang, X.; Zhu, H.; Yang, C.; Yang, R.; Liu, C. Molecular aptamers for drug delivery.Trends Biotechnol. 2011, 29, 634–640.

44. Famulok, M.; Hartig, J.S.; Mayer, G. Functional Aptamers and Aptazymes in Biotechnology, Diagnostics, and Therapy. Chem. Rev. 2007, 107, 3715–3743.

45. Gold, L.; Polisky, B.; Uhlenbeck, O.; Yarus, M. Diversity of Oligonucleotide Functions. Annu. Rev. Biochem. 1995, 64, 763–767.

46. Burmeister, P.E.; Lewis, S.D.; Silva, R.F.; Preiss, J.R.; Horwitz, L.R.; Pendergrast, P.S.; McCauley, T.G.; Kurz, J.C.; Epstein, D.M.; Wilson, C.; et al. Direct in Vitro Selection of a 2′-O-Methyl Aptamer to VEGF. Chem. Biol. 2005, 12, 25–33.

47. Bates, P.J.; Kahlon, J.B.; Thomas, S.D.; Trent, J.O.; Miller, D.M. Antiproliferative activity of G-rich oligonucleotides correlates with protein binding. J. Biol. Chem. 1999, 274, 26369–26377.

48. Dapić, V.; Abdomerović, V.; Marrington, R.; Peberdy, J.; Rodger, A.; Trent, J.O.; Bates, P.J. Biophysical and biological properties of quadruplex oligodeoxyribonucle otides. Nucleic Acids Res. 2003, 31, 2097–2107.

49. Dapić, V.; Bates, P.J.; Trent, J.O.; Rodger, A.; Thomas, S.D.; Miller, D.M. Antiproliferative activity of G-quartetforming oligonucleotides with backbone and sugar modifications. Biochemistry 2002, 41, 3676–3685.

50. Reyes-Reyes, E.M.; Teng, Y.; Bates, P.J. A new paradigm for aptamer therapeutic AS1411 action: Uptake by macropinocytosis and its stimulation by a nucleolin-dependent mechanism. Cancer Res. 2010, 70, 8617–8629.

51. Huang, Y.F.; Shangguan, D.; Liu, H.; Phillips, J.A.; Zhang, X.; Chen, Y.; Tan, W. Molecular assembly of an aptamer-drug conjugate for targeted drug delivery to tumor cells. ChemBioChem 2009, 10, 862–868.

52. Wang, R.W.; Zhu, G.; Mei, L.; Xie, Y.; Ma, H.; Ye, M.; Qing, F.L.; Tan, W. Automated modular synthesis of aptamer-drug conjugates for targeted drug delivery. J. Am. Chem. Soc. 2014, 136, 2731–2734.

53. Senter, P.D. Potent antibody drug conjugates for cancer therapy. Curr. Opin. Chem. Biol. 2009, 13, 235–244.

54. Teicher, B.A.; Chari, R.V.J. Antibody conjugate therapeutics: Challenges and potential. Clin. Cancer Res. 2011, 17, 6389–6397.

55. Lee, M.D.; Dunne, T.S.; Siegel, M.M.; Chang, C.C.; Morton, G.O.; Borders, D.B. Calichemicins, a novel family of antitumor antibiotics. 1. Chemistry and partial structure of calichemicin γ1. J. Am. Chem. Soc. 1987, 109, 3464–3466.

56. Kupchan, S.M.; Komoda, Y.; Court, W.A.; Thomas, G.J.; Smith, R.M.; Karim, A.; Gilmore, C.J.; Haltiwanger, R.C.; Bryan, R.F. Maytansine, a novel antileukemic ansa macrolide from Maytenus ovatus. J. Am. Chem. Soc. 1972, 94, 1354–1356.

57. Pettit, G.R.; Kamano, Y.; Herald, C.L.; Tuinman, A.A.; Boettner, F.E.; Kizu, H.; Schmidt, J.M.; Baczynskyj, L.; Tomer, K.B.; Bontems, R.J. The isolation and structure of a remarkable marine animal antineoplastic constituent: Dolastatin 10. J. Am. Chem. Soc. 1987, 109, 6883–6885.

58. Bai, R.; Petit, G.R.; Hame, E. Dolastatin 10, a powerful cytostatic peptide derived from a marine animal: Inhibition of tubulin polymerization mediated through the vinca alkaloid binding domain. Biochem. Pharmacol. 1990, 39, 1941–1949.

59. Hartley, J.A.; Hamaguchi, A.; Suggitt, M.; Gregson, S.J.; Thurston, D.E.; Howard, P.W. DNA inter strand cross-linking and in vivo antitumor activity of the extended pyrrolo [2,1-c][1,4]benzodiazepine dimer SG2057. Investig. New Drugs2012, 30, 950–958.

60. Hartley, J.A. The development of pyrrolobenzodiazepines as antitumour agents. Expert Opin. Investig. Drugs 2011, 20, 733–744.

61. Moldenhauer, G.; Salnikov, A.V.; Luttgau, S.; Herr, I.; Anderl, J.; Faulstich, H. Therapeutic potential of amanitin-conjugated anti-epithelial cell adhesionmolecule monoclonal antibody against pancreatic carcinoma. J. Natl. Cancer Inst. 2012, 104, 622–634.

62. Chari, R.V.J. Targeted cancer therapy: Conferring specificity to cytotoxic drugs. Acc. Chem. Res. 2008, 41, 98–107.

63. Kemshead, J.T.; Hopkins, K. Uses and limitations of monoclonal antibodies (MoAbs) in the treatment of malignant disease: A review. J. R. Soc. Med. 1993, 86, 219–224.

64. Trail, P.A.; Willner, D.; Lasch, S.J.; Henderson, A.J.; Hofstead, S.; Casazza, A.M.; Firestone, R.A.; Hellstrom, K.E. Cure of xenografted human carcinomas by BR96-doxorubicin immunoconjugates. Science 1993, 261, 212–215.

65. Casi, G.; Neri, D. Antibody-drug conjugates: Basic concepts, examples and future perspectives. J. Control. Release2012, 161, 422–428.

66. DiJoseph, J.F.; Dougher, M.M.; Kalyandrug, L.B.; Armellino, D.C.; Boghaert, E.R.; Hamann, P.R.; Moran, J.K.; Damle, N.K. Antitumor efficacy of a combination of CMC-544 (inotuzumab ozogamicin), a CD22-targeted cytotoxic immunoconjugate of calicheamicin, and rituximab against non-Hodgkin's B-cell lymphoma. Clin. Cancer Res. 2006,12, 242–249.

67. De Groot, F.M.; Damen, E.W.; Scheeren, H.W. Anticancer prodrugs for application in monotherapy: Targeting hypoxia, tumor-associated enzymes, and receptors. Curr. Med. Chem. 2001, 8, 1093–1122.

68. Singh, Y.; Palombo, M.; Sinko, P.J. Recent trends in targeted anticancer prodrug and conjugate design. Curr. Med. Chem. 2008, 15, 1802–1826.

69. Ranson, M.; Sliwkowski, M.X. Perspectives on anti-HER monoclonal antibodies. Oncology 2002, 63, 17–24.

70. Erickson, H.K.; Widdison, W.C.; Mayo, M.F.; Whiteman, K.; Audette, C.; Wilhelm, S.D.; Singh, R. Tumor delivery andin vivo processing of disulfide-linked and thioether-linked antibody-maytansinoid conjugates. Bioconjug. Chem. 2010,21, 84–92.

71. Erickson, H.K.; Park, P.U.; Widdison, W.C.; Kovtun, Y.V.; Garrett, L.M.; Hoffman, K.; Lutz, R.J.; Goldmacher, V.S.; Blättler, W.A. Antibody-maytansinoid conjugates are activated in targeted cancer cells by lysosomal degradation and linker-dependent intracellular processing. Cancer Res. 2006, 66, 4426–4433.

72. Kovtun, Y.V.; Audette, C.A.; Ye, Y.; Xie, H.; Ruberti, M.F.; Phinney, S.J.; Leece, B.A.; Chittenden, T.; Blättler, W.A.; Goldmacher, V.S. Antibody-drug conjugates designed to eradicate tumors with homogeneous and heterogeneous expression of the target antigen. Cancer Res. 2006, 66, 3214–3221.

73. Xie, H.; Audette, C.; Hoffee, M.; Lambert, J.M.; Blättler, W.A. Pharmacokinetics and biodistribution of the antitumor immunoconjugate, cantuzumab mertansine (huC242-DM1), and its two components in mice. J. Pharmacol. Exp. Ther.2004, 308, 1073–1082.

74. Sanderson, R.J.; Hering, M.A.; James, S.F.; Sun, M.M. C.; Doronina, S.O.; Siadak, A.W.; Senter, P.D.; Wahl, A.F. In vivodrug-linker stability of an anti-CD30 dipeptide-linked auristatin immunoconjugate. Clin. Cancer Res. 2005, 11, 843–852.

75. Doronina, S.O.; Toki, B.E.; Torgov, M.Y.; Mendelsohn, B.A.; Cerveny, C.G.; Chace, D.F.; DeBlanc, R.L.; Gearing, R.P.; Bovee, T.D.; Siegall, C.B.; et al. Development of potent monoclonal antibody auristatin conjugates for cancer therapy.Nat. Biotechnol. 2003, 21, 778–784.

76. Francisco, J.A.; Cerveny, C.G.; Meyer, D.L.; Mixan, B.J.; Klussman, K.; Chace, D.F.; Rejniak, S.X.; Gordon, K.A.; DeBlanc, R.; Toki, B.E.; et al. cAC10-vcMMAE, an anti-CD30-monomethyl auristatin E conjugate with potent and selective antitumor activity. Blood 2003, 102, 1458–1465.

77. Jeffrey, S.C.; Andreyka, J.B.; Bernhardt, S.X.; Kissler, K.M.; Kline, T.; Lenox, J.S.; Moser, R.F.; Nguyen, M.T.; Okeley, N.M.; Stone, I.J.; et al. Development and properties of β-glucuronide linkers for monoclonal antibody-drug conjugates. Bioconjug. Chem. 2006, 17, 831–840.

78. De Graaf, M.; Boven, E.; Scheeren, H.W.; Haisma, H.J.; Pinedo, H.M. Beta-glucuronidase-mediated drug release. Curr. Pharm. Des. 2002, 8, 1391–1403.

79. Jeffrey, S.C.; Nguyen, M.T.; Moser, R.F.; Meyer, D.L.; Miyamoto, J.B.; Senter, P.D. Minor groove binder antibody conjugates employing a water soluble β-glucuronide linker. Bioorg. Med. Chem. Lett. 2007, 17, 2278–2280.

80. Burke, P.J.; Senter, P.D.; Meyer, D.W.; Miyamoto, J.B.; Anderson, M.; Toki, B.E.; Manikumar, G.; Wani, M.C.; Kroll, D.J.; Jeffrey, S.C. Design, synthesis, and biological evaluation of antibody-drug conjugates comprised of potent camptothecin analogues. Bioconjug. Chem. 2009, 20, 1242–1250.

81. Alley, S.C.; Benjamin, D.R.; Jeffrey, S.C.; Okeley, N.M.; Meyer, D.L.; Sanderson, R.J.; Senter, P.D. Contribution of linker stability to the activities of anticancer immunoconjugates. Bioconjug. Chem. 2008, 19, 759–765.

82. Doronina, S.O.; Mendelsohn, B.A.; Bovee, T.D.; Cerveny, C.G.; Alley, S.C.; Meyer, D.L.; Oflazoglu, E.; Toki, B.E.; Sanderson, R.J.; Zabinski, R.F.; et al. Enhanced activity of monomethylauristatin F through monoclonal antibody delivery: Effects of linker technology on efficacy and toxicity. Bioconjug. Chem. 2006, 17, 114–124.

83. Phillips, G.D.L.L.; Blättler, W.A.; Lambert, J.M.; Chari, R.V.J.; Lutz, R.J.; Wong, W.L.T.; Jacobson, F.S.; Koeppen, H.; Schwall, R.H.; Kenkare-Mitra, S.R.; et al. Targeting HER2-positive breast cancer with trastuzumab-DM1, an antibody-cytotoxic drug conjugate. Cancer Res. 2008, 68, 9280–9290.

84. Cohen, R.; Vugts, D.J.; Visser, G.W.; Stigter-van, W.M.; Bolijn, M.; Spiga, M.; Lazzari, P.; Shankar, S.; Sani, M.; Zanda, M.; et al. Development of novel ADCs: Conjugation of tubulysin analogues to trastuzumab monitored by dual radiolabeling. Cancer Res. 2014, 74, 5700–5710.

85. Pillow, T.H.; Tien, J.; Parsons-Reponte, K.L.; Bhakta, S.; Li, H.; Staben, L.R.; Li, G.; Chuh, J.; Fourie-O Donohue, A.; Darwish, M.; et al. Site-specific trastuzumab maytansinoid antibody–drug conjugates with improved therapeutic activity through linker and antibody engineering. J. Med. Chem. 2014, 57, 7890–7899.

86. Drake, P.M.; Albers, A.E.; Baker, J.; Banas, S.; Barfield, R.M.; Bhat, A.S.; de Hart, G.W.; Garofalo, A.W.; Holder, P.; Jones, L.C.; et al. Aldehyde tag coupled with HIPS chemistry enables the production of ADCs conjugated site-specifically to different antibody regions with distinct in vivo efficacy and PK outcomes. Bioconjug. Chem. 2014, 25, 1331–1341.

87. Kolodych, S.; Koniev, O.; Baatarkhuu, Z.; Bonnefoy, J.Y.; Debaene, F.; Cianférani, S.; Dorsselaer, A.V.; Wagner, A. CBTF: New amine-to-thiol coupling reagent for preparation of antibody conjugates with increased plasma stability.Bioconjug. Chem. 2015, 26, 197–200.

88. Zhao, N.; Pei, S.-N.; Qi, J.; Zeng, Z.; Iyer, S.P.; Lin, P.; Tung, C.-H.; Zu, Y. Oligonucleotide aptamer-drug conjugates for targeted therapy of acute myeloid leukemia. Biomaterials 2015, 67, 42–51.

89. Haley, B.; Frenkel, E. Nanoparticles for drug delivery in cancer treatment. Urol. Oncol. 2008, 26, 57–64.

90. Vigderman, L.; Zubarev, E.R. Therapeutic platforms based on gold nanoparticles and their covalent conjugates with drug molecules. Adv. Drug Deliv. Rev. 2013, 65, 663–676.

91. Subinoy, R.; Avinash, B.; Rubul, M.; Vincent, M.R. Monolayer coated gold nanoparticles for delivery applications.Adv. Drug Deliv. Rev. 2012, 64, 200–216.

92. Luo, Y.L.; Shiao, Y.S.; Huang, Y.F. Release of photoactivatable drugs from plasmonic nanoparticles for targeted cancer therapy. ACS Nano 2011, 5, 7796–7804.

93. Yang, X.; Liu, X.; Liu, Z.; Pu, F.; Ren, J.; Qu, X. Near-infrared light-triggered, targeted drug delivery to cancer cells by aptamer gated nanovehicles. Adv. Mater. 2012, 24, 2890–2895.

94. Wang, J.; Zhu, G.; You, M.; Song, E.; Shukoor, M.I.; Zhang, K.; Altman, M.B.; Chen, Y.; Zhu, Z.; Huang, C.Z.; et al. Assembly of aptamer switch probes and photosensitizer on gold nanorods for targeted photothermal and photodynamic cancer therapy. ACS Nano 2012, 6, 5070–5077.

95. Wang, J.; Sefah, K.; Altman, M.B.; Chen, T.; You, M.; Zhao, Z.; Huang, C.Z.; Tan, W. Aptamer-conjugated nanorods for targeted photothermal therapy of prostate cancer stem cells. Chem. Asian J. 2013, 8, 2417–2422.

96. Dreaden, E.C.; Austin, L.A.; Mackey, M.A.; El-Sayed, M.A. Size matters: Gold nanoparticles in targeted cancer drug delivery. Ther. Deliv. 2012, 3, 457–478.

97. Mohapatra, M.; Anand, S. Synthesis and applications of nano-structured iron oxides/hydroxides—A review. Int. J. Eng. Sci. Technol. 2010, 2, 127–146.

98. Chen, T.; Shukoor, M.I.; Wang, R.; Zhao, Z.; Yuan, Q.; Bamrungsap, S.; Xiong, X.; Wei, T. Smart multifunctional nanostructure for targeted cancer chemotherapy and magnetic resonance imaging. ACS Nano 2011, 5, 7866–7873.

99. Zhang, L.; Dong, W.F.; Sun, H.B. Multifunctional superparamagnetic iron oxide nanoparticles: Design, synthesis and biomedical photonic applications. Nanoscale 2013, 5, 7664–7684.

100. Mahmoudi, M.; Sant, S.; Wang, B.; Laurent, S.; Sen, T. Superparamagnetic iron oxide nanoparticles (SPIONs): Development, surface modification and applications in chemotherapy. Adv. Drug Deliv. Rev. 2011, 63, 24–46.

101. Wang, Y.X.; Hussain, S.M.; Krestin, G.P. Superparamagnetic iron oxide contrast agents: Physicochemical characteristics and applications in MR imaging. Eur. Radiol. 2001, 11, 2319–2331.

102. Weissleder, R.; Elizondo, G.; Wittenberg, J.; Rabito, C.A.; Bengele, H.H.; Josephson, L. Ultrasmall superparamagnetic iron oxide: Characterization of a new class of contrast agents for MR imaging. Radiology 1990, 175, 489–493.

103. Leenders, W. Ferumoxtran-10 advanced magnetics. IDrugs 2003, 6, 987–993.

104. Wang, A.Z.; Bagalkot, V.; Vasilliou, C.C.; Gu, F.; Alexis, F.; Zhang, L.; Shaikh, M.; Yuet, K.; Cima, M.J.; Langer, R.; et al. Superparamagnetic iron oxide nanoparticle-aptamer bioconjugates for combined prostate cancer imaging and therapy. ChemMedChem 2008, 3, 1311–1315.

105. Yu, M.K.; Kim, D.; Lee, I.H.; So, J.S.; Jeong, Y.Y.; Jon, S. Image-guided prostate cancer therapy using aptamer-functionalized thermally cross-linked superparamagnetic iron oxide nanoparticles. Small 2011, 7, 2241–2249.

106. Laurent, S.; Saei, A.A.; Behzadi, S.; Panahifar, A.; Mahmoudi, M. Superparamagnetic iron oxide nanoparticles for delivery of therapeutic agents: Opportunities and challenges. Expert Opin. Drug Deliv. 2014, 11, 1449–1470.

107. Wahajuddin; Arora, S. Superparamagnetic iron oxide nanoparticles: Magnetic nanoplatforms as drug carriers. Int. J. Nanomed. 2012, 7, 3445–3471.

108. Meng, L.; Zhang, X.; Lu, Q.; Fei, Z.; Dyson, P.J. Single walled carbon nanotubes as drug delivery vehicles: Targeting DOXorubicin to tumors. Biomaterials 2012, 33, 1689–1698.

109. Zhu, Z.; Tang, Z.; Phillips, J.A.; Yang, R.; Wang, H.; Tan, W. Regulation of singlet oxygen generation using single-walled carbon nanotubes. J. Am. Chem. Soc. 2008, 130, 10856–10857.

110. Zhang, H.; Hou, L.; Jiao, X.; Ji, Y.; Zhu, X.; Li, H.; Chen, X.; Ren, J.; Xia, Y.; Zhang, Z. In vitro and in vivo evaluation of antitumor drug-loaded aptamer targeted single-walled carbon nanotubes system. Curr. Pharm. Biotechnol. 2014, 14, 1105–1117.

111. Lu, Y.; Wu, P.; Yin, Y.; Zhang, H.; Cai, C. Aptamer-functionalized graphene oxide for highly efficient loading and cancer cell-specific delivery of antitumor drug. J. Mater. Chem. B 2014, 2, 3849–3859.

112. Rodriguesand, D.; Elimelech, M. Toxic effects of single-walled carbon nanotubes in the development of E. coli biofilm. Environ. Sci. Technol. 2010, 44, 4583–4589.

113. Chen, Y.-C.; Huang, X.-C.; Luo, Y.-L.; Chang, Y.-C.; Hsieh, Y.-Z.; Hsu, H.-Y. Non-metallic nanomaterials in cancer theranostics: A review of silica- and carbon-based drug delivery systems. Sci. Technol. Adv. Mater. 2013, 14, 044407.

114. Slowing, I.I.; Vivero-Escoto, J.L.; Wu, C.W.; Lin, V.S.Y. Mesoporous silica nanoparticles as controlled release drug delivery and gene transfection carriers. Adv. Drug Deliv. Rev. 2008, 60, 1278–1288.

115. Zhu, C.L.; Song, X.Y.; Zhou, W.H.; Yang, H.H.; Wen, Y.H.; Wang, X.R. An efficient cell-targeting and intracellular controlled-release

drug delivery system based on MSN-PEM-aptamer conjugates. J. Mater. Chem. 2009, 19, 7765–7770.

116. Li, L.L.; Yin, Q.; Cheng, J.; Lu, Y. Polyvalent Mesoporous Silica Nanoparticle-Aptamer Bioconjugates Target Breast Cancer Cells. Adv. Healthc. Mater. 2012, 1, 567–572.

117. Chen, Y.; Chen, H.; Shi, J. In vivo bio-safety evaluations and diagnostic/therapeutic applications of chemically designed mesoporous silica nanoparticles. Adv. Mater. 2013, 25, 3144–3176.

118. Dey, N.S.; Rao, M.E.B. Quantum dot: Novel carrier for drug delivery. Int. J. Res. Pharm. Biomed. Sci. 2011, 2, 448–458.

119. Qi, L.; Gao, X. Emerging application of quantum dots for drug delivery and therapy. Expert Opin. Drug Deliv. 2008, 5, 263–267.

120. Bagalkot, V.; Zhang, L.; Levy-Nissenbaum, E.; Jon, S.; Kantoff, P.W.; Langer, R.; Farokhzad, O.C. Quantum dot-aptamer conjugates for synchronous cancer imaging, therapy, and sensing of drug delivery based on bi-fluorescence resonance energy transfer. Nano Lett. 2007, 7, 3065–3070.

121. Savla, R.; Taratula, O.; Garbuzenko, O.; Minko, T. Tumor targeted quantum dot-mucin 1 aptamer-doxorubicin conjugate for imaging and treatment of cancer. J. Control. Release 2011, 153, 16–22.

122. Winnik, F.M.; Maysinger, D. Quantum dot cytotoxicity and ways to reduce it. Acc. Chem. Res. 2013, 46, 672–680.

123. Ghaderi, S.; Ramesh, B.; Seifalian, A.M. Fluorescence nanoparticles “quantum dots” as drug delivery system and their toxicity: A review. J. Drug Target. 2011, 19, 475–486.

124. Hardman, R. A toxicologic review of quantum dots: Toxicity depends on physicochemical and environmental factors. Environ. Health Perspect. 2006, 114, 165–172.

125. Akbarzadeh, A.; Rezaei-Sadabady, R.; Davaran, S.; Joo, S.W.; Zarghami, N.; Hanifehpour, Y.; Samiei, M.; Kouhi, M.; Nejati-Koshki, K. Liposome: Classification, preparation, and Applications. Nanoscale Res. Lett. 2013, 8, 102–110.

126. Wagner, A.; Platzgummer, M.; Kreismayr, G.; Quendler, H.; Stiegler, G.; Ferko, B.; Vecera, G.; Vorauer-Uhl, K.; Katinger, H.

GMP production of liposomes—a new industrial approach. J. Liposome Res. 2006, 16, 311–319.

127. Kaneda, Y. Virosomes: Evolution of the liposome as a targeted drug delivery system. Adv. Drug Deliv. Rev. 2000, 43, 197–205.

128. Cao, Z.; Tong, R.; Mishra, A.; Xu, W.; Wong, G.C.; Cheng, J.; Lu, Y. Reversible cell-specific drug delivery with aptamer-functionalized liposomes. Angew. Chem. Int. Ed. 2009, 48, 6494–6498.

129. Kang, H.; O'Donoghue, M.B.; Liu, H.; Tan, W. A liposome-based nanostructure for aptamer directed delivery. Chem. Commun. 2010, 46, 249–251.

130. Xing, H.; Tang, L.; Yang, X.; Hwang, K.; Wang, W.; Yin, Q.; Wong, N.Y.; Dobrucki, L.W.; Yasui, N.; Katzenellenbogen, J.A.; et al. Selective delivery of an anticancer drug with aptamer-functionalized liposomes to breast cancer cells in vitro and in vivo. J. Mater. Chem. B 2013, 1, 5288–5297.

131. Baek, S.E.; Lee, K.H.; Park, Y.S.; Oh, D.K.; Oh, S.; Kim, K.S.; Kim, D.E. RNA aptamer-conjugated liposome as an efficient anticancer drug delivery vehicle targeting cancer cells in vivo. J. Control. Release 2014, 196, 234–242.

132. Alshaer, W.; Hillaireau, H.; Vergnaud, J.; Ismail, S.; Fattal, E. Functionalizing liposomes with anti-CD44 aptamer for selective targeting of cancer cells. Bioconjug. Chem. 2015, 26, 1307–1313.

133. Langer, R. Drug delivery and targeting. Nature 1998, 392, 5–10.

134. Alexis, F.; Pridgen, E.; Molnar, L.K.; Farokhzad, O.C. Factors affecting the clearance and biodistribution of polymeric nanoparticles. Mol. Pharm. 2008, 5, 505–515.

135. Gref, R.; Minamitake, Y.; Peracchia, M.T.; Trubetskoy, V.; Torchilin, V.; Langer, R. Biodegradable long-circulating polymer nanospheres. Science 1994, 263, 1600–1603.

136. Farokhzad, O.C.; Cheng, J.; Teply, B.A.; Sherifi, I.; Jon, S.; Kantoff, P.W.; Richie, J.P.; Langer, R. Targeted nanoparticle-aptamer bioconjugates for cancer chemotherapy in vivo. Proc. Natl. Acad. Sci. USA 2006, 103, 6315–6320.

137. Dhar, S.; Gu, F.X.; Langer, R.; Farokhzad, O.C.; Lippard, S.J. Targeted delivery of cisplatin to prostate cancer cells by aptamer

functionalized Pt (IV) prodrug-PLGA-PEG nanoparticles. Proc. Natl. Acad. Sci. USA 2008, 105, 17356–17361.

138. Dhar, S.; Kolishettib, N.; Lipparda, S.J.; Farokhzad, O.C. Targeted delivery of a cisplatin prodrug for safer and more effective prostate cancer therapy in vivo. Proc. Natl. Acad. Sci. USA 2011, 108, 1850–1855.

139. Gu, F.; Zhang, L.F.; Teply, B.A.; Mann, N.; Wang, A.; Radovic-Moreno, A.F.; Langer, R.; Farokhzad, O.C. Precise engineering of targeted nanoparticles by using self-assembled biointegrated block copolymers. Proc. Natl. Acad. Sci. USA 2008, 105, 2586–2591.

140. Huang, F.; You, M.; Chen, T.; Zhu, G.; Liang, H.; Tan, W. Self-assembled hybrid nanoparticles for targeted co-delivery of two drugs into cancer cells. Chem. Commun. 2014, 50, 3103–3105.

141. Kedar, U.; Phutane, P.; Shidhaye, S.; Kadam, V. Advances in polymeric micelles for drug delivery and tumor targeting. Nanomedicine 2010, 6, 714–729.

142. Wu, Y.R.; Sefah, K.; Liu, H.P.; Wang, R.W.; Tan, W.H. DNA aptamer-micelle as an efficient detection/delivery vehicle toward cancer cells. Proc. Natl. Acad. Sci. USA 2010, 107, 5–10.

143. Mu, C.; Dave, N.; Hu, J.; Desai, P.; Pauletti, G.; Bai, S.; Hao, J. Solubilization of flurbiprofen into aptamer-modified PEG-PLA micelles for targeted delivery to brain-derived endothelial cells in vitro. J. Microencapsul. 2013, 30, 701–708.

144. Xu, W.; Siddiqui, I.A.; Nihal, M.; Pilla, S.; Rosenthal, K.; Mukhtar, H.; Gong, S. Aptamer-conjugated and doxorubicin-loaded unimolecular micelles for targeted therapy of prostate cancer. Biomaterials 2013, 34, 5244–5253.

145. Li, X.; Yu, Y.; Ji, Q.; Qiu, L. Targeted delivery of anticancer drugs by aptamer AS1411 mediated Pluronic F127/cyclodextrin-linked polymer composite micelles. Nanomedicine 2015, 11, 175–184.

146. Zhang, J.; Chen, R.; Fang, X.; Chen, F.; Wang, Y.; Chen, M. Nucleolin targeting AS1411 aptamer modified pH-sensitive micelles for enhanced delivery and antitumor efficacy of paclitaxel. Nano Res. 2015, 8, 201–218.

147. Deng, C.; Jiang, Y.; Cheng, R.; Meng, F.; Zhong, Z. Biodegradable polymeric micelles for targeted and controlled anticancer drug

delivery: Promises, progress and prospects. Nano Today 2012, 7, 467–480.

148. Tripathy, S.; Das, M.K. Dendrimers and their applications as novel drug delivery carriers. J. Appl. Pharm. Sci. 2013, 3, 142–149.

149. Nanjwadea, B.K.; Bechra, H.M.; Derkar, G.K.; Manvi, F.V.; Nanjwade, V.K. Dendrimers: Emerging polymers for drug-delivery systems. Eur. J. Pharm. Sci. 2009, 38, 185–196.

150. Pednekar, P.P.; Jadhav, K.R.; Kadam, V.J. Aptamer-dendrimer bioconjugate: A nanotool for therapeutics, diagnosis, and imaging. Expert Opin. Drug Deliv. 2012, 9, 1273–1288.

151. Danhier, F.; Feron, O.; Préat, V. To exploit the tumor microenvironment: Passive and active tumor targeting of nanocarriers for anti-cancer drug delivery. J. Control. Release 2010, 148, 135–146.

152. Lee, I.H.; An, S.; Yu, M.K.; Kwon, H.K.; Im, S.H.; Jon, S. Targeted chemoimmunotherapy using drug-loaded aptamer-dendrimer bioconjugates. J. Control. Release 2011, 155, 435–441.

153. MacAdam, A.B.; Shafi, Z.B.; James, S.L.; Marriott, C.; Martin, G.P. Preparation of hydrophobic and hydrophilic albumin microspheres and determination of surface carboxylic acid and amino residues. Int. J. Pharm. 1997, 151, 47–55.

154. Wang, G.; Uludag, H. Recent developments in nanoparticle-based drug delivery and targeting systems with emphasis on protein-based nanoparticles. Expert Opin. Drug Deliv. 2008, 5, 499–515.

155. Xie, L.; Tong, W.; Yu, D.; Xu, J.; Li, J.; Gao, C. Bovine serum albumin nanoparticles modified with multilayers and aptamers for pH-responsive and targeted anti-cancer drug delivery. J. Mater. Chem. 2012, 22, 6053–6060.

156. Yang, L.; Meng, L.; Zhang, X.; Chen, Y.; Zhu, G.; Liu, H.; Xiong, X.; Sefah, K.; Tan, W. Engineering polymeric aptamers for selective cytotoxicity. J. Am. Chem. Soc. 2011, 133, 13380–13386.

157. Wu, C.; Han, D.; Chen, T.; Peng, L.; Zhu, G.; You, M.; Qiu, L.; Sefah, K.; Zhang, X.; Tan, W. Building a multifunctional aptamer-based DNA nanoassembly for targeted cancer therapy. J. Am. Chem. Soc. 2013, 135, 18644–18650.

158. Hu, R.; Zhang, X.; Zhao, Z.; Zhu, G.; Chen, T.; Fu, T.; Tan, W. DNA nanoflowers for multiplexed cellular imaging and traceable targeted drug delivery. Angew. Chem. Int. Ed. 2014, 53, 5821–5826.

159. Zhu, G.; Hu, R.; Zhao, Z.; Chen, Z.; Zhang, X.; Tan, W. Noncanonical self-assembly of multifunctional DNA nanoflowers for biomedical applications. J. Am. Chem. Soc. 2013, 135, 16438–16445.

160. Zhu, G.; Zheng, J.; Song, E.; Donovana, M.; Zhang, K.; Liu, C.; Tan, W. Self-assembled, aptamer-tethered DNA nanotrains for targeted transport of molecular drugs in cancer theranostics. Proc. Natl. Acad. Sci. USA 2013, 110, 7998–8003.

161. Zhou, J.; Li, H.; Li, S.; Zaia, J.; Rossi, J.J. Novel Dual Inhibitory Function Aptamer–siRNA Delivery System for HIV-1 Therapy. Mol. Ther. 2008, 16, 1481–1489.

162. Kim, E.; Jung, Y.; Choi, H.; Yang, J.; Suh, J.-S.; Huh, Y.-M; Kim, K.; Haam, S. Prostate cancer cell death produced by the co-delivery of Bcl-xL shRNA and doxorubicin using an aptamer-conjugated polyplex. Biomaterials 2010, 31, 4592–4599.

163. Herrmann, A.; Priceman, S.J.; Kujawski, M.; Xin, H.; Cherryholmes, G.A.; Zhang, W.; Zhang, C.; Lahtz, C.; Kowolik, C.; Forman, S.J.; et al. CTLA4 aptamer delivers STAT3 siRNA to tumor-associated and malignant T cells. J. Clin. Investig. 2014, 124, 2977–2987.

Index